T0249495

WIRELESS SENSOR AND ACTUATOR NETWORKS

Academic Press is an imprint of Elsevier
84 Theobald's Road, London WC1X 8RR, UK
Linacre House, Jordan Hill, Oxford OX2 8DP, UK
Radarweg 29, PO Box 211, 1000 AE Amsterdam, The Netherlands
30 Corporate Drive, Suite 400, Burlington, MA 01803, USA
525 B Street, Suite 1900, San Diego, CA 92101-4495, USA

**British Library Cataloguing in Publication Data**
A catalogue record for this book is available from the British Library

**Library of Congress Cataloging-in-Publication Data**
A catalog record for this book is available from the Library of Congress

ISBN:   978-0-12-372539-4

For information on all Academic Press publications
visit our website at books.elsevier.com

Working together to grow
libraries in developing countries

www.elsevier.com | www.bookaid.org | www.sabre.org

ELSEVIER      BOOK AID
              International      Sabre Foundation

# Contents

## Part 3 From theory to practice: case studies     277

## 10 From theory to practice: case studies     279

# List of figures

# List of tables

# Preface

## Aim of this book

Wireless sensor and actuator networks (WSANs) are among the most addressed research fields in the area of information and communication technologies (ICT) these days, in the US, Europe and Asia. WSANs are composed of possibly a large number of tiny, autonomous sensor devices and actuators equipped with wireless communication capabilities. One of the most relevant aspects of this research field stands in its multidisciplinarity and the broad range of skills that are needed to approach their design. Theory of control systems is involved, networking, middleware, application layer issues are relevant, joint consideration of hardware and software aspects is needed, and their use can range from biomedical to industrial or automotive applications, from military to civil environments, etc.

This book mainly covers wireless networking and design issues of WSANs with applications.

This research field attracted enormous and ever increasing attention in the past years. However, by looking, for example, at the IEEE literature, the first paper having 'wireless sensor network' in the title in the online IEEE database of scientific papers Xplore, dates back to the year 2000. A query on the ACM database brings us to the same outcome. So, this is a new research field that only very recently attracted the interest of many scientists worldwide. On the other hand, the number of papers in the open literature increased exponentially after the year 2000 (e.g. with a similar query IEEE Xplore shows 3 papers for 2000, 20 in 2001, 34 in 2002, 98 in 2003, 289 in 2004, 622 in 2005, 952 in 2006): this clearly testifies to the relevance of the research field on the one hand; on the other, owing to the chaotic distribution of effort provided by thousands of separate research groups worldwide, a consensus on major design rules of WSANs is still lacking, and it is not unusual to find recent papers using model assumptions which have been proven to be not realistic by others. This book also aims at defining some general design rules for WSANs and a common set of model assumptions that are real-world-proof. Some myths will be destroyed.

## Why a new book on WSANs

As anticipated by the title, this book covers aspects of WSANs, ranging from channel modelling, transmission techniques, communication protocols, localization and signal processing issues. Some of these aspects have already been covered by previously published books, by this and other publishers. The rationale for providing a new book on WSANs is the following. The majority of available books provide extensive descriptions of algorithms and protocols learnt from the literature, while minor relevance is given to their performance evaluation, and to the description of tools, techniques and methodologies needed to set the most important parameters of such algorithms and protocols. For instance, how the node density is related to the estimation error of a given spatial random process, or how network reachability is linked to the lifetime of nodes. Moreover, such books do not emphasize the practical issues related to the development of WSANs, being mainly based on theoretical results published over the scientific literature.

This book intends to provide methodologies and tools to design WSANs for real applications, in real environments. It will describe algorithms and protocols, with the aim of assessing their performance against the most relevant parameters. The design paradigm used comes out from the traditional communication society approach: first, derive charts describing the link between performance metrics and system parameters, then, by fixing requirements on the former, the latter are set as design constraints. In other words, contrary to best effort approaches, we design WSAN-related techniques starting from fixed requirements (on network connectivity, or lifetime, on the desired precision of position estimates, etc.). The tools that can be used to derive the performance metrics are basically of two types: either mathematical or simulation approaches can be applied. In both cases proper modelling of basic aspects like, for example, wireless channel characterization and physical layer techniques, is needed. This can be achieved from the literature, or by conducting on-the-field experiments based on available hardware/software platforms. In this book all these approaches are used: many experimental results are reported, which are useful to the comprehension of basic aspects related to the communications between wireless sensor nodes; also, a careful selection of results taken from the available literature is presented. Many simulation outcomes are discussed based on various types of simulation frameworks. Finally, the largest possible emphasis is given to the use of mathematical and formal descriptions of algorithms and their performance. In fact, we believe that mathematical models allow best comprehension of the relations between system parameters, provided that the models are suitably tuned to the real world. As a result of this approach, the book contains many performance charts.

The general approach used in writing this book is thus oriented to algorithm and protocol design and performance assessment. This is testified by the subtitle: 'Technologies, analysis and design.' Another option was: 'From theory to practice'.

This book is for PhD students and researchers, who aim at creating a solid scientific background about WSANs. It is also intended for engineers who need to design WSANs and want to understand the basic rules underlying their performance. Even if less importance is given to an exhaustive description of the available literature, the table of contents is also designed in order to provide a book useful for beginners.

## About the contents of this book

Apart from the first chapter, introducing the main definitions related to and features of WSANs, the book is composed of three parts.

The first part covers fundamental issues. A description of some the most interesting applications for WSANs is given in Chapter 2. This part is taken from the outcomes of the most relevant European projects on WSANs; moreover, a discussion on how these applications fall into few categories useful for partitioning the set of design guidelines is given. Then Chapter 3 deals with an analysis of the characteristics of the wireless channel which have an impact over link and network performance. The reason for starting with these two chapters is in the usual approach that engineers apply when designing a telecommunication system, that is, building a protocol stack: applications, on top of the stack, set requirements that drive the selection of protocols and transmission techniques; at the other end, the wireless channel poses constraints to the communication capabilities and performance. As a result of the encounter between the requirements set by applications and the constraints posed by the wireless channel, the communication protocols and techniques are selected. Therefore, after these two chapters we can start analyzing the link and network performance of WSANs under several viewpoints. Chapters 4 and 5 deal with network performance in terms of connectivity and lifetime. The impact of node density, transmission ranges and transmit power levels on such aspects are discussed. This part is not technology related. Then Chapter 6 introduces the transmission techniques (at physical and data link layer) that allow achievement of such transmission ranges and network performance. Emphasis is given to IEEE802.15.4 standards, as they represent an almost de facto standard for many applications of WSANs. This first part reports many results of experimental activities performed in order to measure wireless channel characteristics at 2.4 GHz, the band used by IEEE802.15.4 and other transmission techniques like, for example, Bluetooth, which are also suitable for some applications of WSANs. Then both mathematical and simulation models are used to study network performance.

The second part of the book covers issues related to access control and routing (Chapter 7) localisation and time synchronisation (Chapter 8) and signal processing (Chapter 9) for WSANs. Many algorithms and protocols presented in the literature are reported and described and some of them are evaluated through the use of suitable frameworks. This part of the book provides more detailed insight on some of the key functionalities of WSANs. The scientific literature is reviewed and some protocols and algorithms are numerically evaluated. This part is based on theoretical analyses.

The final part of this book (Chapter 10) is dedicated to some case studies: real-world applications of WSANs which are the result of experimental activities performed by the authors in the context of industrial contracts or large cooperative projects. The aim of this part is to clarify how the design guidelines provided in previous chapters can be useful when designing real-world networks. As usual when dealing with field trials and in general with experimental activities, the results are sometimes difficult to interpret and discuss. However, this also provides suggestions on how to build a field trial and how to anticipate the behaviour of a network prototype. Mention of the industrial contexts of the experimental activities is given as a necessary tribute to the availability of results.

In summary, the book tries to follow a path. First, the fundamentals are given in order to establish a sufficient technical background useful to the comprehension of basic aspects.

Since in this phase the pillars of the network are created, validation of assumptions is a key issue, and experiments are conducted under such a view. The design of algorithms and protocols can then be safely based on such pillars, and their performance evaluation can be realized through the use of theoretical approaches. Finally, the techniques designed must be proven on the real world, and the case studies discussed here report some examples.

## Is it possible to skip some chapters?

Of course, it is. However, we dare to suggest you read the first chapter, because it provides some clear introductory statements about WSANs, and the whole first part, in sequential order. Then the chapters of the second part might be read or not depending on specific interests. Concerning the third part, we assume that if you bought the book then you will read chapter 10.

In any case, all chapters are introduced by a short paragraph describing its scope and content. Take a minute to read it before deciding whether to skip the chapter.

## Acknowledgements

Some young scientists (PhD and Master students, or post-docs) directly contributed to this book under the supervision of the authors; they merit explicit acknowledgement as the book would have not been produced without their effort. Chiara Buratti and Flavio Fabbri were very active in the preparation of Chapter 4; Chiara also took care of the first draft of Chapter 2. Moreover, both Chiara and Flavio assisted the authors with corrections, inclusion of references, etc in the whole book. Chiara Taddia was involved in writing Chapter 7. Virginia Corvino was in charge of part of Chapter 6. Enrica Salbaroli and Raffaele Rugin were involved in writing Chapter 10. Thank you very much for your precious help.

Gratitude has also to be expressed to Alberto Zanella, senior researchers at WiLAB, who provided useful comments for the realisation of the second part of the book.

The authors are also very thankful to John Orriss from the University of Manchester, UK, who collaborated for years on connectivity issues with the group of authors, and generated some of the models used in this book. Davide Dardari and Andrea Conti would like to thank Professor Marco Chiani as well as Professor Moe Z. Win and his staff from the Massachusetts Institute of Technology (MIT), Cambridge, USA, for giving them a unique opportunity to improve their expertise in advanced wireless systems and for the strong collaboration which is still fertile both from the professional and personal point of view.

Mention should also go to all those at WiLAB who indirectly provided inputs to this book simply because they were involved in the past years in common research projects with the authors, and as such contributed to the creation of a common scientific background on WSANs at WiLAB: Velio Tralli, Gianni Pasolini, Andrea Giorgetti, Alessandro Bazzi, Barbara Masini, and the younger ones.

Finally, the authors wish to express their immense gratitude to Professor Oreste Andrisano, head of WiLAB: since the beginning of the 1990s he started believing in the potential of a small group of young scientists showing an enthusiastic approach to the field of wireless networks, including the authors of this book.

## Authors' personal notes

When I started thinking about a book on wireless sensor and actuator networks and I had the first talks with Tim Pitts, commissioning editor of Elsevier, it was a long time ago. Both my parents were alive and healthy. Now both are gone. This led to two consequences. First, despite the continuous push from my colleagues and friends, Gianluca, Davide and Andrea, whom I had invited since the beginning to collaborate in the preparation of the book, it resulted in a significant delay. However, it also had positive effects: many interesting scientific results have been published in the last year and it was possible to include the latest achievements; also, some relevant European projects dealing with wireless sensor networks reached their maturity during this period, which means that part of their results are reported in the book. Second, I want to thank who shined upon my life during the past two years, my family, and I can't refrain from dedicating this book to my Mother and my Father who largely contributed to it, through me.

**Roberto Verdone**

I dedicate this book to my wife Paola, son Damiano and daughter Alessia for the support and understanding they provided me with throughout the preparation of this manuscript. A special thanks to my father from whom I inherited the enthusiasm for the wireless world and my mother for her devotion and dedication.

**Davide Dardari**

Writing a book is a nice adventure. It is a trip that you know when it starts but you have no idea when it will be finished. Writing a scientific book is more complex, simply because researchers are working while you are writing, and it could become a race without a finish. Nothing is possible without students who every day ask me new questions that force me to study, friends that stimulate me with continuous discussion and family who support me in any new trip. I dedicate this book to my students, to my friends and to my family.

**Gianluca Mazzini**

I dedicate this book to my family and to persons who have allowed me to grow and cultivate the passion for doing research.

**Andrea Conti**

# About the authors

This book is the outcome of the effort provided by several young and senior scientists working in the area of WSANs for many years in Italy. Roberto Verdone and Davide Dardari are with the University of Bologna, while Gianluca Mazzini and Andrea Conti are with the University of Ferrara; the four authors cooperate in large projects within the framework of WiLAB, the wireless communication laboratory located in Bologna and in putting together scientists working in the field of wireless systems. WiLAB is an organization born under the auspices of the University of Bologna, the National Research Council and CNIT (the National Inter-University Consortium for Telecommunications). The research experience accumulated by the authors in the field of WSANs is the result of several projects carried out both at national and European level at WiLAB. Participation to either long-term research frameworks like the European networks of excellence, NEWCOM and CRUISE, or more industry-oriented contexts, has provided a range of skills that deal with both theoretical aspects and practical implementations. Such diversified experiences appear in the following chapters.

# List of acronyms

| | |
|---|---|
| ACK | acknowledge |
| AcR | autocorrelation receiver |
| AFL | anchor-free localization |
| AoA | angle-of-arrival |
| APTC | adaptive power topology control |
| AWGN | additive white Gaussian noise |
| BEP | bit error probability |
| BLEP | block error probability |
| BSN | body sensor network |
| BPSK | binary phase shift keying |
| BPZF | band-pass zonal filter |
| BSC | binary symmetric channels |
| BT | Bluetooth |
| CAP | contention access period |
| c.d.f. | cumulative distribution function |
| CDMA | code division multiple access |
| CEP | codeword error probability |
| CFA | cooperative fusion architecture |
| CFP | contention free period |
| CH | cluster head |
| ch.f. | characteristic function |
| CG | communication graph |
| CRC | cyclic redundancy check |
| CRLB | Cramér-Rao lower bound |
| CSI | channel state information |
| CSMA | carrier sensing multiple access |
| CSMA/CA | carrier sensing multiple access with collision avoidance |
| CSS | chirp spread spectrum |
| CTR | critical transmission range |
| DDSP | distributed digital signal processing |
| DP | direct path |
| DS-SS | direct sequence spread spectrum |
| ED | energy detector |
| EvD | event detection |
| ECG | eco cardiogram |
| EGC | equal gain combining |
| EIRP | effective isotropic radiated power |
| e.m. | electro-magnetic |
| EMST | Euclidean minimum spanning tree |

| FAD | fixed assessment delay |
|-----|------------------------|
| FC | fusion center |
| FCC | Federal Communications Commission |
| FDMA | frequency division multiple access |
| FEC | forward error correction |
| FFD | full function device |
| FH | frequency hopping |
| FIM | Fisher information matrix |
| GC | giant component |
| GDOP | geometric dilution of precision |
| GFSK | Gaussian-shaped binary frequency shift keying |
| GG | geometric graph |
| GLRT | generalized likelihood ratio test |
| GPRS | general packet radio service |
| GPS | global positioning system |
| GRG | geometric random graph |
| GTS | granted time slot |
| GWBP | Galton-Watson branching process |
| HHA | hybrid hierarchical architecture |
| ICT | information and communication technologies |
| IF | intermediated frequency |
| i.i.d. | independent, identically distributed |
| IM | instant messaging |
| IR-UWB | impulse radio UWB |
| ISM | industrial scientific medical |
| IST | information society technologies |
| LLC | logical link control |
| LOS | line-of-sight |
| LRT | likelihood ratio test |
| LS | least squares |
| MAC | medium access control |
| MB-UWB | multiband UWB |
| MF | matched filter |
| MFR | MAC footer |
| MGF | moment generating function |
| MHR | MAC header |
| ML | maximum likelihood |
| MPDU | MAC payload data unit |
| MRC | maximal ratio combining |
| MSDU | MAC service data unit |
| MSE | mean square error |
| MST | minimum spanning tree |
| MUI | multi-user interference |
| MVE | monitoring volcanic eruptions |

| NBI | narrowband interference |
|---|---|
| NLOS | non-line-of-sight |
| NTP | network time orotocol |
| OFDM | orthogonal frequency division multiplexing |
| OOK | on-off keying |
| O-QPSK | offset quadrature shift keying |
| PAM | pulse amplitude modulation |
| PAN | personal area network |
| p.d.f. | probability distribution function |
| PDP | power delay profile |
| PE | spatial and time random process estimation |
| PEB | position error bound |
| PER | packet error rate |
| PEP | packet error probability |
| PFA | parallel fusion architecture |
| PHR | physical header |
| PHY | physical |
| PoI | phenomenon of interest |
| PPDU | physical protocol data unit |
| ppm | part-per-million |
| PPM | pulse position modulation |
| PPP | Poisson point process |
| PSDU | physical service data unit |
| RA | range assignment |
| RAN | radio access network |
| RBS | reference broadcast synchronisation |
| REI | residual energy information |
| RF | radio frequency |
| RFD | reduced function device |
| RFID | radio frequency identification |
| RG | random graph |
| RMSE | root mean square error |
| RSS | received signal strength |
| RTT | round-trip time |
| r.v. | random variable |
| SD | selection diversity |
| SHR | synchronization header |
| SIM | structural integrity monitoring |
| SIR | signal-to-interference ratio |
| SMS | short message service |
| SN | sensor node |
| SNR | signal-to-noise ratio |
| SNIR | signal-to-noise-plus-interference ratio |
| SP | signal processing |

| | |
|---|---|
| SS | spread spectrum |
| S-V | Saleh-Valenzuela |
| TCS | tracking and communication system |
| TDD | time division duplexing |
| TDMA | time division multiple access |
| TDoA | time difference-of-arrival |
| ToA | time-of-arrival |
| ToF | time-of-flight |
| TH | time-hopping |
| TNR | normalized threshold |
| TR | transmission range |
| TSPN | time-synch protocol for sensor networks |
| UMTS | universal mobile telecommunications system |
| UWB | ultrawide bandwidth |
| WAF | wall attenuation factor |
| WBI | wideband interference |
| WED | wall extra delay |
| WHN | wireless hybrid network |
| Wi-Fi | wireless fidelity |
| WiMAX | worldwide interoperability for microwave access |
| WLAN | wireless local area network |
| WPAN | wireless personal area network |
| WSAN | wireless sensor and actuator network |
| WSN | wireless sensor network |
| ZZLB | Ziv-Zakai lower bound |

# 1

# Introduction

## 1.1  Introduction

This chapter is the only one in this book intended for beginners who need to start from scratch and learn what a wireless sensor and actuator network (WSAN) is and what are its main characteristics. However, even among the category of scientists working in the area of WSANs, sometimes statements like 'WSAN are a specific type of ad hoc network' or 'Power control is an efficient way to increase network lifetime' are found. They are not true. If you want to learn why, read this short chapter. It will provide a definition of WSANs first, and then the main features of WSANs will be highlighted. Some practical aspects concerning energy management, a relevant matter for most WSAN applications, will be discussed. Technical issues that research is still addressing are also described. Finally, the research agendas of some fora that recently defined the goals of research on embedded systems and WSANs for the next years are summarized, to ensure better understanding of what are future expectations.

## 1.2  What is a WSAN?

A wireless sensor network (WSN) in its simplest form can be defined as (Chong & Kumar, 2003; Akyildiz, Su, Sankarasubramaniam & Cayirci, 2002; Culler, Estrin & Srivastava, 2004) a network of (possibly low-size and low-complex) devices denoted as *nodes* that can sense the environment and communicate the information gathered from the monitored field (e.g., an area or volume) through wireless links; the data is forwarded, possibly via multiple hops relaying, to a *sink* (sometimes denoted as *controller* or *monitor*) that can use it locally, or is connected to other networks (e.g., the Internet) through a gateway. The nodes can be stationary or moving. They can be aware of their location or not. They can be homogeneous or not.

This is a traditional single-sink WSN (see Figure 1.1). Almost all scientific papers in the literature deal with such a definition. This single-sink scenario suffers from the lack of scalability: by increasing the number of nodes the amount of data gathered by the sink increases and once its capacity is reached the network size can not be augmented.

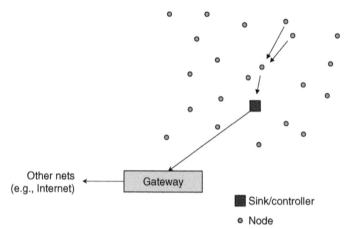

**Figure 1.1**   *Traditional single-sink WSN*

Moreover, for reasons related to medium access control (MAC) and routing aspects, network performance cannot be considered independent from the network size.

Let us give a simple and approximate evaluation of the capacity of a single-sink WSN, defined in terms of maximum number of nodes that can be attached to the sink. We consider a WSN where nodes are requested to send their samples (composed of $D$ bytes each) taken from the monitored space every $T_R$ seconds. Let us start by assuming that all nodes can directly send their data to the sink (single-hop network with star topology).

### 1.2.1   Single-sink single-hop WSN

Denote as $N$ the number of nodes, $R_b$ the channel bit rate. Then we define a factor, $\alpha_A \le 1$, taking account of the overhead introduced by all protocol stack layers: if $S_A$ is the maximum data throughput measured at the application layer, it is given by $S_A = R_b \cdot \alpha_A$. The smaller $\alpha_A$, the lower is the throughput even if the channel bit rate is unchanged. All protocol layers contribute to lower $\alpha_A$; the MAC sub-layer is often the main contributor when random channel access schemes are used. In modern communication systems $\alpha_A$ typically takes values between 0.5 and 0.1.

Under such assumptions, the application throughput will be approximately equal to $N D$ $8/T_R$. Then, we reach the following inequality: $ND8/T_R \le R_b\alpha_A$; therefore,

$$N \le R_b\alpha_A T_R/(8D). \tag{1.1}$$

This equation provides an approximate estimation of the number of nodes that can be part of a single-sink single-hop WSN. To give a numerical example, assume $R_b = 250\,\text{Kbit/s}$, $T_R = 1\,\text{s}$, $\alpha_A = 0.1$, $D = 3$; then the maximum number of nodes is approximately 1000. On the other hand, if $T_R = 10\,\text{ms}$, then $N$ can not exceed 10. It is clear that the requirements

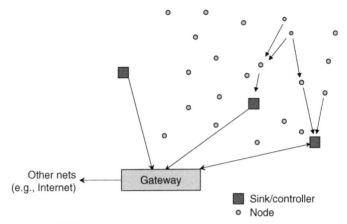

**Figure 1.2**   *Multi-sink WSN*

set by the application scenario play a very relevant role when defining the capacity of a single-sink WSN.

Note also that the protocol overhead can play a significant role, through $\alpha_A$.

In the case discussed above, the $N$ nodes are all within range of the sink. If the transmission range of links between sink and nodes is $R$, then the density of nodes is (no smaller than) $N/\pi R^2$.

## 1.2.2   Single-sink multi-hop WSN

If we now assume that the $N$ nodes are distributed according to a smaller density, then some of them must reach the sink through multiple hops. If a node can send its sample to the sink through $h$ hops, then the delivery of the data sample requires $h$ transmissions. Let us denote by $h_m$ the average number of hops per data sample taken from the field; if no smart reuse of radio resources is introduced, then we have for a single-sink multi-hop WSN:

$$N \leqslant R_b \alpha_A T_R/(8Dh_m). \tag{1.2}$$

Therefore, the capacity of the network is reduced by a factor of $h_m$.

## 1.2.3   Multi-sink multi-hop WSN

A more general scenario includes multiple sinks in the network (see Figure 1.2). Given a level of node density, a larger number of sinks will decrease the probability of isolated clusters of nodes that cannot deliver their data owing to unfortunate signal propagation conditions. In principle, a multiple-sink WSN can be scalable (i.e., the same performance can be achieved even by increasing the number of nodes), while this is clearly not true for a single-sink network. However, a multi-sink WSN does not represent a trivial extension of a single-sink case for the network engineer. There might be mainly two different cases: (1) all sinks are connected through a separate network (either wired or wireless), or (2) the sinks are disconnected. In the former case, a node needs to forward the data collected to any element in the set of sinks. From the protocol viewpoint, this means that a selection can be done based on a suitable criterion (e.g., minimum delay, maximum throughput, minimum number of hops, etc.). The presence of multiple sinks in this case ensures better network performance with respect to the single-sink case (assuming the same number of nodes is deployed over the same area), but the communication protocols must be more complex and should be designed according to suitable criteria. In the second case, when the sinks are not connected, the presence of multiple sinks tends to partition the monitored field into smaller areas; however from the communication protocols viewpoint no significant changes must be included, apart from simple sink discovery mechanisms. Clearly, the most general and interesting case (because of the better potential performance) is the first one, with the sinks connected through any type of mesh network, or via direct links with a common gateway.

If we now want to provide a simple and approximate evaluation of the capacity of a multi-sink WSN, we can assume that each sink (denoting as $N_S$ their overall number in the network) can serve up to $N$ nodes with $N$ limited by expressions (1.1) and (1.2). Therefore, we can write:

$$N \leq N_S R_b \alpha_A T_R / (8 D h_m),\tag{1.3}$$

assuming that clusters of nodes attached to a given sink do not interfere with those attached to any other sinks. To give a numerical example, assume $R_b = 250\,\text{Kbit/s}$, $T_R = 10\,\text{ms}$, $\alpha_A = 0.1$, $D = 3$; then, if there are $N_S = 5$ sinks in the network, the maximum number of nodes is approximately 50.

### 1.2.4   The presence of actuators

Both the single-sink and multiple-sink networks introduced above do not include the presence of actuators, that is, devices able to manipulate the environment rather than observe it. WSANs are composed of both sensing nodes and actuators (see Figure 1.3). Once more, the inclusion of actuators does not represent a simple extension of a WSN from the communication protocol viewpoint. In fact the information flow must be reversed in this case: the protocols should be able to manage many-to-one communications when sensors provide data, and one-to-many flows when the actuators need to be addressed, or even one-to-one links if a specific actuator has to be reached. The complexity of the protocols in this case is even larger.

Given the very large number of nodes that can constitute a WSAN (more than hundreds sometimes), it is clear that MAC and the network layer are very relevant parts of the

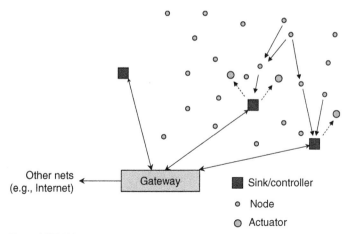

**Figure 1.3** *Typical WSAN*

protocol stack. Tens of proposals specifically designed for WSANs have been made in the past few years. The communication protocols of a WSAN should also allow an easy deployment of nodes; the network must be able to self-organize and self-heal when some local failures are encountered.

## 1.2.5 The nodes' architecture

The basic elements of a WSAN are the *nodes* (either sensors or actuators), the sinks and the *gateways*. Sinks, gateways and even actuator nodes are usually more complex devices than the sensor nodes, because of the functionalities they need to provide, or in some cases owing to the type of actuation mechanisms implied (e.g., mechanical actions). The sensor node is the simplest device in the network, and in most applications the number of sensor nodes is much larger than the number of sinks, or actuators. Therefore, their cost and size must be kept as low as possible. Also, in most applications the use of battery-powered devices is very convenient to make the deployment of such nodes easier. To let the network work under specified performance requirements for a sufficient time, denoted as network lifetime, the nodes must be capable of playing their role for a sufficiently long period using the energy provided by their battery, which in many applications should be not renewed for years. Thus, energy efficiency of all tasks performed by a node is a must for the WSAN design.

The traditional architecture of a sensor node is reported in Figure 1.4. A microprocessor manages all tasks; one or more sensors are used to take data from the environment; a memory is included over the board which is used to store temporary data, or during its processing; a radio transceiver (with the antenna) is also present. All these devices are powered by a battery. Traditional batteries can provide initial charges in the order of 10,000 joules and they should be parsimoniously used for the whole duration of the network lifetime by all these devices. In some cases energy scavenging techniques can

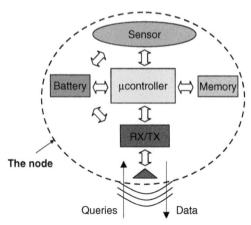

**Figure 1.4**  *Architecture of a sensor node*

be introduced to enlarge lifetime of nodes, but in few applications can this be really considered as a viable technique.

As a result of this need to have energy-efficient techniques implemented over the board,
all data processing tasks are normally distributed over the network; therefore, the nodes
cooperate to provide the data to the sinks. This is also because of the low complexity
that is accepted for the architecture of such nodes.

In conclusion, a WSAN can be generally described as a network of nodes that cooperatively sense the environment and may control it, enabling interaction between people or
computers and the surrounding environment.

The density of nodes and sinks is a very relevant parameter for WSANs: the density of
sensor nodes defines the level of coverage of the monitored space (i.e., what percentage
is such that if an event happens inside it is detected by at least one node); however, it also
defines the degree of connectivity, or reachability, that is a relevant issue as described in
Chapter 4. On the other hand, the density of sinks plays a significant role in defining the
performance of the network in terms of success rate of data transmissions, etc.

## 1.3   Main features of WSANs

The general description given in the previous sections already introduced the main
features of a WSAN: scalability with respect to the number of nodes in the network,
self-organization, self-healing, energy efficiency, a sufficient degree of connectivity
among nodes, low-complexity, low cost and size of nodes are all very relevant features
of WSANs; those protocol architectures and technical solutions providing such features
can be considered as a potential framework for the creation of networks able to implement several types of applications. Unfortunately, the definition of such a protocol architecture and technical solution is not simple, and the research still needs to work on it.

The massive research on WSANs started after the year 2000 (Chong & Kumar, 2003; Akyilxiz, Su, Sankarasubramaniam & Cayirci, 2002; Culler, Estrin & Srivastava, 2004). However, it took advantage of the outcome of the research on wireless networks performed since the second half of the previous century. In particular, the study of ad hoc networks attracted a lot of attention for several decades, and some researchers tried to report their skills acquired in the field of ad hoc networks to the study of WSANs.

According to some general definitions, wireless ad hoc networks are formed dynamically by an autonomous system of nodes connected via wireless links without using an existing network infrastructure or centralized administration. Nodes are connected through 'ad hoc' topologies, set up and cleared according to user needs and temporary conditions. Apparently, this definition can include WSANs. However, this is not true. This is the list of main features for wireless ad hoc networks: unplanned and highly dynamical; nodes are 'smart' terminals (laptops, etc.); typical applications include real-time or non-real-time data, multimedia, voice; every node can be either source or destination of information; every node can be a router toward other nodes; energy is not the most relevant matter; capacity is the most relevant matter (Gupta & Kumar, 2000).

Apart from the very first item, which is common to WSANs, in all other cases there is a clear distinction between WSANs and wireless ad hoc networks. In WSANs nodes are simple and low-complexity devices; the typical applications require few bytes sent periodically or upon request or according to some external event; every node can be either source or destination of information, not both; some nodes do not play the role of routers; energy efficiency is a very relevant matter, while capacity is not for most applications. Therefore, WSANs are not a special case of wireless ad hoc networks. Thus, a lot of care must be used when taking protocols and algorithms which are good for ad hoc networks and using them in the context of WSANs.

## 1.4 Practical issues of WSANs related to energy management

As stated above, energy efficiency is a key issue for most WSAN applications. Network lifetime must be kept as long as possible. Clearly, it depends on how long can be the period of time starting with network deployment and ending when the battery of sensor nodes is no more able to provide the energy needed for communication, sensing or processing. Chapter 5 will deal with this issue. However, it is convenient to start here with a brief discussion about some aspects of the energy management which might be non intuitive for engineers and researchers that did not make practice over real WSAN test beds. Let us assume that sinks, gateways and actuators are plugged: so, energy efficiency is a must only for sensor nodes.

First, what part of a sensor node is responsible for energy consumption?

As shown in Figure 1.4, a node is basically composed of a battery, a microprocessor, a memory, the sensors and the transceiver. Normally, when in transmit mode the transceiver drains much more current from the battery than the microprocessor in active state, or the sensors and the memory chip. The ratio between the energy needed for transmitting and for processing a bit of information is usually assumed to be much larger than one (more than one hundred or one thousand in most commercial platforms). For this reason, the communication protocols need to be designed according to energy-efficient paradigms, while processing tasks are not, usually. The design of energy-efficient protocols is thus a very peculiar issue of WSANs, with no significant precedent in wireless network history. The vast majority of articles on WSANs in the literature deal with the design of energy-efficient protocols, neglecting the role of the energy consumed when processing data inside the node (Verdone, 2004; Xu, Heidemann, Estrin, 2001; Ye, Heidemann & Estrin, 2002).

As a conclusion, the transceiver is the part responsible for the consumption of most energy.

On the other hand, sometimes data-processing techniques are implemented in WSANs that can require long processing tasks to be performed at the microprocessor, much longer than the actual amount of time a transceiver spends in transmit mode. This can cause significant energy consumption by the microprocessor, even comparable to the energy consumed during transmission, or reception, by the transceiver. So, the general rule that communication protocol design is much more important than a careful design of the processing task scheduling can not be considered always true.

Second, what are the transceiver states that require more current drain from the battery?

Intuitively, the transmit state is very energy expensive, as both the baseband and radio frequency (RF) part of the transceiver are active. However, the same is true for the receive state! Therefore, the receive state can consume as much energy as the transmit does. Owing to the hardware design principles, sometimes in the receive state the transceiver can consume even more energy than in the transmit state.

Receive and transmit states are both very energy consuming, and the transceiver must be kept in those two states for the shortest possible percentage of time.

Clearly, permanence in the transmit state is needed only when a data burst needs to be transmitted. The shorter is the data burst to be transmitted, the longer is node life. This suggests to avoid using protocols based on complex handshakes. However, a transceiver might need to stay in receive mode for longer periods of time, if proper scheduling of transmit times is not performed. Protocols should avoid a phenomenon, called 'overhearing' (Ye, Heidemann & Estrin, 2002), such that nodes need to stay in receive time for long periods waiting for a packet while listening to many data bursts sent to other nodes.

However, this is not enough. In fact, many MAC protocols consider channel sensing mechanisms: the transceiver senses the wireless channel for some periods of time in order to determine whether it is busy or free. Depending on the specific hardware platform, channel sensing can be very energy consuming, almost as the transmit and receive states. So, protocols must not abuse of the channel sensing mechanism.

Thus, carrier sensing multiple access (CSMA) is potentially very energy inefficient.

Typically, the transceiver drains much smaller currents from the battery when in sleep or off mode. For this reason, proper use of the energy provided by the battery often requires long (in the order of 95%–99% of time) intervals of time with the transceiver in sleep state. During such periods, a data burst sent to the node can not be detected. Therefore, the management of sleep mode is a very relevant issue for WSANs.

A final consideration needs to be reported in this introductory part, related to the use of power control. This technique, setting the transmit power at the minimum level needed to allow signal correct detection at the receiver, is often used in wireless networks to reduce the interference impact of transmissions and the useless emission of radiowaves with large power. However, setting a proper power level requires information on the channel gain, which might be difficult to achieve in applications with very bursty data transmissions. Therefore it is worthwhile wondering whether power control is a useful technique for WSANs.

Looking at the data reported on the datasheet (Chipcon Products) of a sample transceiver used in many commercial platforms, such as CHIPCON CC2420, one can derive an interesting conclusion. When transmitting at the largest power level (0 dBm), about 17 mA are drained from the battery. At minimum transmit power ($-25$ dBm), the current drained is 8.5 mA, about halved! There is no relevant energy saving when decreasing the power level of transmission by 25 dB. Even if this example is given with reference to a specific chip, there are reasons to state that the conclusion is general.

The energy consumed in transmission state is not proportional to the transmit power level used, and therefore power control is not an efficient technique to reduce energy consumption.

## 1.5   Current and future research on WSANs

Many technical topics of WSANs are still considered by research as the current solutions are known to be non-optimised, or too much constrained.

From the physical layer viewpoint, standardization is a key issue for success of WSAN markets. Currently the basic options for building HW/SW platforms for WSANs are Bluetooth, IEEE 802.15.4 and 802.15.4a. At least, most commercially available platforms use these three standards for the air interface. For low data rate applications (250 Kbit/s on the air), IEEE 802.15.4 seems to be the most flexible technology currently available. Clearly, the need to have low-complexity and low-cost devices does not push research in the direction of advanced transmission techniques.

MAC and network layer have attracted a lot of attention in the past years and still deserve investigation. In particular, combined approaches that jointly consider MAC and routing seem to be very successful.

Topology creation, control and maintenance are very hot topics. Especially with IEEE 802.15.4, which allows creation of several types of topologies (stars, mesh, trees, cluster-trees), these issues play a very significant role.

Transport protocols are needed for WSANs depending on the specific type of application.

However, some of the most relevant issues investigated by research in WSANs are cross-layer, dealing with vertical functionalities: security, localization, time synchronization.

Basically, the research in the field of WSANs started very recently with respect to other areas of the wireless communication society, as broadcasting or cellular networks. The first IEEE papers on WSANs were published after the turn of the Millennium. The first European projects on WSANs were financed after year 2001. In the US the research on WSANs was boosted a few years before. Many theoretical issues still need a lot of investments. Europe will finance projects having WSANs as core technologies for at least the next seven years, within the Seventh Framework Programme. This book does not aim at providing a crystal ball able to announce next year's achievements. However, a look into the future can be launched by reading the research agendas of some research fora, or projects, participated in by the main actors in the research field.

In Europe, during the Sixth Framework Programme, three projects were financed by the European Commission, with explicit activities dedicated to the provision of research agendas, showing the needs, the main application areas, and the gaps to be filled: WISENTS, E-SENSE and CRUISE. The outcome of their work can be found on the official websites. Then, it is also relevant to mention the viewpoint of the main European technology platforms, gathering all stakeholders in the field, related to the area of WSANs: e-Mobility and ARTEMIS. They have drawn research agendas that will drive the selection of large cooperative projects in future years in Europe.

The e-Mobility technology platform gathers all major players in the area of wireless and mobile communications. A strategic research agenda was released and updated in 2006. According to their views, by the year 2020 mobile and wireless communications will play a central role in all aspects of European citizens' lives, not just telephony, and will be a major influence on Europe's economy, wirelessly enabling every conceivable business endeavour and personal lifestyle. The aim of research in the field can be summarized as follows: 'The improvement of the individual's quality of life, achieved through the availability of an environment for instant provision and access to meaningful, multi-sensory information and content.' 'Environment' means that the users will strongly interact with the environment that surrounds them, for example by using devices for personal use, or by having the location as a basis for many of the services to be used. This implies a totally different structure for the networks. Also, the context recognized by the system and it acting dynamically on the information is a major enabler for intelligent applications and services. This also means that sensor networks and radio frequency identifications (RFIDs) are increasingly important. 'Multi-sensory' is related to all the users' devices, and also to the fact that the environment will be capable of sensing the users' presence. Also, virtual presence may be considered, implying more sensory information being communicated, and an ideal of a rich communication close to the quality achieved in interpersonal communications or direct communications with another environment; this could also include non-invasive and context-aware communication characterizing polite human interactions. Therefore, this stretches mobile and wireless communications beyond radio and computer science into new areas of science, like biology, medicine, psychology, sociology, and nano-technologies, and also requires full cooperation with other industries not traditionally associated with communications. Finally, the information should be multi-sensory and multi-modal, making use of all human basic senses to properly capture context, mood, state of mind, and, for example, one's state of health. Clearly, the realization of this vision of mobile and wireless communications demands multi-disciplinary research and development, crossing the boundaries of the above sciences and different industries. Also, the number of electronic sensors and RFIDs surrounding us is quickly increasing. This will increase the amount of data traffic.

The future system will be complex, consisting of a multitude of service and network types ranging across wireless sensor networks, personal area, local area, home networks, moving networks to wide area networks. Therefore, the e-Mobility vision emphasizes the key role played by WSANs as elements of a more complex system linking different types of access technologies.

ARTEMIS (advanced research & technology for embedded intelligence and systems) is the technology platform for embedded systems. The term 'embedded systems' describes electronic products, equipment or more complex systems, where the embedded computing devices are not visible from the outside and are generally inaccessible by the user. The sensor and actuator nodes of WSANs are embedded systems. According to the ARTEMIS strategic research agenda, intelligent functions embedded in components and devices will be a key factor in revolutionizing industrial production processes, from design to manufacturing and distribution, particularly in the traditional sectors. These technologies add intelligence to the control processes in manufacturing shop floors and improve the logistic and distribution chains, resulting in an increasing productivity in a wide range of industrial processes. The grand challenge in the area of sensors and actuators relates to the support of huge amounts of input and output data envisaged in the application contexts with minimal power requirements and fail-safe operation.

# Part 1

# Fundamentals of WSANs design

This part of the book covers fundamental issues for the design of WSANs. A description of some of the most interesting applications for WSANs is given in Chapter 2. This part is taken from the outcomes of the most relevant European projects on WSANs; moreover, a discussion on how these applications fall into few categories useful for partitioning the set of design guidelines is given. Chapter 3 deals with an analysis of the characteristics of the wireless channel which have an impact over link and network performance. The reason for starting with these two chapters is in the usual approach that engineers apply when designing a telecommunication system, building a protocol stack: applications, on top of the stack, set requirements that drive the selection of protocols and transmission techniques; on the other hand, the wireless channel poses constraints to the communication capabilities and performance. As a result of the encounter between the requirements set by applications and the constraints posed by the wireless channel, the communication protocols and techniques are selected. Therefore, after these two chapters we can start analyzing the link and network performance of WSANs under several viewpoints. Chapters 4 and 5 deal with network performance in terms of connectivity and lifetime. The impact of node density, transmission ranges and transmit power levels on such aspects are discussed. This part is not technology related. Chapter 6 introduces the transmission techniques (at physical and data link layer) that allow achievement of such transmission ranges and network performance. Emphasis is given to IEEE 802.15.4 standard, as it represents an almost de facto standard for many applications of WSANs. This first part of the chapter reports many results of experimental activities performed in order to measure wireless channel characteristics at 2.4 GHz, the band used by IEEE 802.15.4 and other transmission techniques, such as Bluetooth and UWB, also suitable for some applications of WSANs. Then, both mathematical and simulation model used to study network performance, are introduced.

# 2

# Applications of WSANs

This chapter provides an overview of the major applications for which WSANs are envisaged. The main application areas are categorized according to the type of information measured or carried by the network; for each application area some scenarios, which depict practical examples of applications, are provided. The main European projects funded by the European Commission dealing with WSANs, such as Creating Ubiquitous Intelligent Sensing Environments (CRUISE), a network of excellence funded by the European Commission through the Sixth Framework Programme,[1] e-SENSE[2] and WiSeNts[3] (Marron, Minder & the Embedded WiSeNts Consortium, 2006), are taken into consideration for the definition of this list of applications and scenarios. Also, other projects run in North America, Europe and Asia have been considered to create this list. Therefore, what is reported in this chapter is the merging of experiences gathered in very diverse environments.

After this first overview, we classify the different scenarios identified in two categories: event detection and estimation of spatial (and temporal) random processes applications. In the first case the aim of the network is to detect an event (a fire in a forest, a earthquake, etc.), whereas in the second case the aim is to estimate a random spatial or temporal process. This classification is meaningful as it can provide a first idea of the type of application requirements set by the scenarios.

Finally, since in the past years the development of new technologies and the standardization of new air interfaces has increased the interest of researchers towards the concept of wireless hybrid networks (WHNs), where separate network paradigms (ad hoc or infrastructure-based networks, etc.) converge and interact, we consider a challenging network architecture, namely the hybrid hierarchical architecture (HHA), which was investigated within CRUISE. The HHA represents a reference model that is recognized as of particular value for WSANs. It is a particular case of WHN, obtained by the merging of infrastructure-based and infrastructure-less networks. After a brief description and analysis of this architecture, a discussion is presented about how the coherence of the different scenarios with the HHA can be evaluated.

---

[1] See the CRUISE website: http://www.ist-cruise.eu
[2] See the e-SENSE IST project website: http://www.ist-esense.org
[3] See the WiSeNts website: http://www.embedded-wisents.org

## 2.1   Application areas and scenarios

This chapter provides an overview of the major applications for which WSANs are provided. The application areas considered are the following:

- Environmental monitoring
- Health care
- Mood-based services
- Positioning and animals tracking
- Entertainment
- Logistics
- Transportation
- Home and office
- Industrial applications.

For each application area, we provide some scenarios which depict a situation where a character (or characters) makes use of different sensors to monitor an environment (environmental monitoring, positioning and tracking), for provision of health services (health care), to enhance his/her experience (mood-based services), to enjoy his/herself (entertainment), to improve an industrial activity (industrial application, logistic and transportation), etc.

For each scenario we provide a description which gives a reader the basic concept about the scenario without going into the details. The description could be a small story describing the scenario: the story highlights the objective of the application, who are the users, what are the users' requirements, what are the objects involved in the scenarios etc. Otherwise, a simple description of the scenario, or a reference to projects, dealing with that specific application area, could be provided.

In the following the different application areas and scenarios are introduced.

### 2.1.1   Environmental monitoring

Environmental monitoring applications have crucial importance for scientific communities and society as a whole (Cardell-Oliver et al., 2004; Szewczyk et al., 2004; Santoni et al., 2006). Those applications may monitor indoor or outdoor environments. Supervized areas may be thousands of square kilometres and the duration of the supervision may last years. Networked microsensors make it possible to obtain localized measurements and detailed information about natural spaces where it is not possible to do this through known methods. Often in these applications, the network provides solutions for security and surveillance concerns; natural disasters such as floods and earthquakes may be perceived earlier by installing networked embedded systems closer to places where these phenomena may occur. The system should respond to the changes of the environment as quick as possible; therefore these applications require real-time monitoring technologies with high security requirements. Not only communications but also cooperation such as statistical sampling and data aggregation are possible between nodes. An environmental monitoring application may be used in either a small or a wide area for the same purposes. One of the first ideas of a WSN concept is to design it to use the system to monitor environments where humans cannot be present all the time. Such systems have to be infrastructure-less and very robust, because of the inevitable challenges in nature, such as living things or atmospheric events. Since the nodes are untethered and unattended in this class of applications, the system must be power efficient and fault tolerant. Long lifetimes of the network must be preserved while the scale increases in the order of tens or hundreds nodes. Environmental monitoring for emergency services is a typical domain which can benefit from networked tiny sensors.

The following scenarios are related to this area:

- Forest fire detection
- Flood detection
- Structural integrity monitoring
- Glacsweb
- Monitoring volcanic eruptions (MVE)-WSN.

### Forest fire detection

In recent years, Portugal has had serious problems with forest fires (CRUISE, WP112, D112.1, 2006). One of the main problems is that when the fire becomes large it also becomes very difficult to put out. In these cases, a WSN could be deployed to detect a forest fire in its early stages. A number of nodes need to be pre-deployed in a forest. Each node can gather different types of information from sensors, such as temperature, humidity, pressure and position. All sensing data is sent by multi-hop communication to the control centre via a number of gateway devices distributed throughout the forest. The gateways will be connected to mobile networks (e.g., Universal Mobile Telecommunications System – UMTS) and will be positioned so as to reduce the number of hops from source of fire detection to the control centre. The gateways will also reduce network congestion in large-scale deployments by extracting data from the network at pre-determined points. It may also be possible in this scenario that some mobile forest patron units act as mobile gateways, collecting environmental data as they traverse through the forest. As soon as a fire-related event is detected, such as sudden temperature rise, the control centre will be alarmed immediately. Operators in the control centre can judge if it is a false alarm by either using the data collected from other sensors or dispatching a team to check the situation locally. Then both fire-fighters and helicopters can be sent to put out the fire before it grows to a severe forest fire.

### Flood detection

Marka and Javier are a couple and they work in different companies (CRUISE, WP112, D112.1, 2006). Both must arrive very early in the morning at work. Besides, they have two children, aged seven and twelve. Every morning Marka and Javier leave prepared the breakfast for their children and their food for noon. However, the children must self-prepare every morning before going to school. The older one helps his brother, but sometimes he forgets some of parents' recommendations such as making sure all the lights are switched off before leaving, or making sure that all the taps have been turned off. Today, his young brother got his hands messy at the last moment and he had to clean them before going to school. They were in a hurry and he forgot to turn off the tap. In a short time the water starts to flow over the sink. One of the flood sensors detects the water and the WSN, since there is not too much water so far, chooses to inform the neighbour, and, of course, sends an SMS to Javier and Marka informing that there is water where there should not be. The neighbour fails to check the situation and the water continues to flood, so the WSN closes the main valve and if the water continues to flow it will inform the fire department. When Marka and Javier arrive home there is only a small amount of water in the toilet floor, but it can be removed by using a mop.

## Structural integrity monitoring
Wireless sensors can be used to detect and localize damages in buildings, bridges, ships, aircraft, etc. In Structural Integrity Monitoring (SIM) (CRUISE, WP112, D112.1, 2006) systems the wireless nodes can infer the existence and the location of damage by measuring structural responses to ambient or forced excitation causes, for example by earthquakes, wind, or vehicles, or when triggered by the monitoring system. The main applications of the SIM system are disaster response monitoring, which checks the health of a structure after an earthquake or an explosion, and continuous health monitoring aimed to assess the effects of the ambient vibrations or wind, etc. After a disaster, for example, a direct damage detection to evaluate the damages can be used by means of a visual inspection, X-ray or mobile video-camera. On the other hand, to evaluate the damage of a structure in the course of time, it is necessary to record the change in structural properties/behaviour, especially through vibration.

## Glacsweb
Glacsweb[4] is a project with the aim of monitoring glacier behaviour via different sensors and linking them together into an intelligent web of resources. Glacsweb aims at the development of a low-power wireless sensor network node capable of surviving for several years, which gathers the data autonomously into a web-accessible database. Probes are placed on and under glaciers and data collected from them by a case station on the surface. Measurements include temperature, pressure and subglacial movement; the aim is to understand what happens beneath glaciers and how they are affected by climate. The data gathered is important in understanding the dynamics of glaciers as well as global warming.

## MVE-WSN
MVE-WSN (Werner-Allen et al., 2005) is a wireless sensor network to monitor volcanic eruptions with low-frequency acoustic sensors developed by the universities

---

[4] See the Glacsweb website: http://www.envisense.org

of Harvard and North California. The WSN was deployed in July 2004 at the volcano Tungurahua, an active volcano in central Ecuador. The network collected infrasonic (low-frequency acoustic) signals at 102 Hz, transmitting data over a 9 km wireless link to a remote base station. During the deployment over 54 hours of continuous data, which included at least nine large explosions, have been collected. Nodes were time-synchronized using a separate global positioning system (GPS) receiver, and the data was later correlated with that acquired at a nearby wired sensor array. In addition to continuous sampling, a distributed event detector that automatically triggers data transmission when a well-correlated signal is received by multiple nodes has been developed. MVE-WSN has a wide range of goals related to both scientific studies and hazard monitoring. Traditionally, dispersed networks of seismographs, which record ground propagating elastic energy, are used to locate, determine the size of, and assess focal mechanisms (source motions) of earthquakes occurring within a volcanic edifice. Another use of seismic networks is the imaging of the internal structure of a volcano through tomographic inversion. Earthquakes recorded by spatially-distributed seismometers provide information about propagation velocities between a particular source and receiver.

### 2.1.2 Health care

These applications include telemonitoring of human physiological data, tracking and monitoring of doctors and patients inside an hospital, drug administrator in hospitals, etc. (Scott, 1998; Stankovic et al., 2005) Merging wireless sensor technology into health and medicine applications will make life much easier for doctors, disabled people and patients. They will also make diagnosis and consultancy processes regardless of location and transition automatically from one network in a clinic to the other installed in patient's home. As a result, high-quality health care services will get closer to the patients. Health applications are critical, since vital events of humans must be monitored. Heterogeneity is an issue because the sensed materials will be various. Localization is important because it is critical to determine where exactly the person is; if he carries a heart rate control device and it detects a sudden heart attack, there must be no mistake or no incapability for finding his location. However, since in most cases single-hop networks will be used and neither topology nor the routing will be changed, mobility is not considered to be a challenging issue for this kind of application. The delay between the source of the event and the other end-point of the system is also important. The data has to be conserved as original, which points to reliability of transmission. Although the idea of embedding wireless biomedical sensors inside the human body is promising, many additional challenges exist; the system must be safe and reliable; require minimal maintenance; and energy-harnessing from body heat. With more researches and progresses in this field better quality of life can be achieved and medical cost can be reduced.

The following scenarios are related to this area:

- Night shift assistant
- Backup shift assistant

- Acute patient monitoring
- Continuous care.

### Night shift assistant

The night shift assistant supports the health care work in a standard situation in which many patients have to be managed by drastically reduced staff (e-SENSE, 2006). This situation also holds true for nursing homes. Being a nurse, Anna frequently has to work night shifts, being responsible for a whole ward by herself. All patients are equipped with sensors that measure their vital functions while at the same time keeping the patients mobile so they can, depending on their health condition, move freely without additional equipment. Anna can observe the patients' data on different screens in her room. In case of a dangerous change in a patient's condition, an alarm sounds and pays special attention to the data of the specific patient. Anna receives an alarm from the third floor where a patient has fallen while leaving the bathroom. His blood pressure has decreased rapidly, so she hurries to the scene. As she discovers that she cannot handle the situation by herself she triggers an instant message for the doctor who is on duty tonight. He usually spends the night shift in a room in a different part of the hospital and is responsible for several wards at a time. While waiting for the doctor to answer her message, Anna pushes a button on her handheld device that will transfer the data of the fallen patient to the doctor's handheld device. He has a quick look at the data and lets her know that he will be at the scene right away. Anna is relieved and knows now that the patient will be taken care of.

### Backup shift assistant

The backup shift assistant is also helping in a standard situation in the hospital and simplifies the support of young assistant physicians by experienced colleagues (e-SENSE, 2006). At the hospital in which Anna works, the physicians differ very much in their age and work experience. Today Robert, a very young doctor, is on duty while a severe patient condition arises. He is as yet inexperienced and does not feel comfortable with handling the situation by himself. Since the patient's condition is critical, he decides to contact a more experienced colleague who is available in the hospital. Robert sends an instant message to the closest available physician, who is doing administrative duties in another building of the hospital. The physician quickly gives feedback via instant message that he will support Robert and receives on his handheld Robert's location and navigation information for the fastest route. Robert also sends him the vital data of the patient. Robert is relieved when the more experienced physician arrives. Together they stabilize the patient's condition.

### Acute patient monitoring

David had no serious trouble with his state of health, but one day he is taken to hospital because of a chest pain (e-SENSE, 2006). The diagnosis is a slight heart attack. David does not need surgery, but he will have to receive medical observation for a few weeks. He does not feel comfortable staying in the hospital for such a long time and would rather recover at home in familiar surroundings. The doctor introduces a new system for medical observation and after-care to him, consisting of a mobile phone and various physiological sensors. He explains that the sensors are able to communicate attained data to the phone and the phone will process this sensor information. In case of a dangerous change in David's echocardiogram (ECG) or breathing rate, an alarm will be sent to the hospital and from there to the on-duty physician's handheld device, and the hospital will get back to David by phone to make sure it is not a false alarm. The phone can be operated manually, transmitting recorded data once an hour and sending an alarm if needed, but David prefers to put it in automatic mode so he does not have to take care of the data transmission himself; the application will do this automatically and continuously. The doctor explains to him that during the first 48 hours of his recovery at home, his data will constantly be transmitted to the nurses' office where it can be monitored continuously and afterwards it will be checked just once a day. David now feels comfortable and secure, knowing that in case of an emergency he will be helped quickly. He programmes the mobile so it will inform his wife by sending a short message service (SMS) to her mobile phone if the hospital is alarmed at his condition. In addition, the application also reminds David to take his medication on time and when his supply of pills is running low, so he can arrange to renew his supply at the pharmacy in advance.

### Continuous care

Mr Brain has had surgery (CRUISE, WP112, D112.1, 2006). He is in the extensive care unit, his vital life parameters are being closely watched by doctors, nurses and medical personnel. When his condition improves in a few days he is moved out of the extensive care into the brain surgery ward. However, continuous surveillance is far from ending. Besides the nurses and the doctors directly involved, specialists from other departments are watching the recovery of his body functions after the heavy surgery and long-time anesthesia. In accordance with their own daily schedules different doctors can log on to the data system remotely and get information in both real and past time collected from the different, hardly noticeable wireless sensors that Mr Brain has attached to his body. Data from his blood tests is loaded on to his mobile device and made available to the requesting doctors. At the same time Mr Brain is steadily getting better and is allowed to walk around in the ward. The wireless sensors and the sink device do not restrict his comfort and his high spirits about a quick recovery. The continuous data flow from the sensors of his body network is being transferred to multiple receivers when requested or saved in the main database for future checks.

## 2.1.3 Mood-based services

These services enable new applications by intelligently processing and providing user states, namely mood. The services give users the possibility to share their emotions with friends and relatives in various situations (sports, leisure, family events, etc.), receive prolonged or repeated provision of emotional events, and manage and tackle negative emotional situations.

The following scenarios are related to this area:

- Tom's ceremony
- Tom's friends

- Danger's warning
- Personal coaching
- Mental counselling
- Daily travel.

### Tom's ceremony

In this scenario, the application enables people to share live or pre-recorded emotions about a real event even when they are separated (e-SENSE, 2006). David is on a business consultancy assignment in a foreign country. His son, Tom, is going to graduate while his father is away. During Tom's graduation ceremony, his mother records the ceremony with her mobile phone and streams it to her husband. David is overwhelmed with joy as he is able to follow the graduation ceremony from his office computer using an instant messaging (IM) service. They also exchange textual messages using that service. The WSN captures the physiological (via a body sensor network – BSN) and physical characteristics (webcam, voice recorder) of David, to infer his emotional state. The WSN is interfaced with the IM service, so that the conversation is augmented and enhanced with automatic replies such as animations or simple smiles, which are sent to his wife. This in turn will increase her happiness in knowing that Tom's father is sharing the family's happy moments. Later, Tom will be able to relive this moment by visualizing his father's emotions recorded on the video.

### Tom's friends

In this scenario, the network enables communities to share emotions when they exchange multimedia content (e-SENSE, 2006). Senders can receive spontaneous feedback of the emotions felt by the recipient of specified content. Each one can participate more actively to the other's experience. Tom is chatting with a couple of friends using an IM application. During the conversation, they inform him that they are enjoying their trip to Africa and Asia. While they are still chatting, pre-recorded pictures and videos of the trip are sent out to Tom, who is so excited that he keeps asking more and more questions. The positive emotion is picked up by the BSN. The IM application inserts smiles and short audio-clips which he had already configured in the system, or graphic overlays on the video feed (from the webcam) so that his friends can get visual feedback of Tom's feelings.

### Danger warning

In this scenario, alert messages are sent to the required service providers when the user feels he/she is in a dangerous and urgent situation in order to ensure a mobile security monitoring (e-SENSE, 2006). After having worked late at the hospital, Anna returns home at a late night hour. She takes her usual route by underground. She is alone in the train when at one of the underground stops a frightening person gets into her carriage. He starts shouting and behaving erratically. She is so terrified that she cannot use her phone to ring somebody. However, her BSN detects her fear and instructs her mobile phone (acting as a gateway) to send a message to the security officers located at the next train stop, because the embedded location sensor infers Anna location as being on a specific underground train and carriage.

## Personal coaching

Personal coaching addresses self-improvement (e-SENSE, 2006). The application enables the users to get information and advice about the subconscious aspects of his/her own behaviour. David is very interested in his personal development and always wants to learn more about his behaviour. That is why he decides to subscribe to a new service that is keeping an eye on his daily life. David physiological parameters and gestures are monitored constantly. These data are loaded up to a server and stored there together with his time schedule information. All data are analyzed and mapped to make inferences concerning unwanted behaviour during certain situations. At the end of each week the system provides David with a report. This way he can see how his behaviour and mood varies with the situation or people involved. David subscribes to monthly meetings with a coach to improve on the basis of these reports. The system can act as a virtual coach, for example could notify David every time he is tapping his foot in meetings. He can change this annoying behaviour within days and feel more relaxed because his colleagues do not nag him about his tapping anymore.

## Mental counselling

This option of sensitive counselling is dedicated to an increasing mental health problem, that is, depression (e-SENSE, 2006). Mood-sensing here is deployed to support people with minor mental disorders and helps family and friends to actively participate in mental support of the main user of the service. Linda, a friend of Anna, suffers from major depression. When she is feeling bad, she is hardly active, does not go out any more and cries and sleeps a lot. At her last session her psychologist suggested an application to her that will measure her physiological data, and told her that using it could assist her in dealing with the disease. Today she feels bad. Her body sensors detect an unusual level and the system automatically alert closer friends as well as George, her husband, who then gives her a call. She asks him to come back home as soon as possible and he promises to do so. On his way home he gets caught in a traffic jam. Linda's situation gets worse and the system automatically informs her psychotherapist. The psychotherapist calls her immediately to evaluate the situation and tries to console her or give advice on what is to be done. He asks her to take some of her prescribed medication and makes an appointment for a therapy session in the afternoon. Linda just took her pills when Anna stops by. George receives another message that her state is improving.

## Daily travel

In this scenario, a service is provided for the daily commute between workplaces according to the variation of user situations and conditions (e-SENSE, 2006). David is on his way home from work and arrives at the bus stop. He has just missed the bus, and while consulting the schedule via his mobile phone he determines that his next bus is in 10 minutes. He sits down at the bus stop and waits. The e-SENSE system detects his inactivity and suggests some activities for David. He has recently received three e-mails which he could read or he can continue watching the movie he started watching in the morning on his way to work. David decides to use his spare time more efficiently and chooses to check his e-mails.

## 2.1.4  Positioning and animals tracking

Another important application area is related to positioning (Wong et al., 2005): the WSAN could be used to evaluate the exact position, for example, of people or objects in a room.

Moreover, related to this area there is another interesting application, that of animal monitoring (Guo et al., 2006). The use of sensor nodes on the animals could facilitate other monitoring activities: detecting the heat period (missing the day where a sow can become pregnant has a major impact on pig production) and possibly detecting illness (such as a broken leg) or detecting the start of farrowing (turning on the heating system for newborns when farrowing starts).

The following scenarios are related to this area:

- Immersive room
- Real-time relative positioning system
- Hogthrob.

### Immersive room

The LEMe room (CRUISE, WP112, D112.1, 2006) is an intelligent environment with a multi-projection display called LEMeWall which is designed to support the research on innovative 2D/3D mixed reality interaction. The LEMeWall is a tiled display system compound of twelve projectors providing a high-resolution visualization of digital content on a large area screen. To increase the interaction immersion, this setup is complemented by a network of five cameras for body gesture tracking and microphones for voice interaction. This room presents an excellent immersive environment for collaborative design review where several users can interact naturally with digital content in front of the wall. Sensors and actuators can play an important role in the interaction capabilities with the LEMeWall, and one of the first implementations is focused on tracking the position of users within the room so that position-dependent-user interaction with the system may be possible. In the LEMe room intelligent environment, the users are supposed to move freely within the room while they interact with the system. To do so, tracking user position is important to identify how and which users are interacting through speech or gesture commands.

### Real-time relative positioning system

People or object localization is important in a given number of scenarios and applications (CRUISE, WP112, D112.1, 2006). Technologies using GPS have existed from some time now but most of the solutions only work in outdoor scenarios. Localization indoors involves many more difficulties and represents an unsolved problem in many cases, especially when relative positioning to others and to objects is required while in movement. As an example, suppose a group of people, is searching inside a building; it would be most helpful to the group commander to monitor the movements of his(her) (wo)men, along with other relevant parameters, and specially their relative positions. If measures could be made regarding the distance between team members a calculation of their relative positions could be achieved.

### Hogthrob

Hogthrob[5] aims at monitoring sensor-wearing sows. A new law requires pregnant sows to move freely in a large pen. This is a challenge for farmers. The farmer needs to identify the sows that should be placed in the smaller pens among all the pregnant sows. Sows are nowadays equipped with RFID tags. The farmer needs to use a tag reader (possibly physically applied to the RFID tag) to identify the sows. This solution is not practical in large pens. A sensor node with an integrated radio placed on each sow could transmit the sow identification to the farmer's handheld PC, thus alleviating the need for a tag reader. Software running on the sensor nodes could also alert the farmer to a sow entering its heat period (there is a correlation between the movement of the sows and their heat period). Nowadays, sows wear tags but farmers need physical contact with those tags to identify the animal; in addition, farmers are on their own to monitor heat.

## 2.1.5  Entertainment

The entertainment application area provides a wide range of heterogeneous services for leisure activities to generally enrich the respective situations for the users.

The following scenarios are related to this area:

- Lea game show
- At the nightclub
- Virtual mood.

### Lea game show

In this scenario, the application enables the actors of live TV shows to react immediately to user emotional feedback (e-SENSE, 2006). The added value for the audience is that the variation of satisfaction is instantly considered. It enables the audience to interact more intuitively with the content of the programme. Lea and her playmates watch an interactive children's show. They wear a sensor kit measuring their mood, movement and excitement during the course of the show. In the show several sketches and scenes are shown that are varied according to the viewer's moods. Today it is a mystery story and the players are currently deciding how to find out how to catch the thief of an ancient mystical book. The scene is rather long, so the kids get bored. Their bored state and non-excitement with two of Lea's friends going to the kitchen to get some cookies and juice, is constantly sent to the show and the actors receive hints about the overall state of their audience. The actors finish the current scene quickly and start to explore a cave where the thief could be hiding. The kids watching now get really excited and the new scene is enhanced so that Lea and her friends enjoy the thrilling show.

---

[5] See the Hogthrob project website: http://www.hogthrob.dk

### At the nightclub

In this scenario, the application enables the manager of entertainment establishments to promptly adapt the ambiance and animation according to the permanent monitoring of the audience mood (e-SENSE, 2006). Personalized additional services are offered to the clients based on the clients' experiences. Tom spends his night with friends at a nightclub. Each nightclub visitor wears some sensor gear, which is able to infer the mood and situation of a person. When the DJ plays a song, he receives feedback from the crowd in the form of a mood distribution (mood map) and is able to make ambient adjustments of audio and visual effects to enhance the users' experiences. Other ambient adjustments to room temperature and noise level can be made automatically by the nightclub system, based on current user perception. Based on the mood level, the club can decide to give several incentives in order to make people happier (e.g., promotional drink offer). The nightclub also offers a service which automatically remembers a playlist of songs Tom was really enjoying. During or after the nightclub visit, Tom is able to view the playlist and even purchase the digital rights of the songs. While dancing and enjoying himself with his friends, the nightclub system offers him the opportunity via his mobile phone to activate the recording of a personalized video sequence. He follows the suggestion and a close-by camera records a short sequence, which he takes home after the club visit to share this memorable experience with other friends.

### Virtual mood

In this gaming option, the BSN enables the user to project himself into the virtual world of the game and to meet virtual characters that are also animated by online friends (e-SENSE, 2006). Their virtual experience corresponds accurately to their gesture and mood variations. Lea is playing a video game. This is a role-playing game in which she has created a character named Elea who evolves in a virtual world. When she plays, Lea wears a set of motion and physiological sensors that capture her gestures and her mood variations. It enables Lea to animate Elea movements and behaviour according to her real gestures and mental condition. Tonight Lea is experiencing virtual relationships in a realistic world in which she meets other virtual characters that are animated by other online players. She is in a bad mood because of a bad day at school. Elea mimics this mood state and also looks sad in the virtual world. Another player from her list of virtual friends can see that Elea is sad. He proposes to her that they go to the virtual discotheque. In this virtual place, he makes her listen to music he likes and selects the music according to the mood variations of Elea (looks happier when Lea likes the music, sadder when she dislikes it). When she likes an item, the virtual friend starts to dance and suggests Elea join him. At home, Lea follows the virtual friend and dances in reality.

## 2.1.6  Logistics

Logistics is a hot research field for new micro and nano technologies. Using WSANs, a product can readily be followed from production setup until it is delivered to the end user. For this purpose, it may cover a large geographical area and many different entities in order to establish communication between entities involved in the application. Therefore, it may require a high degree of distribution. Also, a manufacturer or company would want to minimize its expenses in order to increase its profit. For a corporation the maintenance cost of a logistics application is an especially significant parameter for deciding to establish it or not. In logistic applications it is not easy to let people interfere with individual components of the system at any time for maintenance purposes. Therefore, providing a fault tolerance system is important. The system must also be able to serve to a scale of hundreds or thousands of inventories as well as to tens of them. Mobility and localization of the components are basic requirements of the applications in this category.

The following scenarios are related to this area:

- Target tracking
- Warehouse tracking
- Management at the department store
- Smart storage.

### Target tracking

The aim of this application scenario is the monitoring of trolleys for baggage in a railway station (CRUISE, WP112, D112.1, 2006). The trolleys are located in some fixed locations distributed in the area. When they are used by passengers and not replaced in the appropriate locations, in general they are distributed over the whole area. Moreover, some trolleys could be taken outside the railway station. The problem we have to solve is to locate the different trolleys and to get an idea if there are some zones in which they are too many or not present. Moreover, we want to know if some of them are taken outside the station. To do this we equip each trolley with an active RFID tag and each fixed location and entrance or exit with a sensor node equipped with an RFID reader. Sensors periodically collect the position information coming from the RFID tags, organize themselves in a WSN and transmit all the information to the final user, the master of the network, located, for example, in an office of the station. If we consider, for example, the Roma Termini railway station, we could install three different fixed locations for trolleys in each platform and one fixed location at each entrance or exit and the master could be located in an office.

### Warehouse tracking

A big company has a warehouse building close to its foundry (CRUISE, WP112, D112.1, 2006). It is used for storing the final products before their delivery to the end customers. The standard packaging used to be containers and drums, the warehouse being able to store up to 100,000 containers. This large storage capacity causes problems to Mr Banfield, the warehouse manager, and to the entire handling department, especially taking into account that when the stock of one particular product increases, it has to be stored out of its corresponding area. Fortunately, they have solved this problem. Thanks to wireless (sensor) ID tags, each container has an electronic reconfigurable identifier, able to transmit the product code, date of production, date of storage and a number of other pieces of information useful to the warehouse manager. Moreover, the WSN employed allows locating the position of the container with a precision of centimetres. This quality and accuracy helps the distribution department to easily classify and check the status and location of every individual drum or container.

### Management at the department store

Tom is working in a supermarket. He is in charge of the management of a department (e.g., drinks) (e-SENSE, 2006). Thanks to WSN he is able to monitor: the information about the products (type, variety, state, condition of the storage, expiry date, quantity of stock in the line); the changes in the products location (time of moves, location, position, etc.); the consumers (profile, time spent in the area or in front of a product, products they are interested in, etc.). He infers the flow of the goods in the supermarket section, the efficiency of his marketing strategy in the department, and he learns about the behaviour satisfaction of the consumers according to the supply he designed. In real time, he observes the way the products impact on the consumer behaviour. Then he adapts his strategy and product display according to shopper behaviour. He learns about the best place for each product and about the most attractive areas in his department.

### Smart storage

Anna's sister Ada manages a storage department in a food shipping company (e-SENSE, 2006). The most critical issues for her is to manage the storage duration of the perishable food stocks and the storage incompatibility between products, to make inventories and to avoid thefts from the storage buildings. Thanks to a WSN-based system and RFID-type tags, Ada's tasks are supported. When a new container arrives in Ada's department, she is easily informed about its content as about the storage constraints related to this content by reading the RFID attached to the container. Also, Ada is informed about the history of the enclosed products by the producer packaging (path, duration of transport and intermediary storage, incidents in cold chain). Then Ada can adapt her own storage strategy according to the condition of the products. Due to important sanitary constraint and new consumer needs, Ada must seal the storage of the product in order to ensure that incompatible products are not placed in immediate contact. Ada can automatically identify the appropriate storage location in the building in order to avoid the risk of contact between products, especially during handling and packaging tasks before dispatching to the clients. Ada is informed in real time about the appropriate containers to be exported in priority considering the sell-by date of perishable foodstuffs. Thanks to geo positioning she can locate the right container to be sent in order to ensure a logical flow of containers and avoid unintentional long-term storage (first-in first-out). Thanks to a direct access to her database, Ada gets real time and accurate inventory levels of the stocks. She can manage proactively the needs of her department without spending time manually assessing her inventory. She can easily prevent potential shortages in stock or overstocking products.

## 2.1.7  Transportation

Applications of this class aim at providing people with more comfortable and safer transportation conditions. They offer valuable real-time data for a variety of governmental or commercial services. It is possible to design different scenarios of transportation applications. For instance, in one scenario cars may communicate with each other in order to organize traffic, whereas in another one traffic conditions of roads may be monitored by installing static entities along highways. The main aim of this kind of applications is to obtain an autonomous transportation system. The nature of transportation system starts from being ad hoc, hence infrastructure-less and mobile. One vital issue for those mobile components is localization. These applications are highly required to run in real time; therefore, the synchronization of the components and the end-to-end delay of the whole system are quite critical for such systems.

Two scenarios are related to this area:

- CORTEX
- Safe traffic.

### CORTEX

In CORTEX car control project (Sihavaran et al., 2004) the implementation system will automatically select the optimal route according to desired time for reaching the destination, distance, current and predicted traffic, weather conditions, and any other information that will be necessary for the purpose. Cars cooperate with each other to move safely on the road, reduce traffic conditions and reach their destinations. Cars slow down automatically if there are some obstacles or they are approaching other cars, speed up if there are no cars or obstacles. Cars automatically obey traffic lights. The principal target of this application scenario is to present the sentient object paradigm for real-time and ad hoc environments. It needs decentralized (distributed) algorithms.

### Safe traffic

Safe traffic project (Svensoon, 2005) aims at the implementation of an intelligent communication infrastructure. This communication system would provide all vehicles, persons and other objects located on or near a road with the necessary information needed to make traffic safer. In addition, all road users should be provided with an accurate positioning device. This idea is simple: reducing the number of traffic-related deaths and injuries.

## 2.1.8 Homes and office

### Smart home

A wonderful idea for home automation is using the ability to turn lights on and off remotely, monitor a sleeping baby without being in the room and having a fresh cup of hot coffee in the kitchen for breakfast (CRUISE, WP112, D112.1, 2006). Smart homes have the capacity to acquire and apply knowledge about human surroundings and adapt in order to improve human experience. It is saturated with computing and communication capabilities to make intelligent decisions in an automated manner. Its intelligent assistants provide interaction with the information web. Its advance electronics enable early detection of possible problems and emergency situations.

A smart home works similarly to an ordinary home. However, it is operated in a way that is more useful or appropriate for the people living there. Where it differs is in the installed communication infrastructure which allows various devices and systems in the home to communicate with each other. It provides an adaptive control of home environment such as heating, lighting and ventilation. It creates an interactive, sustainable and adaptive environment to satisfy the needs of people of all ages. It contributes to a better quality of life by increasing self-control and self-fulfillment. In short, its main objective is to provide easy living and to improve the social environment.

### Smart office

We consider an area covered by some indoor UMTS stations, and the employees working in the offices carrying UMTS mobile devices also equipped with Zigbee air interfaces; these mobile terminals can interact with Zigbee-enabled small devices distributed over the corridors and inside the offices (CRUISE, WP112, D112.2, 2006). Such devices might provide localization and logistic information, and are also able to detect the presence in their immediate neighbourhood of objects such as laptops, printers, pieces of equipment, etc. which have Zigbee-enabled, low-cost devices that communicate with the nodes distributed in the environment. In this scenario every employee can scan the environment to get the information on the localization of movable objects. This can also be done through web services implemented in the intranet serving the building: the user sitting in his/her office will get the requested information through a sequence of links from the lower level (the objects) to the upper level (the access ports bringing the information to the infrastructure and the Internet or intranet).

## 2.1.9   Industrial applications

The following scenarios are related to this area:

- Shopping at the store
- Smart shopping list
- Smart factory.

### Shopping at the store

Anna is going to shop in the department store (e-SENSE, 2006). As she enters the store her intelligent shopping application starts. The shopping list she made during the previous days is uploaded and displayed on her terminal. The system added some items to the shopping list that are needed at home and that Anna did not check (automatic update of the elementary ingredients for cooking at home, updating of products in the fridge and usual products for housekeeping, etc.). The system guides Anna through the store aisles to help her locate the products. The system considers the proximity of the products first and also the freshness constraints: fresh products are collected at the end of shopping. While Anna approaches the products, she gets a ranking of the comparable products that are in the section according to her profile preferences (with criteria like: price, quality, fat-free, organic and allergies to ingredients). It will also include information about the products' prices in other stores. Pointing the RFID reader (integrated in the terminal) at the products she gets additional information about the products such as origin and expiry date. If she takes a product that was not on the list, the device alerts her, for example that it contains nut traces and that it is not suitable for her daughter Lea because of her allergies. The check out and payment are automatic, therefore avoiding the lengthy queues at the check out point.

## Smart shopping list

David's family is equipped with a system that monitors the family consumption of products (fridge, housekeeping products) thanks to a RFID-based system across their home (e-SENSE, 2006). The system gathers information about the products' usage at home, the family preferences and behaviour (at least a geo localization system, but also body sensors enabling to monitor physical needs of each family member). The stock level of the food and housekeeping products at home is low. The system infers that a lot of products are needed. The system also notices that the yoghurt has expired. This time fewer yoghurts will be ordered. Lea ate all the peaches in one day. The system infers that she likes them and that peaches are good for her (profile, age, physical condition). This time more peaches will be ordered. The system detected that Tom has a bad cold. Tissues are added to the list. Friends are invited for dinner next Saturday (David and Anna's diary). The quantity of the products is increased proportionally. The relevant shopping list made by Anna will be added to the automatic list. The automatic list of the missing/needed products is uploaded on the Internet via a server that submits the list to an e-shopping browser. This browser checks the best prices for the list among several retailers in the area. The products are ordered from a retailer who has the best offers. The delivery date and hour are planned according to David and Anna's availability.

## Smart factory

The maintenance of equipment and quality control in a factory (e-SENSE, 2006). Ada controls the processes of the food processing factory using a WSN-based system. The sensors are installed on the machines and take data about temperature, humidity, vibrations, lubrication, substance (e.g., moisture sensors) and other relevant parameters of the machines. Each sensor node is able to communicate its observations through other nodes to the gateway destination where data from the network is gathered and processed. Ada is equipped with a mobile device and she has access to the instantaneous values of the sensors. The remote monitoring, remote control, and data exchange is also enabled.

A quick diagnosis of conditions will be realized. In case a relevant parameter on a machine is approaching a critical threshold, Ada will be alerted on her device by an alert message. For example, sensors could detect that the fruit press machine overheats and that the level of lubrication oil is abnormally low. An alert is sent to Ada with the information about the thresholds and the location of the machine.

## 2.2 Event detection and spatial and time random process estimation

With the aim of creating a very general taxonomy of WSN applications, they could be classified into the following two categories: event detection (ED) and spatial and time random process estimation (PE).

In the first type of application, sensors must detect an event, for example a fire in a forest, a quake, etc. Therefore, no periodic monitoring by the sink(s) is required. Rather, the nodes must be able to detect an event that might happen at any time in any location within the monitored space.

In the second case, the WSN aims at periodically estimating a given physical phenomenon (e.g., the atmospheric pressure in a wide area or the ground temperature variations in a small volcanic site) which can be modelled as a bi-dimensional random process (generally non-stationary). If we consider a typical application related to data gathering from an area and forwarding data to a final sink(s), when sensors receive a service request they take a sample from the environment and transmit it by following an appropriate communication protocol to the final sink(s), which is in charge of collecting all information detected by nodes and estimating the process realization.

The two types of applications are characterized by separate user requirements and impose specific characteristics at the WSN.

In the ED applications the signal processing within devices is very simple, owing to the fact that each device has to compare measurement results to a given threshold and send the binary information to the final sink(s). The density of nodes must ensure that the event is detected with given probability; therefore the network must have a sufficient coverage. The network coverage is related to the sensing range of nodes and the event type. Moreover, distributed localization algorithms could be used to identify the exact position in which the event occurs. The report can be received by the sink(s) with given probability; connectivity issues, related to transmission range of nodes, must be kept under control and the communication protocols must be designed so that the alarm arrives at the sink(s) with a high probability and in a short time.

In case the sink(s) periodically queries the nodes, the frequency of queries must be chosen so that the event is detected with a given probability and must also ensure that the reports reach the final sink(s) in time.

Therefore, the main issues which characterize this type of application are:

- coverage
- distributed localization
- connectivity
- communication protocols (they must ensure low losses of packets and low delays).

Typical requirements for ED applications are therefore:

- minimum probability of coverage
- maximum localization error
- minimum probability of connectivity
- maximum packet loss probability
- maximum packet delivery delay
- and so on.

As far as PE applications are concerned, the density of nodes must ensure that the process is accurately estimated; therefore signal processing is one of the main issues. The samples can be received by the sink(s) with given probability; connectivity of the network, related to transmission range of nodes, must be kept under control and the communication protocols must be designed so that the process estimation error is maintained under a given threshold. Here the sampling frequency must be chosen so that the process evolution is tracked.

Therefore the main issues which characterize this type of application are:

- signal processing
- connectivity
- time synchronization
- communication protocols (they must ensure low process estimation error).

Typical requirements for PE applications are therefore:

- maximum estimation error
- maximum localization error
- minimum probability of connectivity
- maximum packet loss
- and so on.

Tracking (of targets: objects, animals, people, etc.) is sometimes considered as a special case of application. Indeed, tracking can be seen as a special case of ED or PE applications, depending on the specific case. If the location of targets to be tracked must be continuously monitored, then tracking is an application of PE type, where the process to be estimated is given by the position of the object in the space, mathematically represented by a moving impulse. If the target must be tracked so that the network must detect when it passes close to some specific points (such as doors, warehouse entrance, etc.), then the application is of ED type.

In the following, the scenarios identified above are assigned to either ED and/or PE types. Since the two classes are characterized by defined sets of requirements, this categorization will help in defining the main types of requirements that will drive the network design. However, in a few cases the applications are not easily classifiable according to this procedure.

## 2.2.1   Environmental monitoring

### Forest fire detection

This scenario belongs to the ED applications. The aim of the network is to detect a fire and to give the alarm. The deployed sensor network should cover the possible fire point in such a way that the sensors can detect a fire event before it becomes uncontrollable. When a fire-related event, such as sudden temperature rise, is detected, the control centre should be warned with the sensed data and the location of the suspected fire. Finally, the system must be real time, otherwise it will be of no use.

### Flood detection

This scenario belongs to the ED applications. The aim of the network is to detect a flood and to give the alarm. It could be also desirable that the WSN is able to stop the water flooding, such as closing taps or doors (in this case we have a WSAN). Nodes must be deployed in the home so that they could detect the flood and the system must be realtime in order to have more time to act and reduce the damage.

### Structural integrity monitoring

This scenario belongs to the ED applications. The aim of the network is estimating the state of structural health for buildings, bridges or in general for large structures. A SIM system can detect the changes in the structure which can modify its physical and dynamic characteristics. From the user perspective, SIM systems have to be able to assess the timescale and the severity of the changes. Timescale depicts how quickly changes occur, and severity represents the degree of the changes. Since the SIM systems have to prevent disastrous situations, they must provide the requirements of low latency and reliability.

### Glacsweb

This scenario belongs to the PE applications. The WSN aims to measure some spatial processes, like temperature, pressure and subglacial movement.

*MVE-WSN*

This scenario belongs to both the ED and PE applications, because it has two goals. On one hand, the WSN has the aim to monitor volcanic eruptions, that is, to determine the source mechanism and location of an earthquake or explosion (ED). On the other hand, the seismometers are distributed to study the interior structure of the volcano, and differentiate true eruptions from noise or other signals (e.g., mining activity) not of volcanological interest (PE).

## 2.2.2 Healthcare

*Night shift assistant*

This scenario belongs to the ED applications. The WSN must be able to detect dangerous changes in a patient's condition and give the alarm. Moreover, the network must provide to the nurse the exact position of the patient.

*Backup shift assistant*

This scenario belongs to the PE applications. The WSN is located in the body of the patient (BSN) and has the aim to periodically measure vital data and transmit them.

*Acute patient monitoring*

This scenario belongs to the ED applications. In the case of a dangerous change in David's ECG or breathing rate, an alarm will be sent to the hospital, therefore the WSN has to detect a dangerous situation.

*Continuous care*

This scenario belongs to the PE applications. Medical WSNs aim at providing surveillance of the vital physical parameters of a patient, therefore the WSN has to periodically estimate some vital physical parameters and transmit them.

## 2.2.3 Mood-based services

All the scenarios related to this application belong to the PE applications. In all cases, in fact, the WSN has to measure physiological data, such as voice carrier frequency, ECG, breathing rate, etc. to deduce the mood of the person.

## 2.2.4 Positioning and animal tracking

The immersive room and real time relative positioning system scenarios do not belong to the ED and PE types of applications. However, the Hogthrob project is an ED application. Here the WSN allows farmers to track sows and to monitor the start of the heat period. Localization and time syncronization are very relevant issues in this type of applications.

## 2.2.5   Entertainment

All the scenarios related to this application belong to the PE applications. In all cases, in fact, the WSN has to measure physiological data, such as voice carrier frequency, ECG, breathing rate, etc. to deduce the mood of the person.

## 2.2.6   Logistics

### Target tracking

This scenario belongs to the class of ED applications. As stated above, tracking can be a particular case of ED applications when the event (the target) moves and its position needs to be tracked with reference to specific locations. In this case we need to locate the position of the trolleys with reference to specific points in the area. The more stringent requirement of the application is related to the delay with which a packet coming from an RFID tag reaches the final master. As far as fault tolerance is concerned it depends on the information transmitted: in case the information is related to the exit of a trolley the tolerance should be low (the application requires to receive all the packets coming from sensors located at the exits); whereas it could be larger in case the information comes from sensors inside the station.

### Warehouse tracking, management at the department store, smart storage

These three scenarios belong to both the types ED and PE. If we have to track a target that can be moved from one location to another, this is an ED application. On the other hand, we may want to monitor the product, or in some cases to locate its position precisely, and therefore the WSN has to periodically transmit some data related to the product, for example product code, date of production, date of storage or other valuable data. According to the above definitions this functionality is related to PE applications.

## 2.2.7   Transportation

These two scenarios belong to the PE applications. The aim of these scenarios is to make the traffic safer. To do this different information between devices located on the vehicles and on the road are transmitted: this information is, for example, related to vehicles speed, the distance between vehicles, etc. However, these applications must satisfy strong real-time requirements that are not, in general, the main issues of PE applications.

## 2.2.8   Homes and office

The smart home scenario belongs to both the applications ED and PE. When the WSN is used, for example to evaluate the temperature of a room, the estimation of a random spatial process (the temperature, in this case) is provided (PE applications). When, instead, the network is used, for example to monitor a baby or to turn lights on and off, an ED functionality is used.

In the smart office scenario the aim of the WSN is to locate objects (books, for example) in an indoor environment. Therefore, we have once again the tracking of a target, that is, the object (we want to know, for example, if a book is taken out of a room, etc.); thus it is an ED application.

### 2.2.9 Industrial applications

*Shopping at the store, smart shopping list*
These scenarios belong to both the applications, ED and PE. When a product in the fridge or some elementary ingredients for cooking are finished, the system automatically updates the shopping list. This functionality is related to ED. Moreover, in the first scenario the system guides Anna through the store to help her locate the products and when she is in proximity of the products it collects some information related to them: freshness, price, quality, etc. This functionality is related to PE. In the second scenario, where the system gathers information about the products' usage at home, the family preferences and behaviour, a geo localization system and body sensors enabling to monitor physical needs are used (PE).

*Smart factory*
These scenario belongs to both the applications, ED and PE. Sensors installed on the machines have to take data about temperature, humidity, vibrations, lubrication, substance and other relevant parameters of the machines; therefore the estimation of some processes is provided (PE). Moreover, in case a relevant parameter on a machine is approaching a critical threshold, an altered message is sent to Ada (ED).

## 2.3 The hybrid hierarchical architecture

In the past few years the development of new technologies and the standardization of new air interfaces both for infrastructure-less and infrastructure-based wireless networks (such as, for example, wireless fidelity (Wi-Fi), worldwide inter-operability for microwave access (WiMAX), Bluetooth, Zigbee, etc.) has increased the interest of researchers towards radio systems composed of sub-parts implementing separate technologies and network paradigms (for instance, ad hoc and cellular networks). We denote these systems as WHNs. They are characterized by the coexistence of several communication technologies and the presence of devices with different functionalities and computational capabilities. Several network architectures can be devised in such context, depending on the type of application area and of converging networks. In this framework we refer in particular to a network architecture denoted as HHA, which in fact is a particular case of WHN. The HHA has been selected as reference architecture in the EC-funded project CRUISE. It represents a particular case of network architecture: it is strictly hierarchical with four levels, where the wireless nodes have some specified (and some unspecified) characteristics depending on the level they belong to. As such, the HHA does not aim at being general: there might be other network architectures based on different paradigms and topologies that deserve similar attention. However, it is recognized that some of its features make it worthwhile considering the HHA as a realistic and innovative reference scenario,

In other words, though specific and therefore limiting the area of investigation, the HHA might represent a reference architecture for many WSAN applications, and might therefore help some standardization process, or support the development of widely accepted concepts and tools.

For this reason, the HHA is discussed in this book.

The HHA is reported in Figure 2.1. At level zero, radio access ports (i.e., fixed stations covering the area through radio access networks (RANs) using air interface standards such as general packet radio service (GPRS) or UMTS or Wi-Fi) provide access to mobile terminals (denoted here as mobile gateways, level one) usually carried by people. These mobile devices can also be connected through a different air interface (e.g., Zigbee or Bluetooth) to a lower level of wireless nodes (level two) with limited energy and processing capabilities which can find access to the fixed network only through the gateways. These wireless nodes are distributed in the environment and provide information taken from it; they might be sensor nodes (SNs), or actuators or beacons providing localization data; moreover they interact through possibly different air interfaces with tiny devices at level

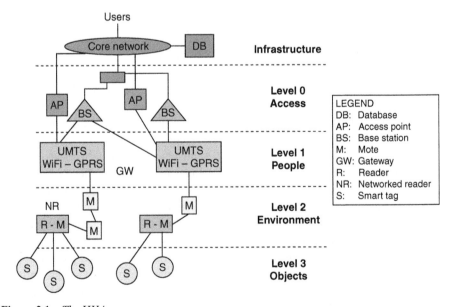

**Figure 2.1**  *The HHA*

three (e.g., smart tags or very-low-cost sensors) which are part of movable objects (e.g., printers, books, tickets, etc.). The hierarchy is thus composed of four levels. Under a network topology perspective, this scenario defines a forest of (possibly disjointed) trees with heterogeneous radio interfaces at the different levels. Note that if the environmental level is connected through a tree-based topology, then the number of levels in the hierarchy further increases as level two is subdivided into sub-levels.

One of the most peculiar aspects of the HHA is in the fact that mobile terminals using wireless local area networks (WLANs) in indoor environments or cellular networks outdoor might be used to gather the data sensed by SNs and to transport them towards the data storage systems and applications servers. The user terminals of these types of networks might therefore act as mobile gateways collecting the data from the WSAN and forwarding it. This scenario can be also considered as an extension of the traditional WSAN scenario where a sink collects information from sensor nodes distributed in the environment, through wireless links. In this case, there are multiple sinks, and their locations are not known. Moreover, the sinks are in fact gateways forwarding the information collected to higher levels through heterogeneous wireless interfaces characterized by different parameters in terms of transmit power, capacities, etc. The interest towards WHN is clearly increasing in the research arena the last few years. For example, the information society technologies (IST) project e-SENSE is working on the integration of WSNs with beyond 3G networks, and the network of excellence CRUISE has selected the HHA as the reference scenario of the project.

Basically, the main features of the HHA which are intrinsically included in its definition are the following:

- The HHA is heterogeneous: different radio communication techniques are involved at the various interfaces between the levels (e.g., UMTS between 0 and 1 and Zigbee between 1 and 2) and also within a given level the nodes might use heterogeneous air interfaces (e.g., Bluetooth or Zigbee at level 2).
- The HHA is hybrid: the air interfaces and devices implement different communication paradigms (e.g., mesh or flat topologies) at the different levels.
- Nodes at level 1 are mobile gateways, such as laptops or cellular phones carried by people and therefore are not specifically deployed for the aim of collecting data from the environment; rather, for cost reasons, these devices are exploited for such aims while used by people for their personal specific use (telephone conversations, web browsing, etc.).
- Multiple nodes with possible overlapping communication zones are present at all levels.
- The HHA is highly dynamic, because of the movement of nodes at two levels of the structure, namely levels 1 and 3.
- The HHA is based on a fixed hierarchy: nodes at a given level can only attach to nodes at the immediate adjacent level. However, in some cases this feature might be removed.

## 2.3.1   Categorization of the application scenarios according to the HHA

The application scenarios described above could be analyzed to evaluate if they some-
how reflect these features and they are coherent with the HHA or at least part of it (some
levels might be absent). As some of the main characteristics of the HHA are not fully
specified, there are many possible variations, or realizations, of the HHA. Let us expand
this concept. We now summarize the main characteristics lying behind the four levels of
the HHA. While the description of all characteristics of the wireless nodes belonging
to each level is neither possible nor meaningful, we refer to those characteristics that
have a significant impact on the requirements posed to the communication algorithms
and protocols implemented in the devices (for instance, the node mobility, its known or
unknown location, etc.).

- Level 0 (Access). Multiple nodes in fixed known locations, selected through an overall
  planning phase, possibly connected with mesh topology, partially overlapped coverage areas; interface to infrastructure with similar communication capabilities as interface towards level 1; no energy constraints; nodes have IP address; no data processing
  capabilities.
- Level 1 (People). Multiple nodes in unknown (or loosely known) locations, mobile
  (low or high speed) or still, uncoordinated, not directly connected through any topology, large potential overlap of communication areas; possible significant difference
  between communication capabilities towards level 0 and level 2 (more limited); weak
  energy constraints; nodes have IP address; strong data processing capabilities.
- Level 2 (Environment). Multiple nodes in known or unknown positions, either planned
  or unplanned, connected through mesh or tree topologies, large potential overlap of
  communication areas; possible significant difference between communication capabilities towards level 1 and level 3 (more limited); strong energy constraints; nodes with
  global or local network address; weak data processing capabilities.
- Level 3 (Objects). Multiple nodes in unknown locations, mobile (low speed) or still,
  uncoordinated, possibly connected through tree topology (objects containing objects);
  large potential overlap of communication areas; strong energy constraints; nodes
  might have or not have global or local address; no data processing capabilities.

Basically, at each level we are recognizing the following main features characterizing or
not the nodes:

1. Mobile or still
2. Known or unknown locations
3. Mesh, or tree, or absent topology connecting nodes within the level
4. Coordinated planning or uncoordinated spatial distribution
5. Strong or limited overlap of communication areas
6. Different or similar communication capabilities at the two interfaces
7. Strong, or weak, or absent energy constraints

8. Global, or local or absent network node address
9. Strong, or weak, or absent data processing capabilities

Clearly such features can significantly influence the selection of communication proto-cols and algorithms at the various air interfaces. To clarify, let us consider the interface between levels 1 and 2 (basically, between sinks and sensor or actuator nodes):

1. Whether the nodes are mobile or still has an influence over the physical layer, since some modulation/demodulation schemes suffer more than others the effects of moving link ends; moreover, if node move, the topology of nodes can be highly dynamical and therefore suitable topology maintenance protocols must be implemented.
2. Whether the nodes are in known or unknown locations deeply impacts the procedures for topology formation, and the selection of routing algorithms.
3. The use of mesh, or tree, or absent topology connecting nodes within the level impacts the addressing mechanisms and routing algorithms to be implemented.
4. Whether the nodes are deployed according to a coordinated planning, or their spatial distribution is the result of an uncoordinated process, can heavily impact the perform-ance of the network protocols.
5. Localization and time synchronization algorithms are deeply influenced by the strong or limited overlap of communication areas among separate nodes; also, different cat-egories of routing algorithms can be chosen depending on this feature.
6. Whether the nodes at a given level have different or similar communication capabil-ities at the two air interfaces can determine the need to properly schedule transmission and reception intervals at the two interfaces.
7. The protocol selection requires knowledge about strong, or weak, or absent energy constraints.
8. Whether global, or local or absent network node address is present plays a key role in the choice of the transport layer paradigms to be used.
9. Clearly, signal processing procedures are deeply influenced by the data processing capabilities of nodes.

The generation of a proper classification of the applications according to such features in the context of the HHA is therefore very relevant.

Let us now create a notation for the possible values taken by the nine features character-izing each level (0 to 3):

1. Node mobility – M/S
2. Locations – K/U
3. Topology – M/T/A
4. Distribution of nodes – C/U
5. Overlap of communication zones S/L
6. Communication capabilities at the two interfaces – D/S

7. Energy constraints – S/W/A
8. Node network address – G/L/A
9. Data processing capabilities – S/W/A

To exemplify for the sake of clarity, node mobility = M means the nodes are mobile, S stands for still, etc.

At any level $j$ ($j = 0$, 1, 2, or 3) nodes characteristics can be thus represented by a sequence of nine alphabetical values (e.g., j-MKAUSDSAS). This sequence of nine alphabetical values can be extended if one wants to take account of other characteristics of the wireless nodes (e.g., hardware limitations). Now, we denote a specific variation of the HHA based on the sequence of features characterizing the nodes at the four levels. As an example, the following sequence denotes a realization of HHA with all four levels represented: 0-SKMCLSAGA 1-MUAUSDWGW 2-SUTUSDSLW 3-SUTUSDSAW. This can also be more efficiently represented through a table reporting inside columns the values of the nine main features described above for the nodes at the various levels, with the rows indicating the level in the hierarchy.

If an application scenario reflects part of the HHA, then it might be classified by defining the proper set of alphabetical values for the levels pertaining to the scenario. By doing so, every scenario would be represented by a set of basic features which are the ones that drive the selection of communication protocols to be used. Therefore, the potential advantage of this would be the creation of a procedure useful to guide the selection of protocols to be implemented in a network, according to application requirements. Clearly, this will be not possible for all scenarios.

Some of the application scenarios described above can be easily categorized according to the HHA reference, provided that the number of levels is reduced, as not all of them are present. In a few cases, all four levels are included in the scenario.

For one of the scenarios described in this chapter, taken as an example, the table representing its features according to the HHA reference is reported in the following.

**Table 2.1**  *Four levels of node features*

| – | I | II | III | IV | V | VI | VII | VIII | IX |
|---|---|----|-----|----|---|----|-----|------|----|
| 0 | S | K | M | C | L | S | A | G | A |
| 1 | M | U | A | U | S | D | W | G | W |
| 2 | S | U | T | U | S | D | S | L | W |
| 3 | S | U | T | U | S | D | S | A | W |

### Forest fire detection

In this scenario, three levels are present: 0 (if sinks are connected to UMTS infrastructure), 1 and 2. The features of nodes at the various levels is reported in the table.

**Table 2.2** *Three levels of node features*

| – | I | II | III | IV | V | VI | VII | VIII | IX |
|---|---|----|-----|----|---|----|-----|------|----|
| 0 | S | K | T | C | L | S | A | G | S |
| 1 | M | U | A | U | L | D | W | G | W |
| 2 | S | U | T | U | S | – | S | L | W |

# 3

# Channel modelling

## 3.1 Introduction

The ultimate performance limits of a wireless communication system, as well as the performance of practical systems, are determined by the channel in which it operates. Realistic channel models are thus of utmost importance for system design and testing.

Many works in the WSANs scientific literature assume deterministic distance-dependent and threshold-based packet capture models. In other words, all nodes within a circle centred at the transmitter, with given radius, can receive a packet sent by the transmitting one; if a receiver is outside the circle, reception is impossible. While the threshold-based capture model, which assumes that a packet is captured if the signal-to-noise ratio (SNR) (in the absence of interference) is above a given threshold, is a good approximation of real capture effects, the deterministic channel model does not represent realistic situations in most cases. The use of realistic channel models is therefore of paramount importance in wireless systems. This aspect will be more evident in Chapter 4 where the impact of channel models on network connectivity is investigated.

This chapter provides a discussion on the narrowband and wideband channel models to be used in WSANs performance assessment. As narrowband channels are regarded, some experimental results performed with nodes using the IEEE 802.15.4 physical (PHY) layer standard at 2.4 GHz industrial scientific medical (ISM) band, deployed in different environments (grass, asphalt, indoor, etc.), are reported. The measurements provide inputs for understanding the basic aspects of narrowband propagation in typical WSANs scenarios at 2.4 GHz.

As will be explained in Section 6.2, the ultrawide bandwidth (UWB) technology is gaining a growing interest for WSN applications due to its capability to achieve high data rates, with potentially low complexity, as well as high precision positioning in harsh propagation environments. UWB propagation channels show fundamental differences from conventional narrowband propagation in many aspects. In Section 3.4, a short overview on UWB channel modelling is presented.

## 3.2 Basics of electromagnetic propagation

The wireless channel, as is well known, often shows unpredictable behaviours. Terrestrial communication links (i.e., at the level of the ground, or underground/underwater) are always affected by the presence of elements that interfere with the propagation of electromagnetic (e.m.) waves. In free-space conditions these effects can be neglected and the well-known Friis formula can be used to express the *power loss* between transmitter and receiver antenna connectors as a function of the distance $d$ and the e.m. wavelength $\lambda$

$$L_{\text{free}}(d) = \frac{1}{G_{\text{t}} \cdot G_{\text{r}}} \cdot \left( \frac{4\pi d}{\lambda} \right)^2, \tag{3.1}$$

where $G_{\text{t}}$ and $G_{\text{r}}$ represent the transmit and receive antenna gain, respectively.

However, in many applications of wireless systems the presence of the elements (the ground, first of all) interfering with the propagation of e.m. waves, can not be neglected. They introduce several effects which make the power loss a random process (Rappaport, 1996).

The *shadowing* or *large-scale* fading effect is caused by the obstruction of the main signal path due to obstacles along the path, either moving or still. In particular, e.m. waves tend to be affected by objects having size at least in the order of the wavelength. For example, any obstacle with size larger than about 10 cm in the 2.4 GHz band can potentially be a source of deviation from the Friis formula if it is located inside the first Fresnel ellipsoid. This happens when the obstacle intersects the ellipsoid whose foci coincide with the transmit and receive devices and whose circular cross-section has radius

$$r_1 = \sqrt{\frac{\lambda d_1 d_2}{d_1 + d_2}}, \tag{3.2}$$

with $\lambda$ being the wavelength and $d_1$, $d_2$ the distances between the obstacle and the two endpoints, respectively. The received power level variations happen in a scale of hundreds of wavelengths.

The *multipath* phenomenon introduces either constructive or destructive interference between separate signal paths thus causing additional signal fading. The time coherence of such a process mainly depends on the obstacles, transmitter and receiver speeds. This effect has a small-scale behaviour (*small-scale fading*), that is, the received power level variations happen in a scale in the order of the wavelength $\lambda$. In addition, the presence of multiple paths can make *frequency selective* effects relevant depending on the signal bandwidth: the signal distortion caused by the relative delays between separate signal paths can play a significant role on receiver performance.

As a rule of thumb, one can assume that no frequency selectivity is induced by the wireless channel if the maximum delay between signal paths is smaller than the inverse of signal bandwidth. For instance, if the latter is 5 MHz, frequency selectivity effects are negligible if the the delay is not larger than 0.2 µs.

## 3.2.1   Narrowband channel models

In a narrowband system the *path loss* at a distance $d$ is conventionally defined as

$$L_0(d) = \frac{P_t}{\mathbb{E}_l\{\mathbb{E}_s\{\widetilde{P}_r(d)\}\}},  \tag{3.3}$$

where $P_t$ is the transmit power, $\widetilde{P}_r(d)$ is the short-term received power,[1] as seen at the antenna connectors of transmitter and receiver, respectively. The statistical expectations $\mathbb{E}_s\{\cdot\}$ and $\mathbb{E}_l\{\cdot\}$ are taken over an area that is large enough to allow averaging out of small-scale and large-scale fading, respectively.[2]

The distance dependence of the path loss is usually modelled by a conventional power law, thus in dB scale

$$L_0(d) = k_0 + 10\,\alpha\log_{10} d,  \tag{3.4}$$

where $k_0$[3] is the path loss in dB at the reference distance of 1 metre evaluated using (3.1). The parameter $\alpha$ is the path loss exponent depending on the environment which typically assumes values in [2, 4] (Rappaport, 1996).

In the 2.4 GHz ISM band, the following path-loss model (*break-point model*) is often adopted

$$L_0(d) = \begin{cases} 40.2 + 20\log(d), & d \le 8\,\text{m}, \\ 58.3 + 33\log(d/8) & d > 8\,\text{m} \end{cases}  \tag{3.5}$$

This model assumes line-of-sight (LOS) propagation for the first 8 metres.

The scientific literature reports many results on wireless channel modelling over decades, because of the large-scale deployment of wireless networks for a long time. However, almost all experiments and models were focused on cellular or broadcast systems, where one node in the link can be placed close to ground, while the other is located further from the ground (e.g., on a mast). The case of WSANs is different: nodes can be deployed on the grass, on asphalt, on the walls. In every link, both ends can be close to ground. As a result, the well-known channel models used in the past for cellular and broadcast systems can not be used. Some papers have been recently published reporting experimental results for WSANs. However, a lot of research is still needed to provide stable models.

---

[1] The short-term received power is often referred to as instantaneous received power.
[2] The large-scale averaging is usually performed in log scale.
[3] Also known as intercept at 1 m in terms of dB.

For example, the effect of foliage has been investigated in Benzair, 1995 where the following model for the path loss in dB is proposed for the 1–4 GHz band

$$L_0(d) = L_e + L_{free}(d) \tag{3.6}$$

where

$$L_e = \alpha_f \cdot d_f$$
$$\alpha_f = a\, f_{[GHz]}^b. \tag{3.7}$$

The $L_e$ is the excess path loss due to the presence of the tree, $d_f$ is the direct distance through the foliage, $\alpha_f$ is the differential attenuation due to the foliage in dB/m (modelled as power dependent on the frequency $f$ expressed in GHz). Constants $a$ and $b$ are given in Benzair, 1995 for different seasons. Other path loss models accounting for the effect of foliage can be found in Batariere, Blankenship, Kepler & Krauss, 2004; Al-Nuaimi & Stephens, 1998; Weissberger, 1981; Dalley, Smith & Adams, 1999.

In indoor environments walls also impact the path loss. This is often modeled through a *multi wall* path loss model, namely wall attenuation factor (WAF), given by

$$L_0^{(wall)}(d) = L_0(d) + \sum_{i=1}^{N_{wall}} A_i^{(wall)}, \tag{3.8}$$

where $N_{wall}$ is the number of walls between transmitter and receiver, and $A_i^{(wall)}$ is the attenuation in dB introduced by the $i$th wall (Motley & Keenan, 1988; Rappaport, 1996). In the case of $A_i^{(wall)} = A^{(wall)}$ for all $i = 1, \ldots, N_{wall}$, the second term of (3.8) becomes $N_{wall}A^{(wall)}$.

Usually, small-scale fading is averaged out by the modulation/coding schemes, hence only the shadowing effect is accounted for explicitly during the wireless network planning. To this purpose it is worthwhile to define the local average received power $P_r(d) = \mathbb{E}_s\{\widetilde{P}_r(d)\}$, where the expectation is taken with respect to the small-scale fading, and introduce the (average) power loss (in linear scale)

$$L(d) = P_t/P_r(d), \tag{3.9}$$

which will be used in the following chapters for network performance characterization and design. Without loss of generality, the power loss, in dB scale, can be expressed in the following form

$$L(d) = L_0(d) + S, \tag{3.10}$$

where $L_0(d)$ is the path loss given, for example by (3.4), (3.5), (3.6) or (3.8), and $S$ is a random variable (r.v.) that accounts for random fluctuations due to large-scale fading (shadowing).

It models the randomness of the geometry (presence of obstacles, etc.) (Fanimokun & Frolik, 2003).

The model in (3.10) can be applied to a wide range of statistical fading models present in the literature (e.g., log-normal, Rayleigh, Rice, or combinations of them). In Section 3.3.7, the widely used log-normal model will be introduced.

## 3.3 Experimental activity aimed at modelling the wireless channel at 2.4 GHz for WSANs

The nodes in WSANs are often deployed in the presence of obstacles, on the ground in outdoor scenarios or on the walls in indoor scenarios. These are clearly non-ideal propagation environments which cause random channel fluctuations and dependence of power loss on direction of transmission. On the other hand, distinction between shadowing and fading effects in WSANs, where links are in most cases not longer than 10–100 metres, is not simple, as separation of multipath and obstructive effects is a complex task. Therefore, in the following measurement results, the randomness of the received power (and hence, the power loss) is characterized through a unique description, without any intention to separate the two components, unlike is often made on other wireless systems, for example cellular networks.

Both Zigbe and Bluetooth devices use the 2.4 GHz ISM band (please refer to Chap. 6 for more details). Owing to the availability of commercial platforms for several years, most results reported in the literature refer to such band, while very little has been done at different frequencies. The following subsections report experimental results performed with nodes working at 2.4 GHz, deployed in different environments (grass, asphalt, indoor, etc.). In particular, the nodes use the IEEE 802.15.4 PHY layer standard. The measurements have been obtained by exploring the 2.4 GHz ISM band with channel bandwidth of 5 MHz, and spanning over 16 channels.

The standard experiments performed are oriented to the measurement of the received power and the packet error rate (PER) of a link where both transmitting and receiving nodes were placed at the same height from the ground, at specified distance, both having omnidirectional antenna (i.e., $G_t = G_r \simeq 1$). Figure 3.1 shows the 16 frequency carriers used in each experiment (channel 1 corresponds to 2405 MHz up to channel 16 which corresponds to 2480 MHz). At each frequency, 1000 packets were transmitted continuously and the results reported refer to such population: the received power has been averaged over the 1000 samples (one per packet), while the PER has been evaluated over the 1000 receptions. Thus, values of PER below 10% are meaningless because we conventionally assume that at least 100 packets must be received in error in order to have a sufficient statistic. Finally, all measurements were repeated with three different values of transmit

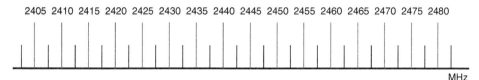

2405 2410 2415 2420 2425 2430 2435 2440 2445 2450 2455 2460 2465 2470 2475 2480

MHz

**Figure 3.1**    *The 16 frequency carriers used in each experiment*

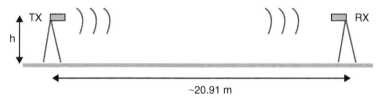

~20.91 m

**Figure 3.2**    *Measurement scenario over the grass*

power: 3.6 dBm, 0 dBm (nominal power) and $-16.6$ dBm, while receiver sensitivity (defined at PER $= 0.01$) was $-92$ dBm.

Conclusions on the measurement campaigns will be given in Section 3.3.7.

## 3.3.1   Measurements over the grass

Figure 3.2 shows the measurement scenario in this case. The devices were located $h = 80$ cm above flat ground. With a link range of about 21 metres, the Fresnel first ellipsoid is not obstructed by the ground and almost LOS communication might be expected.

The PER is reported in Figure 3.3 as a function of the 16 channels used. At minimum power ($-16.6$ dBm) the PER shows values significantly larger than zero. However, these values range from zero to more than 90%. Figure 3.4 shows the values of received power for the same set of experiments.

Having in mind that the receiver sensitivity is equal to $-92$ dBm, the values of average received power at minimum transmit power reported in Fig. 3.4 clearly motivate PER values above 0.01. The reason for values of PER in Fig. 3.3 much above 0.01 are to be related to the non linear relation between PER and the measured received power values which are affected by small errors caused by the measurement equipment.

Let us also note that the values of received power do not change significantly when spanning the 16 channels. This is basically because there are no frequency-selective effects in such an environment and all frequencies behave in the same way. The 16 experiments at fixed transmit power (spanning the 2.4 GHz band) were conducted in a relatively short period of time (about two seconds); therefore, strong time coherence was kept between two measurements at two different frequencies.

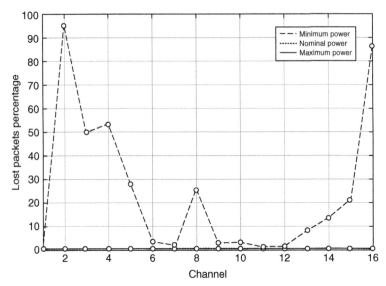

**Figure 3.3**  *PER as a function of the 16 channels over the grass*

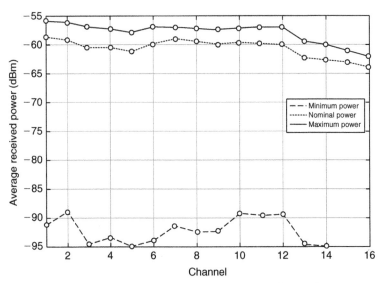

**Figure 3.4**  *Average received power as a function of the 16 channels over the grass*

Finally, it is worth noting that the average (over the 16 channels) values of received power, when changing the transmit power level, change significantly and not proportionally. In other words, the power loss given by the difference between the transmit power and received power in logarithmic scale is not identical in the three cases. For the maximum and nominal powers, it is about 60 dB, while at minimum power it is about 76 dB. The large difference is due to the fact that the three measurements were performed under the same topology, but at different times, with many seconds needed to set a new power level.

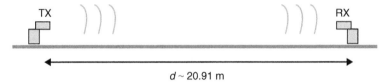

**Figure 3.5**   *Measurement scenario over aspahlt*

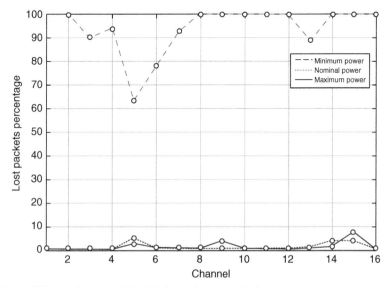

**Figure 3.6**   *PER as a function of the* 16 *channels over asphalt*

Therefore, there is no time coherence between the three measurement campaigns. This shows that random channel fluctuations which are not related to the link distance variations can play a very significant role. In fact, assuming 0 dB gain for both antennas, the free-space loss obtained by applying the Friis formula should be equal to about 67 dB, a value which lies in between those measured. The channel fluctuations in this case might be caused by the wind moving the grass, the tree leaves and the bushes that were located in the close neighbourhood of the communication link.

## 3.3.2   Measurements over asphalt

The next experiments are with nodes placed on flat asphalt, about 20 cm above the ground, as shown in Figure 3.5, while Figures 3.6 and 3.7 show the PER and received power in such condition.

The link performance is generally worse in this case with respect to the previous measurements. This is for two reasons: the nodes are closer to ground and the Fresnel first ellipsoid is not free in this case. Moreover, the asphalt, with a reflection coefficient much

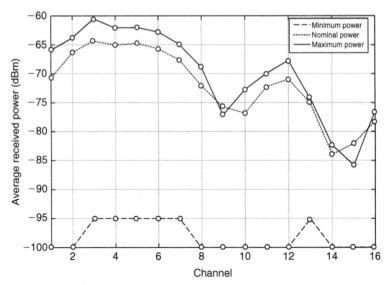

**Figure 3.7**  *Average received power as a function of the 16 channels over asphalt*

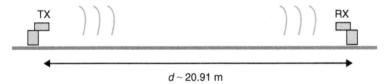

$d \sim 20.91$ m

**Figure 3.8**  *Measurement scenario on grass*

larger than that of the grass, causes stronger multipath components. This is testified also by the much larger variance of results obtained at different frequencies, with the same transmit power: by changing the frequency carrier, the power loss can vary significantly because of a strong path reflected from the ground. In other words, the frequency coherence is much smaller in this situation with respect to the case with grass.

### 3.3.3  Measurements on grass

The next set of experiments over putting the devices much closer to ground, a few cm above (see Figures 3.8, 3.9 and 3.10). A larger power loss is obtained with respect to the case of devices at a higher height from the ground; on average, the attenuation is 5 dB larger. This is the contribution of the ground which largely impacts the e.m. propagation.

### 3.3.4  Measurements over ground at different heights

To reinforce the concept that the height above ground plays a significant role, a set of experiments was performed at different heights above ground, over grass and asphalt. Figures 3.11 and 3.12 show the received power in the two cases, respectively, with both devices placed

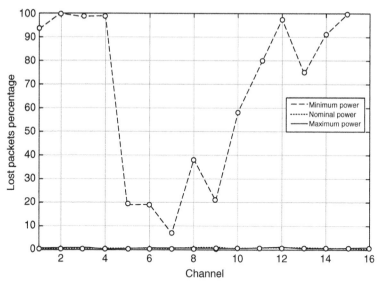

**Figure 3.9** *PER as a function of the* 16 *channels on grass*

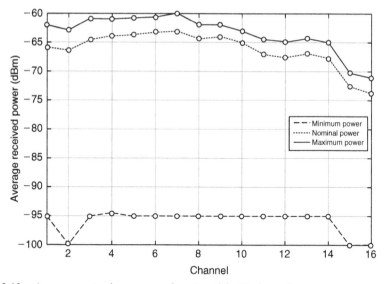

**Figure 3.10** *Average received power as a function of the* 16 *channels on grass*

1 or 0.8 metres above ground. Results clearly show how significant is the impact of the antenna height.

## 3.3.5 Measurements in a parking lot

To gain deeper knowledge of the channel behaviour in urban environments, some measurements were performed in a parking lot, with devices located in the close neighbourhood

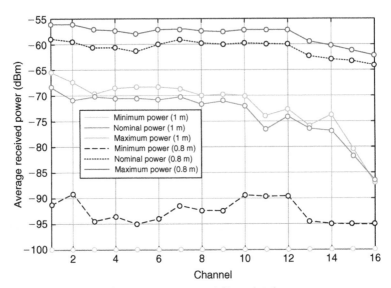

**Figure 3.11**   *Average received power over grass at different height*

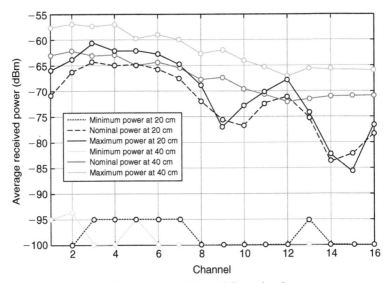

**Figure 3.12**   *Average received power over asphalt at different height*

of cars. In particular, nodes were placed at a height above ground set at 6, 80, 180 cm (see Figure 3.13). In the first and third cases, a direct path is present between transmitter and receiver, while this is not true for the middle case. The experiments were performed with an increasing number of cars between the nodes (see Figure 3.14).

Figures 3.15, 3.16 and 3.17 show the average received power for the three antenna heights, when the number of cars range from 1 to 4. Clearly, when a direct path exists, if the

**Figure 3.13**    *Node placement for urban environment characterization*

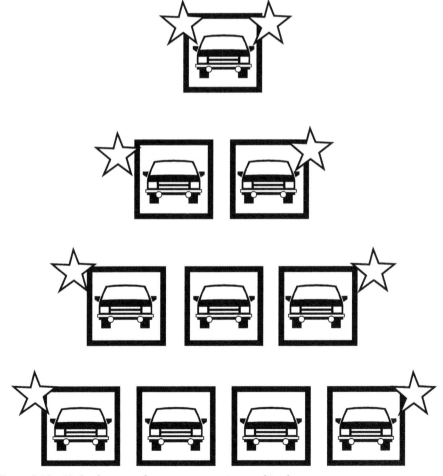

**Figure 3.14**    *Node placement for measurements in a parking lot*

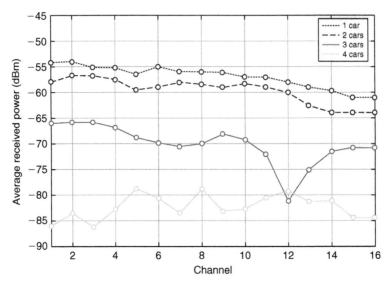

**Figure 3.15**   *Average received power with antenna at 6 cm*

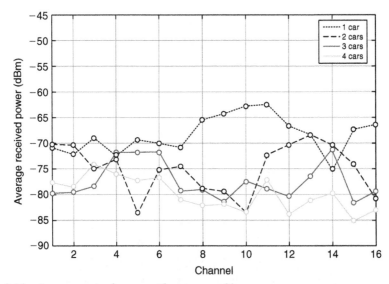

**Figure 3.16**   *Average received power with antenna at 80 cm*

number of cars is larger (then the link distance is larger) the received power is smaller. The same is not true at 80 cm above ground; in this case, the multipath phenomenon plays a role which is even more important than distance.

## 3.3.6   Measurements in indoor environments

To check channel behaviour in indoor environments, some measurements have been performed in an office building, whose map is reported in Figure 3.18, with transmitter and receiver located in a corridor.

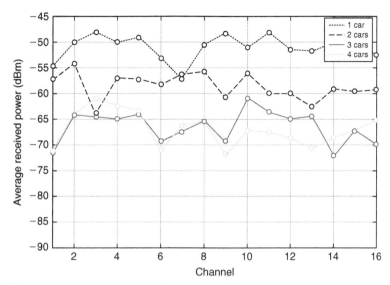

**Figure 3.17**  *Average received power with antenna at* 180 *cm*

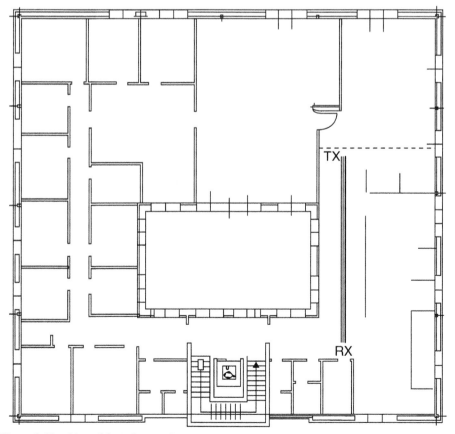

**Figure 3.18**  *Map of the indoor environment*

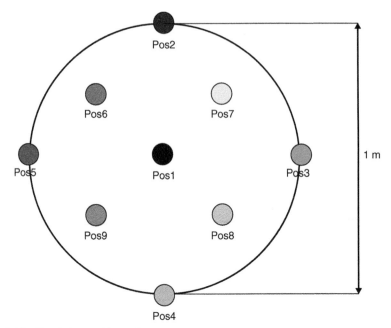

**Figure 3.19**    *Receiving position points*

As in indoor environments the relative distance between objects (walls, doors, etc.) is much smaller than outdoor, so it is important to generate measurements that are representative of an average situation. Thus, it was decided to move the receiving node within a circle having a diameter of 1 metre and centred in 'RX' of Figure 3.18; all measurements have been performed in nine different locations within the circle as shown in Figure 3.19.

The link distance was short enough (about 10 metres) to have all receive power levels above the receiver sensitivity; so the measurements were aiming at deriving a statistics of the received power obtained by varying the multipath contributions. In fact, the movement of the receiver node was 'simulating' the movements of objects (doors opened or closed, people walking, etc.) in the environment. As a result, the power measurements can provide an idea of the amount of randomness generated by the non-ideal effects encountered by the e.m. waves travelling in such a harsh environment.

Figure 3.20 shows the results obtained. More than 20 dB of difference is found between the best and worst result.

### 3.3.7   Conclusions: A narrowband channel model for WSANs

A suitable comparison between the measurements performed and some simple analytical expressions has been conducted for all environments considered: with nodes over asphalt, grass, in a parking lot, and in indoor environments. It was found for the received power in logarithmic scale that in general a Gaussian model can approximate the measurements

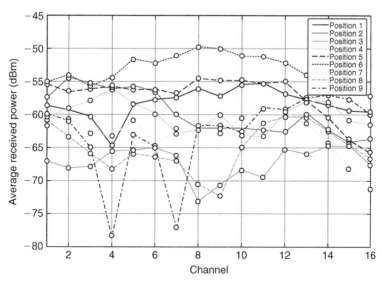

**Figure 3.20** *Average received power for the different positions*

fairly well, with different values of the standard deviation. In general, there was no deterministic dependence on distance as testified by all measurements reported in this chapter.

Some papers in the literature report results achieved in similar environments, and the Gaussian model seems to be accredited. Note that since the scenario is usually stationary, the assumption of a (slow-varying) shadowing environment is acceptable. Therefore, the following *log-normal* model is widely adopted in the literature to describe the power loss, in dB, between two communicating nodes

$$L(d) = k_0 + 10\alpha \log_{10} d + S = k_0 + k_1 \ln d + S, \qquad (3.11)$$

where $k_1 = \alpha 10/\ln 10$ and $S$ is a shadowing sample which is assumed to be Gaussian distributed (i.e., log-normal distributed in linear scale), with zero mean and standard deviation $\sigma$ (*shadowing spread*). The latter usually takes values between 2 and 3 for rural environments (above grass) while indoor it can also take values around 5.

## 3.4 Ultrawide bandwidth channel models

The most widely accepted definition of a UWB signal is a signal with instantaneous spectral occupancy in excess of 500 MHz or a fractional bandwidth of more than 20%. The fractional bandwidth is defined as $B/f_c$, where $B = f_H - f_L$ denotes the $-10$ dB bandwidth and $f_c = (f_H + f_L)/2$ is the centre frequency, with $f_H$ being the upper frequency of the $-10$ dB emission point, and $f_L$ the lower frequency of the $-10$ dB emission point. A way to generate such signals is by driving an antenna with very short electrical pulses with duration in the order of one nanosecond or less. Different from narrowband channels, due to the very large signal bandwidth, the antennas in UWB systems determine a significant pulse-shaping

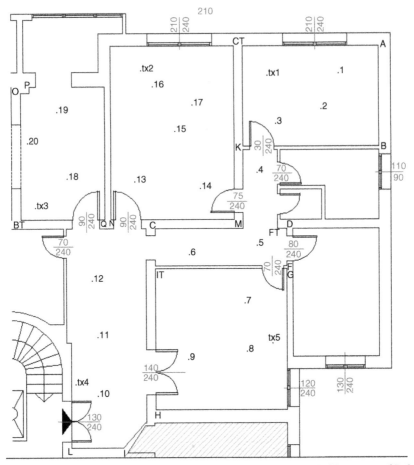

**Figure 3.21**   *The scenario considered for UWB measurements at WiLAB, University of Bologna, Italy*

filtering. The transmission of ultra-short pulses can potentially resolve an extremely large number of paths experienced by the received signal, especially in indoor environments.

A measurement campaign has been performed to characterize the UWB channel behaviour in a typical office indoor environment shown in Figure 3.21 at WiLAB, University of Bologna, in cooperation with the staff of Professor Moe Win from Massachusetts Institute of Technology, Cambridge. UWB pulses are transmitted from location tx1 in the band 3.2–7 GHz using the PulseOn 210 devices (TimeDomain[4]). Figure 3.22 shows a typical measured pulse channel response at LOS receiving location 1. It can be noted the dense multipath present even in strong LOS condition. Figure 3.23 shows the pulse channel response at receiving location 10 which is clearly in non-line-of-sight (NLOS) condition.

---

[4] See the TimeDomain website: http://www.timedomain.com

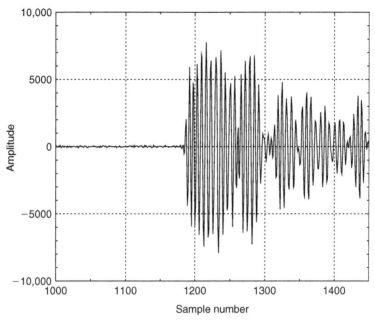

**Figure 3.22** *Example of measured multipath profile at receiving location* 1 *(LOS condition). Sampling interval of* 41.3 *ps*

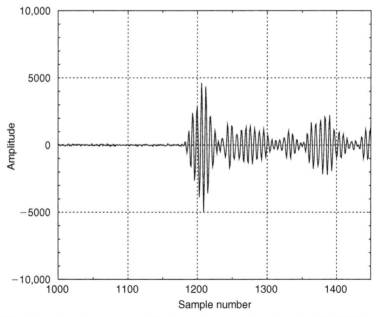

**Figure 3.23** *Example of measured multipath profile at receiving location* 10 *(NLOS condition). Sampling interval of* 41.3 *ps*

A number of UWB channel models have been proposed depending on the frequency range and operating conditions (Cassioli, Win & Molisch, 2002; Molisch, Cassioli, Chong, Emami, Fort, Kannan, Karedal, Kunisch, Schantz, Siwiak & Win, 2006; Chong & Yong, 2005). The IEEE 802.15.3a task group developed a channel model that is valid from 3 to 10 GHz for indoor residential and office environments with restricted distance (<10 m). A more general model valid for the frequency range 3–10 GHz in a number of different environments as well as for the frequency range below 1 GHz in office environments has been presented within the IEEE 802.15.4a task group (Molisch et al., 2005, 2006).

## 3.4.1  Path loss

In a wideband or UWB channel, the wideband path loss is a function of frequency as well as of distance. In this case it can be defined the *frequency-dependent path loss* (Molisch et al., 2005):

$$L_0(f,d) = \delta f \left[ \mathbb{E}_1 \left\{ \mathbb{E}_s \left\{ \int_{f-\delta f/2}^{f+\delta f/2} |H(\tilde{f},d)|^2 d\tilde{f} \right\} \right\} \right]^{-1} \tag{3.12}$$

where $H(f,d)$ is the transfer function from transmitter antenna connector to receiver antenna connector and $\delta f$ is chosen small enough so that e.m. characteristics of materials can be considered constant within the integration bandwidth. The expectation is taken with respect to large-scale as well as small-scale fading. As first approximation, we can factorize $L_0(f,d)$ in two distinct frequency-dependent and distance-dependent terms:

$$L_0(f,d) = L_0^{(\text{freq})}(f) \cdot L_0(d), \tag{3.13}$$

where $L_0(d)$ is given by (3.4) (expressed in linear scale) and

$$L_0^{(\text{freq})}(f) \propto f^{-2\kappa}. \tag{3.14}$$

The parameter $\kappa$ is the frequency dependency decaying factor. In UWB channels, the frequency characteristics of the antennas should also be taken into account in (3.14). In (Molisch et al., 2006) the following frequency-dependent path loss model is given:

$$L_0(f,d) = 2k_0 \frac{(f/f_c)^{2(\kappa+1)}d^{\alpha}}{\eta_{TX}(f)\eta_{RX}(f)}, \tag{3.15}$$

where $k_0$ is the isotropic path loss, in linear scale, at the reference distance of 1 metre and centre frequency $f_c$, $\eta_{TX}(f)$ and $\eta_{RX}(f)$ are, respectively, the transmit and the receive antenna efficiencies and $f_c$ is the centre frequency usually taken equal to 5 GHz. The coefficient 2 approximates the typical loss due to the presence of a person close to the antenna. Shadowing effects can be accounted for similarly to what was done in (3.11).

## 3.4.2 Multipath characterization

Most of the UWB channel models are based on an extended version of the classical Saleh-Valenzuela (S-V) indoor channel model (Saleh & Valenzuela, 1987), where multipath components arrive at the receiver in groups (clusters) following the Poisson distribution. According to the S-V model, the complex baseband channel impulse response is given as

$$h(t) = \sum_{k=0}^{K} \sum_{l=0}^{L} a_{k,l} e^{j\phi_{k,l}} \delta(t - T_k - \tau_{k,l}), \qquad (3.16)$$

where $a_{k,l}$ is the amplitude of the $l$th path in the $k$th cluster, $T_k$ is the delay of the $k$th cluster, $\tau_{k,l}$ is the delay of the $l$th path relative to the $k$th cluster arrival time $T_k$. The phases $\phi_{k,l}$ are uniformly distributed in the range $[0, 2\pi)$. The number of clusters $K$ is assumed to be Poisson distributed. The classical S-V model also considers a Poisson process for the ray arrival times $\tau_{k,l}$. However, to better fit for indoor and outdoor channel measurements, the IEEE 802.15.4a task group proposes to model the ray arrival times with mixtures of two Poisson processes. More details can be found in (Molisch et al., 2006). The power delay profile (PDP) is assumed exponential negative within each cluster:

$$\Lambda_{k,l} = \mathbb{E}\{|a_{k,l}|^2\} \propto \exp(-\tau_{k,l}/\epsilon_k), \qquad (3.17)$$

where $\epsilon_k$ is the intra-cluster decay time constant.

Another widely adopted model is the dense multipath model with a single cluster composed of $L$ independent equally spaced paths and exponential PDP. In this case the (real) baseband impulse response is given by

$$h(t) = \sum_{l=0}^{L} a_l p_l \delta(t - \tau_l), \qquad (3.18)$$

where $a_l$ is the path amplitude and $p_l$ is a r.v. which takes, with equal probability, the values $\{-1, +1\}$. The average path power gains $\Lambda_l$ are given by

$$\Lambda_l = \mathbb{E}\{|a_l|^2\} = \frac{(e^{\Delta/\epsilon} - 1)e^{-\Delta(l-1)/\epsilon}}{e^{\Delta/\epsilon}(1 - e^{L\Delta/\epsilon})} \qquad (3.19)$$

for $l = 1, \ldots, L$ (Cassioli et al., 2002; Dardari & Win, 2006). The parameter $\epsilon$ describes the multipath spread of the channel (decay time constant) and $\Delta$ is the resolvable time interval. In most of the above mentioned models the small-scale fading, which characterizes the path amplitudes $a_l$, follows a Nakagami-*m* distribution:

$$f(x) = \frac{2}{\Gamma(m)} \left(\frac{m}{\Lambda}\right)^m x^{2m-1} \exp\left(-\frac{m}{\Lambda} x^2\right), \qquad (3.20)$$

where $m$ is the severity parameter, depending on the working conditions, and $\Gamma(\cdot)$ is the gamma function.

# 4

# Connectivity and coverage

## 4.1 Introduction

This chapter deals with some basic statistical models to characterize network connectivity and coverage which provide useful general insights on network parameters design rules, such as node density and transmission power, in WSANs. This chapter is mainly based on theoretical work and comparison to simulation.

Some general lessons will be learned that are useful to understand some general principles of WSAN design. In wireless ad hoc networks the best performance is achieved when data generated by a node can flow along the network and reach any possible endpoint. Thus the goal of connectivity is to make it possible for any node to reach any different node, perhaps in a multi-hop fashion. That provided, the network is said to be *fully connected*.

Although WSNs are sometimes thought of as a special case of wireless ad hoc networks, they present a substantial difference, that is, nodes are at least of two different types: sensor and sink nodes. The purpose of this kind of network is to process data originated by sensors, and sinks are in charge of collecting such data. Thus, the goal of connectivity is somewhat different here because it is sufficient for any sensor node to be able to reach at least one sink node, either directly or through other sensor nodes.

## 4.2 Connectivity in wireless ad hoc and sensor networks

The *connectivity theory* studies networks formed by large numbers of nodes distributed according to some statistics over a limited or unlimited region of $\mathbb{R}^d$, with $d = 1, 2, 3$, and aims at describing the potential set of links that can connect nodes to each other, subject to some constraints from the physical viewpoint (power budget, or radio resource limitations).

Connectivity depends on the number of nodes for unit area (nodes' density), and on the transmission power. The choice of an appropriate transmit power level is an important aspect of network design as it affects network connectivity. In fact, with a high transmit power a large number of nodes are expected to be reached via a direct link. On the contrary, a low transmit power would increase the possibility that a given node cannot reach any other node, that is, it is isolated.

In the literature, the connectivity theory typically considers three types of scenarios:

- a Poisson Point Process (PPP) with intensity $\rho$ over an unbounded region
- a square area of side $L$ (or disk of unit radius) with nodes distributed in the region according to a PPP with intensity $\rho$
- a square area of side $L$ (or disk of unit radius) with $N$ (fixed) nodes uniformly distributed at random in the region.

According to the stochastic geometry models, a PPP is characterized by nodes uniformly distributed with density $\rho$ in $\mathbb{R}^d$, with positions independent one from each other.[1] Moreover, the PPP is verified also in a limited region; in fact, given a finite region $A$ of $\mathbb{R}^d$ (see Figure 4.1), the number of nodes in $A$, $n_A$, is once again Poisson distributed, that is,

$$P(n_A = n) = \frac{(\rho|A|)^n}{n!} e^{-\rho|A|}, \tag{4.1}$$

with mean $N_A = \rho|A|$.[2] The probability that the region is empty is $P(n_A = 0) = e^{-N_A} = e^{-\rho A}$. Thus PPPs may be considered either on bounded or unbounded regions.

When dealing with connectivity, one has to think of it as a global attribute, meaning that it is referred to the entire network, not just to a node or set of nodes. This is the reason why connectivity is strictly related to other typical features of the network, such as topology, power consumption, transmission range of the radio transceiver, etc. Therefore, connectivity theory is closely related to topology control theory.

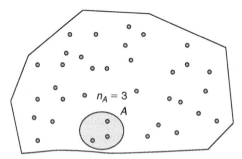

**Figure 4.1** *Poisson Point Process in a finite area* A

---

[1] The unit of measure of $\rho$ depends on the dimensionality $d$ of the region considered, i.e., in [nodes/$m^d$].
[2] The operator $|A|$ indicates the size of $A$ expressed in [$m^d$]. For notation simplicity, in the following we will simply indicate it with $A$.

*Topology control* aims at controlling the set of links that connect couples of nodes, in order to simplify and allow routing of messages between all pairs of nodes.

The *physical* topology of a network can be controlled through the physical layer, in most cases power control techniques are used.

The *logical* topology of a network is controlled by entities working at layer two and three, and is based on a reduced set of links.

Therefore, a *physical* topology is different from a *logical* topology. Connectivity theory deals mainly with physical topologies, but can be extended to some aspects of logical topologies taking non-electromagnetic aspects into account.

The maximum distance at which a node can reach a neighbouring node is called transmission range (TR) of that node. It is worth noting that by varying the transmission range of nodes, a different network is obtained that will show a certain degree of connectivity.

The purpose of topology control is to find the optimum transmission range for the sensor nodes, given some specifications on the whole network (power consumption, capacity, lifetime, . . .). Optimality might refer, for example, to full connectivity with minimum radiated power. If the transmitted power is the same for every node we are dealing with the critical transmission range (CTR). Also the number of sensors and sinks that constitute the network affects the shape of the network: *node density* is also crucial to connectivity as well as the TR.

## 4.3 Link connectivity

Before studying network connectivity, we have to define single link connectivity properties. In the following a short description of the main models present in the literature, all considering narrowband systems, is provided.

In the absence of interference, a receiver node is considered to be connected to a transmitter node, transmitting with power $P_t$, if the received SNR is above a minimum threshold $\gamma$, that is, if the received power $P_r$ is above a minimum threshold $P_{r_{min}}$ which depends on the modulation and receiver characteristics. This is the typical *threshold effect* that distinguishes digital from analogue communication systems. Considering that, in dB scale, $P_r = P_t - L$, where $L$ is the channel attenuation or power loss (including antennas gain) expressed in dB, it results that a direct radio link between nodes exists if $L < L_{th}$, where $L_{th} \triangleq P_t - P_{r_{min}}$. The threshold $L_{th}$ depends on the transmitter power, the receiver sensitivity and antennas gain, and represents the maximum power loss tolerable by the communication system.

In the presence of co-channel interference the situation becomes more complicated. In fact, it is not sufficient, for link connectivity, that the received SNR overcomes $\gamma$, but also the signal-to-interference ratio (SIR) must be above a minimum threshold.

## 4.3.1 Deterministic model: Disk model

This is a simple model often adopted in the literature for analytical convenience, where the connection between nodes is based on pure geometric considerations. According to this model, the power loss $L$ is a deterministic function $L_0(r)$ of the distance $r$ between the nodes. As explained in Chapter 3, typically $L_0(r)$, in dB scale, takes the form

$$L_0(r) = k_0 + k_1 \ln r, \tag{4.2}$$

where $k_0$ is the path loss at the reference distance of 1 metre, which depends on the radio device characteristics as well as the wavelength. In this model the TR is constant and it is simply given by

$$\text{TR} = r_0 = e^{\frac{L_{\text{th}} - k_0}{k_1}}. \tag{4.3}$$

Given a certain node, all nodes located within the disk of radius $r_0$ are connected. For this reason this model is also referred to as *disk model*.

## 4.3.2 Statistical model: Log-normal shadowing

The main limitation of the disk model is that it often leads to inaccurate analysis due to ignorance of the stochastic nature of wireless channel. For example, it will be shown that channel randomness usually improves the connectivity properties of the wireless networks.

Note that since the scenario is usually stationary, the assumption of a (slow-varying) shadowing environment is acceptable. It has been seen in Chapter 3, equation (3.11), that the log-normal model, here reported for reader convenience, is widely adopted in the literature to describe the power loss, in dB, between two communicating nodes

$$L = k_0 + k_1 \cdot \ln r + S, \tag{4.4}$$

where $S$ is a shadowing sample which is assumed to be Gaussian distributed, with zero mean and standard deviation $\sigma$, to model the randomness of the geometry (presence of obstacles, etc.). Generally, link reciprocity is assumed and the different shadowing samples competing to different nodes are considered to be independent.

Recalling (4.3), one could define again the concept of TR that now is affected by fluctuations, namely its value may get larger or smaller depending on the outcome $s$ of $S$

$$\text{TR} = e^{\frac{L_{\text{th}} - k_0 - s}{k_1}}. \tag{4.5}$$

Hence, the number of nodes that are audible by a certain node (those which satisfy $L < L_{\text{th}}$) may vary from one shadowing sample to the other. This simple consideration is based on the assumption that, though random, the TR is equal in all directions for a node

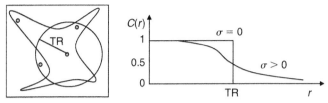

**Figure 4.2** *Link connectivity with or without shadowing effects*

(i.e., $s$ is node-independent). In true situations, $s$ varies from link to link. However, the above consideration still holds on average.

### 4.3.3 Probability of link connection

We have seen that in the presence of propagation fluctuations, for fixed values of distance and transmit power, the connectivity of a pair of nodes cannot be treated deterministically but only in a probabilistic sense.

According to (3.10), the probability $C(r)$ that two nodes are connected, as a function of the distance $r$, is

$$C(r) = \mathbb{P}\{L \le L_{th}\} = \mathbb{P}\{S \le L_{th} - L_0(r)\}, \tag{4.6}$$

From (4.4) it is easy to derive the expression of $C(r)$ for the log-normal model

$$C(r) = 1 - 0.5 \; \text{erfc}\left(\frac{L_{th} - k_0 - k_1 \ln r}{\sqrt{2}\sigma}\right), \tag{4.7}$$

where erfc$(\cdot)$ is the complementary error function. The disk model is obtained by considering $\sigma \to 0$, that is, $S = 0$, thus leading to the following expression for $C(r)$

$$C(r) = \begin{cases} 1 & r \le r_0 \\ 0 & \text{otherwise} \end{cases}, \tag{4.8}$$

where $r_0$ is the TR given by (4.3).

As we can see in Figure 4.2, taking into account a specific transmitting node, the effect of the shadowing is to make audible some nodes that are not reachable when adopting the disk model ($\sigma = 0$) because they are outside the circumference having radius TR; but, on the other hand, it also makes non-audible some nodes which are inside the circumference.

## 4.4 Single-hop link connectivity in WSNs

We consider a single sink scenario with one or more sensor nodes which attempt to establish a connection with the sink. As usual, we assume nodes to be spatially distributed according to a PPP with density $\rho$. We aim to determine the probability distribution of the number $k$ of nodes which are heard by, and hence connected to, the sink. Due to the random nodes position and channel fluctuation effects, the number $k$ of nodes that are directly connected to the sink results to be random.

Given the generic link connection probability function $C(r)$, it can be shown that the probability distribution $g_k$ of $k$ is still Poisson distributed regardless of the specific shape of $C(r)$. The Poisson behaviour does not depend on the channel model but it is inherited by the underlying homogeneous PPP.[3] Hence, the probability distribution of $k$ is

$$g_k = \frac{\mu^k}{k!} e^{-\mu},$$

(4.9)

with the corresponding generating function

$$G_1(s) = \sum_{k=0}^{\infty} g_k s^k = e^{\mu(s-1)},$$

(4.10)

where the mean value $\mu$ represents the average number of neighbours the sink, and hence each node, can hear. In particular, the probability that a node is isolated is simply given by

$$P_{\text{iso}} = \mathbb{P}\{\text{isolated node}\} = g_0 = e^{-\mu}.$$

(4.11)

In general, given a finite or infinite region $A \subseteq \mathbb{R}^d$, the mean value $\mu$ is

$$\mu = \int_A \rho C(\|\mathbf{p}\|) d\mathbf{p},$$

(4.12)

where $\mathbf{p}$ is the generic point in the $\mathbb{R}^d$ space and $\|\cdot\|$ represents the Euclidean norm. One case of interest is to consider $A = \mathbb{R}^d$ (unlimited region), where (4.12) simplifies in (Dardari et al, 2008)

$$\mu = \int_0^{\infty} \rho C(r)(2\pi r)^{d-1} \, dr \qquad \text{for } d = 2,3$$

$$\mu = 2 \int_0^{\infty} \rho C(r) dr \qquad\qquad \text{for } d = 1.$$

(4.13)

When the disk model (4.2) is considered, (4.13) further simplifies in

$$\mu = 2\rho r_0 \qquad \text{for } d = 1$$

$$\mu = \pi \rho r_0^2 \qquad \text{for } d = 2$$

$$\mu = \frac{4\pi}{3} \rho r_0^3 \qquad \text{for } d = 3,$$

(4.14)

---

[3] This result is also known as the marking theorem for point processes.

where $r_0$ is given by (4.3). In Dardari, 2007; Orriss & Barton, 2003; Verdone, 2005 closed-form expressions for the mean value $\mu$ are derived when the log-normal channel model (4.4) is adopted for $d = 1$, $d = 2$ and $d = 3$, respectively. In particular,

$$\mu = 2\rho r_0 \, e^{\frac{\sigma^2}{2k_1^2}} \qquad \text{for } d = 1$$

$$\mu = \pi \rho r_0^2 \, e^{\frac{2\sigma^2}{k_1^2}} \qquad \text{for } d = 2$$

$$\mu = \frac{4\pi}{3} \rho r_0^3 \, e^{\frac{9\sigma^2}{2k_1^2}} \qquad \text{for } d = 3. \tag{4.15}$$

By comparing (4.14) and (4.15), it can be easily verified that the presence of shadowing increases the number of audible nodes with respect to the deterministic case. Contrary to what we would have expected by intuition, we can conclude that channel randomness has a beneficial effect on link connectivity, and hence on network connectivity, as will be shown in the next sections. At least, for unlimited regions.

## 4.5 Multi-hop link connectivity in WSNs

### 4.5.1 *Characterization of the number of connected nodes*

Previous results related to single-hop connectivity suggest the possibility to exploit the Galton-Watson branching process (GWBP) theory to deal with a larger number of hops. In particular, each node connected to the sink is able to connect to a random number of nodes with Poisson distribution according to the result of the previous section. Each node has a random number of connected nodes (children) in the next hop (generation) and so on. These random variables are independent realizations of $k$ and have distribution $g_k$ and generating function $G_1(s)$ given, respectively, in (4.9) and (4.10). We denote with $Z_n$ the number of nodes connected at the $n$th hop, that is, the number of children present in the network at the $n$th generation. Obviously, the generating function of $Z_1$ is $G_1(s)$ and $Z_0 = 1$ (the sink node).

By considering the properties of the generating function, after $n$ hops the generating function of $Z_n$ is

$$G_n(s) = G_1(G_{n-1}(s)). \tag{4.16}$$

Implicitly, we include the possibility that a certain node can be connected at different generations. From the knowledge of $G_n(s)$ it is possible to derive, for example, the mean value of $Z_n$

$$\mathbb{E}\{Z_n\} = \left. \frac{dG_n(s)}{ds} \right|_{s=1} = \mu^n, \tag{4.17}$$

as well as the probability that after $n$ hops no children are present (extinction)[4]

$$\mathbb{P}\{\text{no children after } n \text{ hops}\} = G_n(0). \tag{4.18}$$

If $\mu < 1$, the average total number of nodes connected in the network after an infinite number of hops can be easily evaluated using (4.17)

$$\mathbb{E}\left\{\sum_{n=1}^{\infty} Z_n\right\} = \frac{1}{1-\mu}. \tag{4.19}$$

Since the expectation is finite,

$$\mathbb{P}\left\{\sum_{n=1}^{\infty} Z_n < \infty\right\} = 1, \tag{4.20}$$

that implies

$$\mathbb{P}\{\text{extinction}\} = 1. \tag{4.21}$$

Hence, when $\mu < 1$ the population of connected nodes is finite regardless of the number of hops. In other words, $\mu < 1$ is a sufficient (but not necessary) condition to state that a packet flooded in the network stops propagating (extinct) even for an infinite number of hops.

### 4.5.2  Characterization of the connected nodes positions

In this paragraph the connectivity problem is addressed by evaluating the statistics of the connected nodes position in a multi-hop wireless random sensor/actuator network following the approach proposed in Dardari, 2007.

For simplicity we consider a mono-dimensional scenario with a population of nodes randomly distributed according to a homogeneous PPP with density $\rho$ [nodes/$m$]. The problem we want to solve is the following: given a source node in the origin (e.g., the sink), we are interested in evaluating the probability distribution function (p.d.f.) of the connected positions when the source message is flooded throughout the network with a maximum number $h$ of hops. Starting from that, expressions for other interesting figures of merit such as the *coverage probability* and the *coverage distance* are derived. The statistical channel model (3.10) is considered. As a consequence, the p.d.f. of the position $x$ of each node directly connected to the source is given by

$$f_1(x) = \frac{C(|x|)}{\int_{-\infty}^{\infty} C(|x|)dx}. \tag{4.22}$$

In the presence of $k$ links (i.e., $k + 1$ nodes, the source node and its connected neighbours' nodes), due to the independence assumption of the fading samples, each connected node

---

[4]This is also the probability that a packet flooded in the network stops propagating.

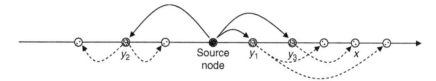

⊕ One hop connected node

⊙ Two hops connected node

**Figure 4.3** *Example of nodes connection after two hops*

experiences the same p.d.f. as in (4.22), hence the overall connected nodes position distribution after one hop is

$$P_1(x) = \sum_{k=0}^{\infty} g_k \frac{\delta(x) + \overbrace{f_1(x) + f_1(x) + \cdots + f_1(x)}^{k}}{k+1}$$

$$= p_0 \delta(x) + (1 - p_0) f_1(x), \tag{4.23}$$

where

$$p_0 = \frac{1 - e^{-\mu}}{\mu} \tag{4.24}$$

and $\mu$ is given in (4.13). The presence of the Dirac pseudo-function $\delta(x)$ accounts for the position p.d.f. of the source node which is located at the origin. Each term in the sum in (4.23) is composed of the factor $1/(k + 1)$, which represents the probability to be one of the $k + 1$ connected nodes, multiplied by the corresponding position p.d.f. $f_1(x)$ (or $\delta(x)$ if the source node).

In order to make clear how the general model is derived, we start with the simple two-hops case (see Figure 4.3). Each node which, during the first hop, established a direct connection to the source node has still, after the second hop, a random number of children nodes distributed according to the Poisson statistics with the same mean $\mu$.

The connected nodes position distribution after two hops can be obtained by considering all possible two-hops link combinations and by averaging their positions according to the statistic (4.22). Specifically, we have

$$P_2(x) = \sum_{k=0}^{\infty} g_k \left[ \delta(x) + \int_{y_1} \cdots \int_{y_k} \frac{P_1(x - y_1) f_1(y_1) + \cdots + P_1(x - y_k) f_1(y_k)}{k+1} dy_1 \cdots dy_k \right]$$

$$= p_0 \delta(x) + \sum_{k=0}^{\infty} g_k \frac{k}{k+1} \int_{y} P_1(x - y) f_1(y) dy$$

$$= p_0 \delta(x) + (1 - p_0) p_0 f_1(x) + (1 - p_0)^2 f_2(x), \tag{4.25}$$

where $f_2(x)$ is defined as

$$f_2(x) \triangleq \int_{\infty}^{\infty} f_1(x - y) f_1(y) dy = f_1(x) \otimes f_1(x). \tag{4.26}$$

The operator $\otimes$ represents the convolution integral.

The general $h$-hops case can be easily deducted by iterating the same procedure followed in (4.25) leading to the p.d.f. of the connected nodes position after $h$ hops

$$P_h(x) = p_0 \delta(x) + \sum_{n=1}^{h-1} (1 - p_0)^n p_0 \, f_n(x) + (1 - p_0)^h \, f_h(x), \tag{4.27}$$

where $f_n(x)$ is defined as the $(n - 1)$-fold convolution integral of $f_1(x)$ with itself.

Even though the model (4.27) is very simple, its numerical evaluation poses some problems when considering a high number of hops, due to the presence of the function $f_n(x)$ which requires the evaluation of $(n - 1)$ multiple convolution integrals. When the disk channel model (4.2) is adopted, a closed form for $f_n(x)$ exists

$$f_n(x) = \begin{cases} \dfrac{n}{2r_0} \displaystyle\sum_{p=0}^{\left\lfloor \frac{nr_0-x}{2r_0} \right\rfloor} \dfrac{(-1)^p \left( \dfrac{nr_0 - x - 2pr_0}{2r_0} \right)^{n-1}}{p!(n - p)!} & |x| < nr_0, \\[4ex] 0 & \text{otherwise,} \end{cases} \tag{4.28}$$

where $\lfloor x \rfloor$ denotes the floor operator. Unfortunately a closed-form solution for the lognormal channel model (4.4) cannot be found and we need to resort to some kind of approximation. In this regard, it is well known that as $n$ grows, the $(n - 1)$-fold convolution integral tends to the Gaussian function. This suggests the approximation of $f_n(x)$, for $n > 1$, with the following Gaussian function

$$f_n(x) \simeq \frac{b}{\sqrt{n\pi}} e^{-\frac{b^2 x^2}{n}} \qquad n > 1, \tag{4.29}$$

where

$$b \triangleq \left( 2 \int_{-\infty}^{\infty} x^2 \, f_1(x) dx \right)^{-\frac{1}{2}}. \tag{4.30}$$

The comparison with simulations confirms the good accuracy of the approximation (4.29) even for small values of $n$.

From (4.27) and by means of the approximation (4.29), it is possible to derive the *probability of coverage* at distance $x$, defined as the probability that the distance $x$ is achieved (or overcome) with a limited number $h$ of hops. Specifically,

$$P_{\text{cov}_h}(x) \triangleq 2\int_x^\infty P_h(x)dx$$

$$\simeq 2(1-p_0)p_0\int_x^\infty f_1(\xi)d\xi + \sum_{n=2}^{h-1}(1-p_0)^n p_0\,\text{erfc}\left(\frac{bx}{\sqrt{n}}\right)$$

$$+ (1-p_0)^h\,\text{erfc}\left(\frac{bx}{\sqrt{h}}\right). \tag{4.31}$$

Some numerical results follow to show the validity of the presented model. We refer to a typical node operating at 2.4 GHz with the following parameters: transmitted power 0 dBm, antennas gain 2 dB, receiver sensibility $-86$ dBm. The corresponding maximum tolerable power loss is $L_{\text{th}} = 90$ dB. The channel model (4.4) is considered with parameters $k_0 = 40$ dB (free space path loss at 1 metre distance), $k_1 = 35$, and $r_0 = 27$ metres.

In Figure 4.4 the connected nodes position distribution $P_h(x)$ for different shadowing spreads $\sigma = 0, 4, 8$ is shown. The maximum number of hops is $h = 4$ and the node density $\rho = 0.0925$ [nodes/m] ($\mu \approx 5$ nodes) is considered. The case $\sigma = 0$ corresponds to the deterministic disk channel model (4.2) which exhibits a step behaviour at distance $r_0$. It can be noted that as the channel randomness increases (higher $\sigma$), the distance distribution spreads in favour of larger distances.

The probability of coverage $P_{\text{cov}_h}(x)$ given by (4.31), for different values of $h$ and shadowing spreads $\sigma$, is plotted in Figure 4.5. It can be observed that both high shadowing

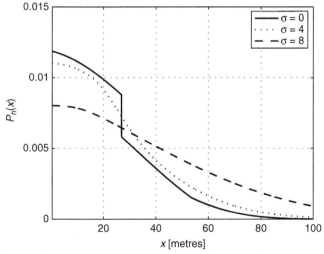

**Figure 4.4** *Connected nodes position distribution for different shadowing conditions. Maximum number of hops* h $= 4$, $\rho = 0.0925$ *nodes/m*

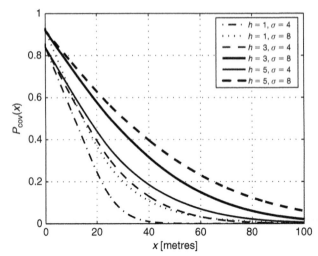

**Figure 4.5**  *Probability of coverage for different number of hops* h *and shadowing conditions* σ

spread and number of hops values have a positive impact on the probability of coverage, especially at large distances.

Starting from (4.31) it is possible to investigate what is the maximum distance $d_{cov}$, namely *coverage distance*, which satisfies a desired probability of coverage $P^*_{cov}$, as a function of system parameters. Figure 4.6 shows the coverage distance as a function of the node density $\rho$, for different number of hops $h$ and a target probability of coverage $P^*_{cov} = 0.1$. It can be observed that, given a fixed maximum number of hops $h$, the coverage distance exhibits a saturation behaviour after a certain value of node density. A further increase of the node density does not correspond to a significant coverage increase. Beyond that saturation point the only way to increase $d_{cov}$ is in adopting higher values of hops.

In Figure 4.7 the coverage distance is plotted for a different number of hops and shadowing conditions $\sigma$ having fixed $P^*_{cov} = 0.5$. It can be concluded that the shadowing spread, that is, the channel randomness, has a strong positive impact on the coverage and that the use of simple disk channel model ($\sigma = 0$) leads to very conservative results.

## 4.6  Characterization of the interference

We consider a single sink two-dimensional scenario with one or more sensor nodes which attempt to establish a connection with the sink. As usual, we assume nodes to be spatially distributed according to a PPP with density $\rho$, but this time they occupy a bounded circular domain of radius $r_T$. A sensor node $N_1$ is placed inside the domain, while a sink node $N_0$ is placed in the centre. The propagation environment is still characterized by distance-dependent loss and log-normal channel fluctuations. However, we recognize that the link connectivity model (4.6) suffers from a serious weakness: the

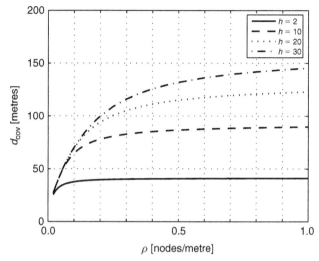

**Figure 4.6** *Coverage distance vs node density* ρ *for different number of hops* h. *Target coverage probability* 0.1, σ = 4

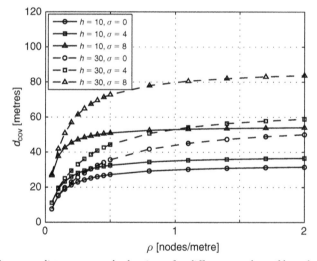

**Figure 4.7** *Coverage distance vs node density* ρ *for different number of hops* h *and shadowing conditions* σ. *Target coverage probability* 0.5

received power alone (or, equivalently, the maximum tolerated loss) does not permit to assess whether a connection has been established or not, but also the interference must be taken into account through the SIR. As an example, a sink that 'hears' a sensor node with a maximum signal strength, might still not be able to 'understand' what it 'says', because other sensor nodes also 'talk'. That said, throughout this section we increasingly move towards the direction of *quality* of connection rather than quantity of received power by taking into account the effect of the interference.

We first derive the distribution of the channel gain between two communicating nodes, then we obtain an expression for the general moment of the total interference received by $N_0$: in this way, the SIR can also be easily computed.

## 4.6.1   Channel gain characterization

We assume that all the nodes in the area transmit by using the same power $P_t$ and that the log-normal propagation model (4.4) is valid. Now consider that the distance between $N_1$ and $N_0$ does not exceed $r_T$ and that the two nodes are within the (random) communication range. Although the unconditioned $S$ in (4.4) is a normal r.v., the distribution of $S$ conditioned on the fact that $N_1$ communicates with $N_0$ is not normal. However, to simplify the analysis, we assume that the distribution of $S$ conditioned on the fact that $N_1$ communicates with $N_0$ is still normal with zero mean and standard deviation $\sigma$. We will show that the approximate distribution we obtain with this assumption is extremely tight.

Let us take a realization $s^*$ of the r.v. $S$ and consider the following two cases:

*Case I:*

$$r_T > e^{\frac{L_{th}-k_0-s^*}{k_1}}.$$

In this case, which occurs when $s^* > L_{th} - k_0 - k_1 \ln r_T$ (with probability $1 - C(r_T)$), the maximum communication distance is less than $r_T$. In such a situation, the probability that the distance $D$ between nodes is less or equal to a certain value $d$ conditioned on $s = s^*$ can be written as

$$\mathbb{P}\{D \le d | s = s^*\} = C^{(I)}d^2, \quad 0 \le d \le e^{\frac{L_{th}-k_0-s^*}{k_1}} \tag{4.32}$$

with $C^{(I)} \triangleq e^{-2\frac{(L_{th}-k_0-s^*)}{k_1}}$. Therefore the joint p.d.f. of the r.v.'s $D$ and $S$ is

$$f_{D,s}^{(I)}(d,s^*) = C^{(I)} \frac{2de^{-2\frac{(L_{th}-k_0-s^*)}{k_1}} e^{\frac{-s^{*2}}{2\sigma^2}}}{\sigma\sqrt{2\pi}} \tag{4.33}$$

for $0 \le d \le e^{\frac{L_{th}-k_0-s^*}{k_1}}$ and $s^* > L_{th} - k_0 - k_1 \ln r_T$. $C^{(I)}$ is a normalizing factor that accounts for the fact that the channel fluctuation is bounded by $L_{th} - k_0 - k_1 \ln r_T$. The range of $d$ and $s^*$ can be re-written as

$$L_{th} - k_0 - k_1 \ln r_T \le s^* \le L_{th} - k_0 - k_1 \ln d$$
$$0 \le d \le r_T \tag{4.34}$$

Starting from (4.33), we make the following change of variables

$$\begin{cases} G = -k_0 - k_1 \ln D - S \triangleq h_1(D,S) \\ Y = S \triangleq h_2(D,S) \end{cases}, \tag{4.35}$$

where $G$ represents the channel gain in dB. The Jacobian of the transformation is given by

$$J(d, s^*) = \begin{vmatrix} \dfrac{\partial h_1}{\partial d} & \dfrac{\partial h_1}{\partial s^*} \\ \dfrac{\partial h_2}{\partial d} & \dfrac{\partial h_2}{\partial s^*} \end{vmatrix} = \begin{pmatrix} -k_1/d & -1 \\ 0 & 1 \end{pmatrix}, \tag{4.36}$$

whose determinant is $-k_1/d$. The joint p.d.f. of the r.v.'s $G$ and $Y$ is therefore given by

$$f_{G,Y}^{(I)}(g, y) = \dfrac{f_{D,s}^{(I)}\left(e^{-\frac{g+k_0+y}{k_1}}, y\right) e^{-\frac{g+k_0+y}{k_1}}}{k_1} = \dfrac{2C^{(I)}e^{-\frac{2}{k_1}(g+L_{\text{th}})}e^{-y^2/(2\sigma^2)}}{\sigma\sqrt{2\pi}k_1} \tag{4.37}$$

for

$$L_{\text{th}} - k_0 - k_1 \ln r_{\text{T}} < y < \infty, \qquad -L_{\text{th}} < g < \infty$$

The p.d.f. of $G$ can be derived by integrating $Y$ over its range of definition

$$f_G^{(I)}(g) = \int_{L_{\text{th}} - k_0 - k_1 \ln r_{\text{T}}}^{\infty} f_{G,Y}^{(I)}(g, y)dy$$

$$= \dfrac{2C^{(I)}\, e^{-\frac{2}{k_1}(g+L_{\text{th}})}}{\sigma\sqrt{2\pi}k_1} \int_{L_{\text{th}} - k_0 - k_1 \ln r_{\text{T}}}^{\infty} e^{-y^2/(2\sigma^2)}\, dy$$

$$= \dfrac{C^{(I)}e^{-\frac{2}{k_1}(g+L_{\text{th}})}}{k_1}\, \text{erfc}\left(\dfrac{L_{\text{th}} - k_0 - k_1 \ln r_{\text{T}}}{\sqrt{2}\sigma}\right) \tag{4.38}$$

*Case II:*

$$r_{\text{T}} \le e^{\frac{L_{\text{th}} - k_0 - s^*}{k_1}}.$$

In this case, which occurs with probability $C(r_{\text{T}})$, the maximum communication distance is greater than $r_{\text{T}}$ and the cumulative distribution function (c.d.f.) of $D$ can be written as

$$\mathbb{P}\{D \le d\} = C^{(II)}d^2 = \dfrac{d^2}{r_{\text{T}}^2}, \qquad 0 < d < r_{\text{T}}, \tag{4.39}$$

where $C^{(II)}$ is the normalizing factor. Therefore, the joint p.d.f. of the r.v.'s $D$ and $S$ is

$$f_{D,s}^{(II)}(d, s^*) = \dfrac{C^{(II)}2de^{-(s^*)^2/2\sigma^2}}{r_{\text{T}}^2\sigma\sqrt{2\pi}} \tag{4.40}$$

for

$$0 \le d \le r_{\text{T}}, \qquad -\infty < s^* \le L_{\text{th}} - k_0 - k_1 \ln r_{\text{T}},$$

where now $C^{(II)}$ accounts for the fact that $s$ is bounded by $-\infty$ and $L_{th} - k_0 - k_1 \ln r_T$. Making the change of variable in (4.35) we obtain

$$
f_{G,Y}^{(II)}(g, y) = \frac{2C^{(II)} e^{-\frac{2}{k_1}(g+k_0+y)} e^{-y^2/(2\sigma^2)}}{r_T^2 \sigma \sqrt{2\pi}}
$$

$$
\times \operatorname{rect}\left( \frac{e^{-\frac{g+k_0+y}{k_1}} - r_T/2}{r_T} \right), \tag{4.41}
$$

for $-\infty < y \leqslant L_{th} - k_0 - k_1 \ln r_T$, where

$$
\operatorname{rect}(x) \triangleq \begin{cases} 1, & |x| < 1/2 \\ 0, & |x| \geq 1/2 \end{cases}. \tag{4.42}
$$

The p.d.f. of $G$ can be obtained as

$$
f_G^{(II)}(g) = \int_{-\infty}^{L_{th} - k - k_1 \ln r_T} f_{G,Y}^{(II)}(g, y) dy
$$

$$
= \frac{C^{(II)} e^{-\frac{2k_1 g + 2k_0 k_1 - 2\sigma^2}{k_1^2}}}{r_T^2 k_1} \left[ \operatorname{erfc}\left( \frac{-gk_1 - k_0 k_1 - k_1^2 \ln r_T + 2\sigma^2}{\sqrt{2}\sigma k_1} \right) \right.
$$

$$
\left. - \operatorname{erfc}\left( \frac{L_{th} k_1 - k_0 k_1 - k_1^2 \ln r_T + 2\sigma^2}{\sqrt{2}\sigma k_1} \right) \right]. \tag{4.43}
$$

Since the two cases are mutually exclusive, the final distribution of $G$ can be written as

$$
f_G(g) = \mathbb{P}\{\text{case I}\} \cdot f_G^{(I)}(g) + \mathbb{P}\{\text{case II}\} \cdot f_G^{(II)}(g). \tag{4.44}
$$

Putting together (4.38), (4.43) and (4.44), we obtain

$$
f_G(g) = \frac{e^{-\frac{2}{k_1}(g+L_{th})}}{k_1} \operatorname{erfc}\left( \frac{L_{th} - k_0 - k_1 \ln r_T}{\sqrt{2}\sigma} \right)
$$

$$
+ \left[ \operatorname{erfc}\left( \frac{-gk_1 - k_0 k_1 - k_1^2 \ln r_T + 2\sigma^2}{\sqrt{2}\sigma k_1} \right) \right.
$$

$$
\left. - \operatorname{erfc}\left( \frac{L_{th} k_1 - k_0 k_1 - k_1^2 \ln r_T + 2\sigma^2}{\sqrt{2}\sigma k_1} \right) \right] \frac{e^{-\frac{2k_1 g + 2k_0 k_1 - 2\sigma^2}{k_1^2}}}{r_T^2 k_1}. \tag{4.45}
$$

If $r_T \to \infty$ (infinite plane), the distribution of $G$ converges to

$$
f_G(g) = \frac{2}{k_1} e^{-\frac{2}{k_1}(g+L_{th})}, \tag{4.46}
$$

which does not depend on the shadowing spread $\sigma$. Although the channel fluctuations affect the number of nodes that can communicate with the receiver and the probability that a node is isolated, the comparison between (4.45) and (4.46) shows that the channel fluctuations do not give any contribution on the distribution of channel gain in case of infinite plane, but play a significant role when a finite area is considered.

## 4.6.2 Interference power characterization

It is worth noting that the previous analysis has been carried out by assuming that a node is able to communicate with $N_0$ if the received power is larger than threshold $P_{r_{min}}$. Unfortunately, in the presence of interference the measure of the received power is not sufficient to state whether the nodes can communicate or not. That means the single link quality characterization cannot be disjoined by a network level analysis. In this regard, it is important to characterize the distribution of the total received power due to the interference.

In case of infinite plane and PPP, when $L_{th}$ approaches infinite the distribution of $G$ tends to zero (see (4.46)). This behaviour can be easily explained by considering that nodes can be arbitrarily far away. In case of a finite plane, the distribution converges to a finite value (see (4.45)). This means each interferer node in the area could be a source of a significant contribution to the total received power. Furthermore, owing to the PPP model we have considered for the nodes' positions, the number of nodes in the area, denoted by $N$, is a random variable. To characterize the total amount of interference, let us assume that there are $n$ nodes in the area and consider the random variable $I_n = P_t(G_1 + \cdots + G_n)$, where $G_i$ represents the channel gain for the $i$th interferer and $P_t$ the transmitted power in linear scale. $I_n$ gives the amount of interference conditioned to the fact that the number of interferers is exactly $n$. If we evaluate the statistical average over $N$, we can evaluate the $p$th moment of the unconditional interference power $I$, that is,

$$\mu_I^{(p)} = \mathbb{E}\{I_n^p\} = \sum_{n=0}^{\infty} \mathbb{P}\{N = n\} \cdot \mu_{n,I}^{(p)} = \sum_{n=0}^{\infty} \frac{e^{-\mu_N}\mu_N^n}{n!}\mu_{n,I}^{(p)}, \tag{4.47}$$

where $\mu_{n,I}^{(p)}$ is the $p$th moment of $I_n$. In (4.47) we have used the property that when the spatial distribution of nodes in a circular area with radius $r_T$ follows a PPP, $N$ is Poisson r.v. with mean $\mu_N = \pi\rho r_T^2$. Recalling that $\mu_{G_i}^{(p)} \triangleq E[G_i^p] = \mu_G^{(p)}$ for each $i$ and $p$, we obtain

$$\mu_{n,I}^{(p)} = \mathbb{E}\{I_n^p\} = P_t^p \sum_{i_1=0}^{p} \cdots \sum_{i_{n-1}=0}^{i_{n-2}} \binom{p}{i_1} \cdots \binom{i_{n-2}}{i_{n-1}} \prod_{\ell=1}^{n} \mu_G^{(i_{\ell-1}-i_\ell)}$$

$$= P_t^p p! \sum_{i_1=0}^{p} \cdots \sum_{i_{n-1}=0}^{i_{n-2}} \frac{\prod_{\ell=1}^{n} \mu_G^{(i_{\ell-1}-i_\ell)}}{\prod_{m=1}^{n}(i_{m-1} - i_m)!}, \tag{4.48}$$

with $i_0 = p$, $i_n = 0$. Note that the propagation model (3.10) is only valid when the distance between the nodes is larger than some wavelengths. To account for this fact, we assume that the attenuation cannot be smaller than $k_0$. In the absence of channel fluctuations, this is equivalent to assuming a 'dead-zone' having a radius of 1 metre which is free from sensors. The expression for $\mu_G^{(p)}$ can be written as

$$\mu_G^{(p)} = \mathbb{E}\{G^p\} = \int_{1/L_{th}}^{1/k_0} x^p f_G(x) dx. \tag{4.49}$$

where $f_G(x)$ is given by (4.46). As special cases we can derive the first two moments of $I$

$$\mu_I \triangleq \mu_I^{(1)} = P_t \sum_{n=0}^{\infty} \frac{e^{-\mu_N} \mu_N^n}{n!} n\mu_G = P_t \mu_N \mu_G \tag{4.50}$$

and

$$\begin{aligned}\mu_I^{(2)} &= P_t^2 \mu_G^{(2)} \mu_N + P_t^2 \mu_G^2 \left[\mu_N^{(2)} - \mu_N\right] \\ &= P_t^2 \mu_N \left[\mu_G^{(2)} + \mu_G^2 \mu_N\right]\end{aligned} \tag{4.51}$$

where we have used the identities $\mu_{n,I}^{(2)} = P_t^2 \left[n\mu_G^{(2)} + n(n-1)\mu_G^2\right]$ and $\mu_N^{(2)} = \mu_N(1 + \mu_N)$.

We now show some numerical results and compare the approximate model with simulations. The following parameters have been fixed: $k_0 = 30\,\text{dB}$, $L_{th} = 104\,\text{dB}$, $\alpha = 4$.

Figure 4.8 shows the p.d.f. of the channel gain in logarithmic scale for different values of the radius $r_T$ (ranging from 0 to 200 m), $\sigma = 5$ dB. The figure shows that for small values of the radius there exists a pick in the distribution; in the case of large values of $r_T$, the distribution tends to be concentrated around the smallest values of $G$. The comparison between analysis and simulations shows an excellent agreement.

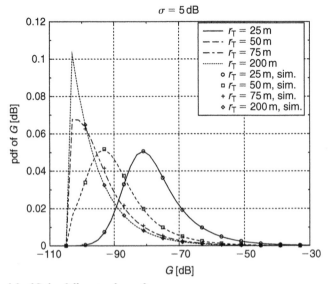

**Figure 4.8**   *p.d.f. of G for different values of* $r_T$

Figures 4.9 and 4.10 show the distribution of the SNR, defined as $\gamma = P_T G / \sigma_N^2$ for different values of the radius $r_T$ and of the standard deviation of the shadowing (ranging from 0 to 7.5), $P_t = 10^{-3}$ W and $\sigma_N^2 = 6.3 \cdot 10^{-10}$ W . $\sigma = 5$ in Figure 4.9 and $r_T = 25$ m in Figure 4.10. The comparison with simulation shows that the model, although approximate, is extremely tight. As expected, when the value of the radius increases, the distance between nodes increases and the values of $\gamma$ tend to become smaller.

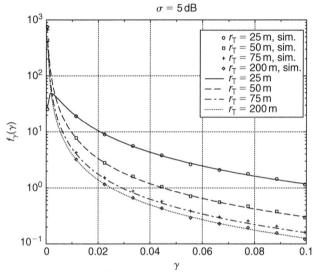

**Figure 4.9** *p.d.f. of $\gamma$ for different values of $r_T$*

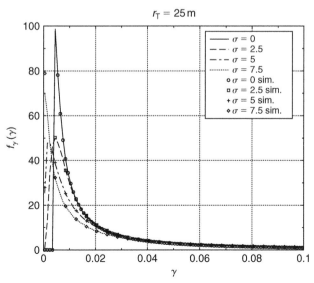

**Figure 4.10** *p.d.f. of $\gamma$ for different values of $\sigma$*

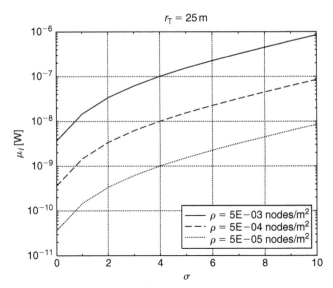

**Figure 4.11**   $\mu_I$ *as a function of $\sigma$ for different values of the nodes' density*

Figure 4.10 shows the role played by shadowing. Unlike bi-dimensional scenarios with infinite area, where shadowing does not influence the p.d.f. of $\gamma$ (Salbaroli & Zanella, 2006), in the case of scenarios with finite dimension the value of $\sigma$ has an impact on the distribution of the channel gain.

Finally, the mean value $\mu_I$ of the total interference $I$ is plotted in Figure 4.11 as a function of $\sigma$, where three different values of the nodes' density are considered and $r_T = 25\,\text{m}$. As expected, shadowing and the nodes' density have a significant influence on the overall interference.

As a concluding remark, note that in Section 4.4 we showed that the presence of log-normal fluctuations is beneficial from the connectivity point of view, because, on average, it increases the number of nodes that are audible from a give node. That does not necessarily imply that the single link (the sensor/sink pair) performs better: in fact the channel fluctuations may raise the number of interferers and hence lower the SIR, which is the real performance indicator at the link level. It is the task of the MAC to mitigate the effect of the interference through suitable channel access techniques.

## 4.7   Network connectivity

While so far we have considered connectivity relating couples of nodes, the aim of the present section is to focus on the entire network. We will start by introducing some general results taken from the literature that regard wireless ad-hoc networks, that is, those networks which contain nodes of the same kind. WSANs present nodes of at least two different types (sensors and sinks) and specific properties: nonetheless, the theorems discussed and the analysis carried out here constitutes a fundamental introduction to the problem of connectivity in WSANs, which will be addressed in Section 4.8.

Usually in the network connectivity theory, the network is modelled as a (random) graph where the vertices represent the nodes and the edges represent the capability of the end nodes to communicate. In an arbitrary network, the presence of an edge connecting two nodes indicates that a radio link is established between them. In this section the basic elements of the *graph theory* are introduced and some important results of graph connectivity are presented. Most of these results derive from the percolation theory which has been successfully developed in the past for completely different applications.

## 4.7.1 Elements of graph theory

In this paragraph a series of definitions related to the graph theory will be introduced, to help the reader in the comprehension of the rest of the chapter.

### Definitions

1. *Graph* A graph $G$ can be defined as a pair $(V, E)$, where $V$ is a set of vertices and $E$ is a set of edges between the vertices $E = (u, v)|u, v$ in $V$. Thus, a graph is a set of items (vertices or nodes) connected by edges, that is, with binary relation between them (adjacency relation). Typically, it is assumed that self-loop (i.e., edges of the form $(u, u)$, for some $u \in V$) are not contained in a graph.
2. *Directed or undirected graph* A graph $G = (V, E)$ is directed if the edge set is composed of ordered node pairs; whereas a graph is undirected if the edge set is composed of unordered node pairs. If the graph is undirected, the adjacency relation defined by the edges is symmetric, or $E = (u, v)$ with $u, v$ in $V$ (sets of vertices rather than ordered pairs).
3. *Geometric graph* (*GG*) A graph where vertices have a geometric location in $\mathbb{R}^d$.
4. *Random graph* (*RG*) A graph where edges between pairs of nodes exist according to random statistics.
5. *Geometric random graph* (*GRG*) An RG where edges exist according to proximity relation between nodes and nodes are in unknown positions. In GRGs, the set of nodes is normally finite, and their number deterministically known.
6. *Complete graph* An undirected graph with an edge between every pair of vertices.
7. *Acyclic graph* A graph with no path that starts and ends at the same vertex.
8. *Connected graph* A graph $G = (V, E)$ is connected if for any two nodes $u, v \in E$ there exists a path from $u$ to $v$ in $G$.
9. *Edge connectivity* The smallest number of edges whose deletion will cause a connected graph to not be connected.
10. *Node connectivity* The smallest number of vertices whose deletion causes a connected graph to not be connected.
11. *Degree of a node* The number of neighbours (connected vertices) of that node.

12. *Tree*   A connected, undirected, acyclic graph. It is a data structure accessed begin-
    ning at the root node, where each node is either a leaf or an internal node. An inter-
    nal node has one or more child nodes and is called the parent of its child nodes. All
    children of the same node are siblings. Contrary to a physical tree, the root is usu-
    ally depicted at the top of the structure, and the leaves are depicted at the bottom.
13. *Weighted graph*   A graph having a weight, or number, associated with each edge.
14. *Euclidean tree*   A tree in a weighted GG where weights are assigned to edges based
    on Euclidean distances.
15. *Spanning tree*   A connected, acyclic subgraph containing all the vertices of a graph.
16. *Minimum spanning tree (MST)*   A minimum-weight tree in a weighted graph which
    contains all of the graph's vertices.
17. *Euclidean minimum spanning tree (EMST)*   A MST where each connection has a
    weight which corresponds with the Euclidean distance between nodes.
18. *Steiner Tree*   A minimum-weight tree connecting a designated set of vertices, called
    terminals, in an undirected, weighted graph. The tree may include non-terminals, which
    are called Steiner vertices.

## 4.7.2   Communication graph and critical transmission range

In graph theory, a graph is an abstract entity. To better fit the real networks, the concept of
*communication graph* can be introduced. In a real network, in fact, the existence of a link
between two entities, $u$ and $v$ (sensors, or ad hoc network nodes) depends on their distance,
the transmission power used to send packets, and on the scenario in which the network is
located. In general, a *network* is a pair $(N, L)$, where $N$ is a set of wireless nodes, having
size $n$, located in a unit square. $L$ is the function mapping every node $u$ to a position $L(u)$.

A *range assignment (RA)* for a network $(N, L)$ is a function assigning to every node $u \in N$
a transmit range $RA(u) \in (0, r]$, where $r$ represents its TR, that is, the radius of the circumference inside which data transmitted from $u$ can be received correctly. Usually the
disk model (4.2) for link connectivity is assumed for mathematical convenience.

The *communication graph (CG)* is the directed graph $(G, E)$ where the directed edge
$(u, v)$ exists if the Euclidean distance between $u$ and $v$ is less or equal than $RA(u)$. In
this case $v$ is a neighbour of $u$. If $u$ is also a neighbour of $v$ for all pairs $(u, v)$, the CG is
undirected and all links are symmetrical. An RA for a network is connecting if the correspondent CG is connected. An RA where all nodes have the same transmit range $r$ is
said to be $r$-homogeneous.

Many studies related to connectivity and to topology control on wireless ad hoc and
sensor networks are focused on the research of the minimum value of the transmission
range which allows the network to be a connected CG.

*Definition – critical transmission range (CTR) for connectivity.* The minimum value $r_c$ of
the transmission range $r$ such that the $r$-homogeneous RA is connected. If the nodes in

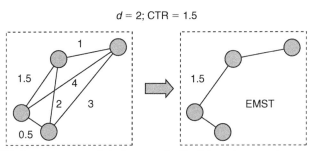

**Figure 4.12** *CTR for connectivity. Labels report the Euclidean distances between nodes*

the network are randomly distributed, the CTR is a random variable. In Santi, 2005 it is shown that the CTR, $r_c$, for connectivity of a network with nodes distributed in a region $R = [0, L]^d$, with $d = 1, 2, 3$, equals the length of the longest edge of the EMST of the corresponding communication graph. In Figure 4.12 an example of CTR required for connectivity is shown. Specifically, the following theorem exists:

**Theorem** – *Let N be a set of n nodes placed in R = $[0, 1]^d$, with d = 1,2 or 3. The CTR for connectivity, $r_c$, of the network composed of nodes in N equals the length of the longest edge of the EMST built on the same set of nodes (Santi, 2005).*

According to the above theorem, the evaluation of the CTR requires to build the EMST of the network and find the longest edge in the EMST. Unfortunately, this way to compute the CTR is not suitable to distributed implementation, since the building of the EMST requires global knowledge (e.g., all nodes' positions), which can be acquired in a distributed setting only by exchanging a considerable amount of information.[5] This is the reason why considerable attention has been devoted to characterize the CTR in the presence of some form of uncertainty about nodes' positions. If nodes' positions are not known, the minimum value of $r$ ensuring connectivity in all possible cases is $r \simeq L\sqrt{d}$, since nodes could be located at the opposite corners of $R$. However, this is very conservative, thus a typical approach is to assume that nodes are distributed in $R$ with a certain distribution $F$ and to study conditions for asymptotically almost certain connectivity.

If the nodes in the network are randomly distributed, the CTR is a random variable. The statistic characterization could be useful to answer fundamental questions that arise during the network design phase, such as: given a number $n$ of nodes uniformly distributed in a region $R$ and given a certain node's spatial distribution $F$, what is the minimum value $r_c(n, F)$ of the transmission range that ensures connectivity of the network with high probability?; and then, given a transmitter technology and a distribution $F$, what is the minimum number $n_c(r, F)$ of nodes to be deployed in order to realize a connected network with high probability?

To answer these questions some numerical results have been obtained through a simulation tool. The reference scenario considered consists of a unit square region, $R = [0, 1]^2$,

---

[5] For example, there exist scenarios in which positions are not known a priori, as in many sensor network application scenarios.

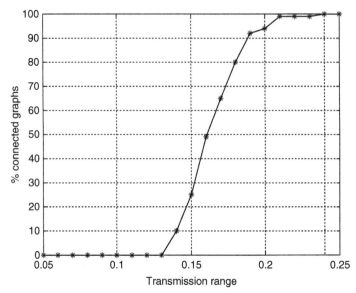

**Figure 4.13** *Percentage of connected graphs as a function of the transmission range for n = 100.*

where a number *n* of nodes are randomly and uniformly distributed. The objective of the study is to evaluate the percentage of graphs connected by varying the value of the transmission range. To this purpose, 100 different network topologies, characterized by different nodes positions in the area, have been considered. For each topology, the network connectivity has been tested.

The existence of a link between nodes has been verified using the Dijkstra (Dijkstra, 1959) routing algorithm which allows the identification of the minimum length (characterized by the minimum Euclidean distance) path between the two nodes.

In Figure 4.13 the percentage of connected graphs, obtained in a network constituted by 100 nodes (*n* = 100), with the procedure described above, as a function of the transmission range, is shown.

As we can see, by increasing the transmission range, the percentage increases. For example, in case we want to obtain a network connected with a probability of 99%, the transmission range value must be about 0.23.

In Figure 4.14 a comparison between the percentage of connected graphs, obtained for different values of number *n* of nodes, is shown. In particular, we consider the cases *n* = 50, *n* = 100, *n* = 200 (once again each simulation has been made by considering 100 different network topologies). As we can see, by increasing the number of nodes in the network, that is, increasing the nodes' density, the transmission range needed to obtain the same percentage of graphs connected decreases. Moreover, the fully connectivity of the network is reached for lower values of the transmission range. Since the CTR is equal to

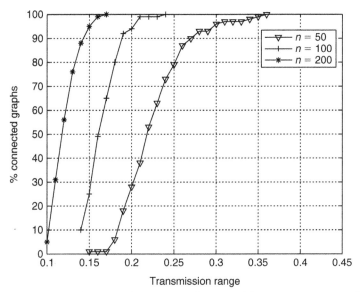

**Figure 4.14** *Percentage of connected graphs as a function of the transmission range for different values of* n

the longest EMST edge, statistical solutions to the CTR problem in dense networks can be derived using results concerning the asymptotic distribution of the longest EMST edge.

**Theorem (Penrose, 1997)** – *Given the unit square* $[0, 1]^2$ *and* n *nodes distributed randomly and uniformly, and being* $M_n$ *the random variable denoting the length of the longest edge of the MST built with the* n *nodes, then*

$$\lim_{n \to \infty} \mathbb{P}\{n\pi(M_n)^2 - \log n \le \beta\} = \frac{1}{\exp(e^{-\beta})}, \tag{4.52}$$

for any $\beta \in \mathbb{R}$ (Penrose, 1997). It is clear that $M_n$ indicates also the r.v. CTR.

**Corollary** – *For* n *approaching infinity and chosen* $\beta = f(n)$*, with* $f(n)$ *being a function that approaches infinity when* n → ∞*, we have*

$$\mathbb{P}\{CTR \le \sqrt{(\beta + \log n)/n\pi}\} = 1. \tag{4.53}$$

In other words, $\sqrt{(\beta + \log n)/n\pi}$ is an upper bound for $r_c$ when $n$ approaches infinity and since it tends to zero for proper selections of $\beta$, it is a very tight upper bound.

For finite values of $n$, (4.53) does not necessarily represent an upper bound. To prove this, a comparison between the CTR value obtained through simulations and the correspondent theoretical value has been made. The asymptotic value of the CTR is obtained by setting $\beta = \log \log n$ in (4.53). Instead, the experimental CTR values are obtained by considering $n$ nodes uniformly distributed at random in $[0, 1]^2$, and by evaluating the transmitting range yielding 99% of connected communication graphs. In Table 4.1 the two CTR values for

**Table 4.1**  *Comparison of the values of* $r_c$ *obtained from the Penrose theorem and through simulations with the 99% of confidence, for different values of* n.

| $n$ | Theoretical $r_c$ (Penrose theorem) | $r_c$ simulated |
|-----|-------------------------------------|-----------------|
| 10 | 0.3160 | 0.6587 |
| 100 | 0.1397 | 0.2349 |
| 1000 | 0.0530 | 0.0773 |

different values of $n$ are shown; as we can see, for low values of $n$ the two values of $r_c$ are very different, but also when $n = 2500$, the relative gap between the theoretical and actual CTR is still in the order of 28%. Thus the problem is that the theorem allows the evaluation of the CTR value, with a negligible error, not for practical values of $n$.

## 4.7.3   The giant component

The *giant component* (*GC*) is a phenomenon appearing also for practical values of $n$. Let us consider a process in which all network nodes have initially transmission range $r = 0$ that is successively increased. As the range increases, new edges arise in the communication graph and some fully connected sub-graphs appear in the network, even if the global network is not fully connected (*clusterization effect*). We consider two events: the event in which the last isolated node disappears from the communication graph and the event in which the communication graph becomes connected. We denote the ranges corresponding to the two events as $r_1$ and $r_c$, respectively. It is clear that $r_1 \leqslant r_c$. The following Penrose theorem states that, for large values of $n$, the condition $r_1 = r_r$ is asymptotical almost surely.

**Theorem (Penrose, 1999a)** – *Assume* n *points are distributed in* R $= [0, 1]^d$ *according to the uniform distribution, with* d $= 2, 3$. *Then,* $\lim_{n \to \infty} \mathbb{P}(r_1 = r_c) = 1$.

An important consequence of the above theorem is that, for low values of $n$, when the last isolated node disappears the communication graph of the network is constituted, with high probability, by a very big cluster ('giant component') and only few other nodes or small clusters. This phenomena is shown in Figures 4.15, 4.16 and 4.17, where the percentage of fully connected graphs and the average percentage of nodes included in the GC are reported for $n = 50$, 100 and 200, respectively. Numerical results are obtained through simulations by considering a network composed of $n$ nodes uniformly distributed in $R = [0, 1]^2$. For example, if we consider a transmission range $r = 0.14$ with $n = 100$, even if the rate of graphs connected is only 0.1, there is an average percentage of nodes included in the GC of 90%. When $n = 200$ there is only the 5% of fully connected graphs whereas the GC is constituted by the 92% of network nodes.

From these results it can be observed that, in general, the network is close to be fully connected even with a transmission range significantly smaller than $r_c$. In case the application can tolerate the presence of some isolated nodes, it could be possible to use a reduced transmission range with a consequent energy consumption saving.

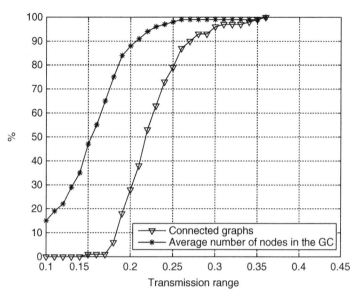

**Figure 4.15** *Giant component in the case* n = 50

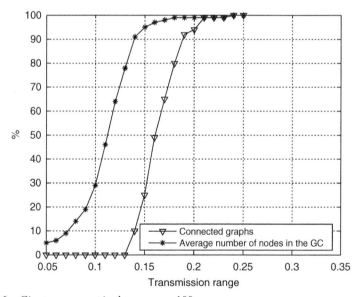

**Figure 4.16** *Giant component in the case* n = 100

## 4.7.4 Probability of node isolation and connectivity

In the previous section we have seen that the condition where the last isolated node disappears is strictly related to graph connectivity. For this reason, the graph connectivity characteristics can be analyzed in a more convenient way through the investigation of the *probability of node isolation*.

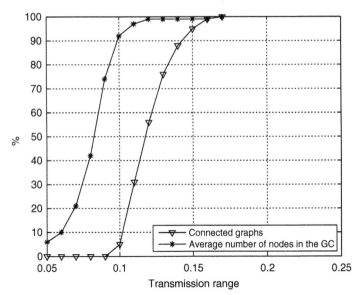

**Figure 4.17**   *Giant component in the case* n = 200

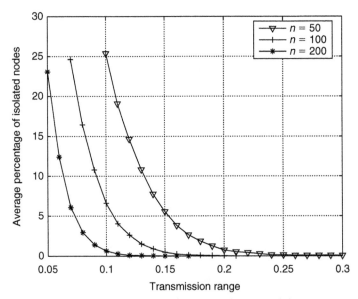

**Figure 4.18**   *Average percentage of isolated nodes as a function of the transmission range for different values of* n

A node is isolated when it experiments, on all the possible links, a loss larger than the threshold $L_{th}$. For example, in Figure 4.18 the average percentage of isolated nodes as a function of the transmission range, in the absence of shadowing, by varying the number of nodes in the network, is shown. The scenario is composed of a fixed number of nodes

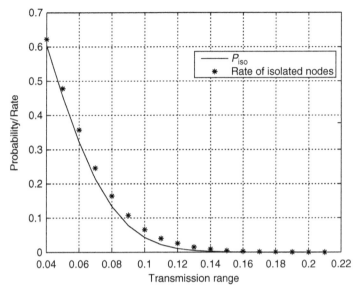

**Figure 4.19** *Rates of isolated nodes by simulation and by theoretical analysis: the effective rate is larger because of border effects*

uniformly distributed in a unitary square area. As we can note, by increasing the network size the performance improves.

In case nodes are distributed in a uniform way in the bidimensional infinite plane, the probability that a node is isolated is given by (4.11) and (4.13). In case, instead, nodes are distributed in a finite region of size $L$ (as in the scenario considered in the following simulation results), (4.11) is accurate only when the average area covered by a single node is much smaller than the region area $L^2$. When border effects are not negligible, (4.11) gives a lower bound of the actual probability of node isolation. This is due to the fact that the nodes close to the border have a greater probability to be isolated because no nodes are present outside the region.

In Figure 4.19, the probability of node isolation, $P_{iso}$, derived using (4.11), and the rate of isolated nodes, obtained through simulations, are shown as a function of the transmission range and in the absence of shadowing and $L = 1$. As we can see simulation results are very close to the theoretical analysis results.

Another important statistic of interest is the probability that there are *no isolated nodes* in the network (i.e., is the probability that all the $N$ nodes composing the network are connected). Considering, as first approximation, the probability that a node is isolated independent from node to node, when $N \gg 1$ and $r_0/L \ll 1$ (border effects negligible), it is

$$\mathbb{P}\{\text{no isolated nodes} \,|\, N = n\} \simeq (1 - P_{iso})^n. \tag{4.54}$$

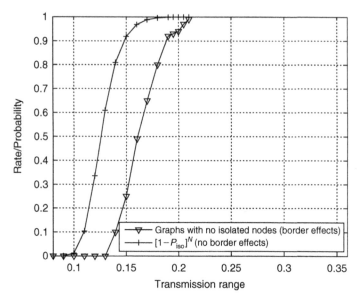

**Figure 4.20** *Rates of non-isolated nodes by simulation and by theoretical analysis: the effective rate is lower because of border effects*

In Figure 4.20 the probability defined by (4.54) and the actual rate of graphs in which no isolated nodes are present, obtained through simulations, are compared. Simulations consider a finite area with $L = 1$ whereas theoretical results do not consider border effects. By fixing a certain transmission range, the rate of graphs where there are no isolated nodes is lower than the theoretical probability that there are not nodes isolated. Thus, when border effects are present, a larger transmission range is required to achieve the same probability obtained by neglecting border effects.

To verify the statistical independence assumed in (4.54) for the different number of nodes, in Figure 4.21 we compare the rate of graphs with no isolated nodes obtained by simulation and the probability of obtaining a network with no isolated nodes calculated as $[1 - \text{(rate of isolated nodes)}]^N$: the independence still holds in the presence of border effects.

Finally, in Figure 4.22 a comparison of the rate of fully connected graphs and the rate of graphs without isolated nodes, is provided. Results are obtained through simulations by considering 100 nodes uniformly and randomly distributed in the unitary square area and considering 100 different network topologies. As stated in Section 4.2, the absence of isolated nodes is a necessary, but not sufficient condition to have a fully connected network. Thus, having fixed the number of nodes in the network, $N$, the transmission range needed to obtain a certain probability $p$ that there are no isolated nodes, is a lower bound for the transmission range needed to obtain a fully connected network with the same probability $p$.

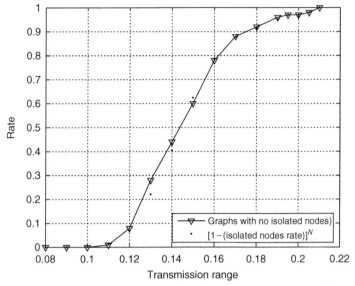

**Figure 4.21**  *Rate of graphs with no isolated nodes obtained by simulation and the probability of obtaining a network with no isolated nodes calculated as* $[1 - (rate\ of\ isolated\ node)]^N$

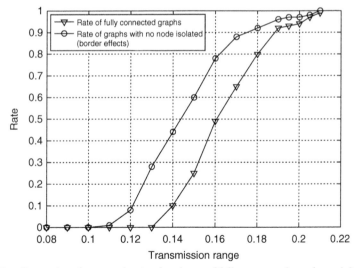

**Figure 4.22**  *Comparison between the simulated rate of fully connected graphs and the simulated rate of graphs with no isolated nodes*

As previously stated, the Penrose theorem (1999) introduced in Section 4.7.3 is veri-fied for dense networks ($N \gg 1$), having high $\mathbb{P}\{$no isolated nodes$\}$; thus, based on that result and on Figure 4.22, we can write the probability of fully connected network

$$\mathbb{P}\{con\} = \mathbb{P}\{no\ isolated\ nodes\} - \epsilon ;$$
$$\epsilon \to 0 \quad if\ \mathbb{P}\{no\ isolated\ nodes\} \to 1 \tag{4.55}$$

From simulation results, and also from the previous theoretical considerations, we can deduce that the probability that a network, with a number $N \gg 1$ of uniformly distributed nodes in a square area having side $L = 1$, is fully connected, has an upper bound given by the probability that there are no isolated nodes in the network. The bound becomes tighter as $\mathbb{P}\{$no isolated nodes$\} \rightarrow 1$.

So far we have considered connectivity issues for a network composed by a fixed number of nodes uniformly distributed in a square area. Here we extend the mathematical model to study the network connectivity in the presence of a random number of nodes uniformly distributed in a finite region $A$. In particular we refer to a PPP distribution of nodes, distributed with density $\rho$ in a square region having side $L$. Thus, we do not know the exact number $N$ of nodes in the area, but we know its statistic; from (4.1) it is

$$\mathbb{P}\{N = n\} = \frac{(\rho A)^n}{n!} e^{-\rho A}. \tag{4.56}$$

We can then evaluate the probability that there are no isolated nodes as

$$\mathbb{P}\{\text{no isolated nodes}\} = \sum_{n=0}^{\infty} P(N = n) \mathbb{P}\{\text{no isolated nodes}| N = n\}$$

$$= \sum_{n=0}^{\infty} \frac{(\rho A)^n}{n!} e^{-\rho A} \cdot (1 - P_{\text{iso}})^n = e^{-\rho A \cdot P_{\text{iso}}}, \tag{4.57}$$

whereas the probability $P_{\text{iso}}$ that a node is isolated is given by (4.11).

Thus the probability that there are no isolated nodes in $A$, is

$$\mathbb{P}\{\text{no isolated nodes}\} = \exp[-\rho A \cdot \exp(-\mu)], \tag{4.58}$$

where $\mu$ can be derived from (4.15). The above probability is exact when border effects are not present, thus when nodes outside $A$ can act as intermediate nodes to connect nodes inside $A$. Moreover, according to the Penrose theorem, (4.58) is an upper bound for the probability in (4.55) that all nodes in $A$ are connected (i.e., there exists a link between each couple of nodes in the area).

To verify this model we have considered a scenario constituted by a simulation square area with side of $L = 2000$ metres, which contains another square area $A$ with side $L_{\text{int}} = 1000$ metres (see Figure 4.24). To minimize border effects, only the nodes within $A$ are accounted for in determining the performance. We have fixed the transmission range in nominal conditions to $r_0 = 160$ [m]; in this way nodes inside $A$ can be connected through nodes outside $A$ (but inside the square of side $L$), being $(L - L_{\text{int}})/2 > r_0$. Simulation results have been obtained by considering 500 different network topologies. In Figure 4.23 the probability that there are no isolated nodes, given by (4.58), as a function of nodes' density for different values of shadowing spread $\sigma$, is shown. As can be seen, in the presence of channel fluctuations (larger $\sigma$), the probability that there are not isolated nodes increases, that is, channel randomness increases the network connectivity, as previously anticipated.

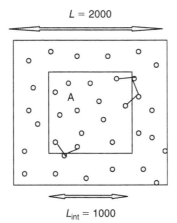

Figure 4.23 (L = 2000 at top, $L_{int}$ = 1000 at bottom, A labeled inside)

**Figure 4.23** *Reference scenario considered in the simulations*

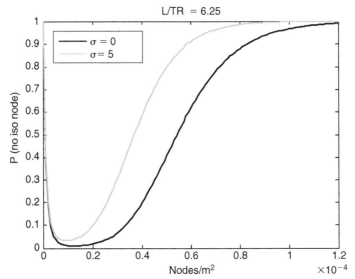

**Figure 4.24** *Probability that there are no isolated nodes as a function of nodes' density, for different values of σ*

In Figures 4.25 and 4.26, a comparison between the rate of graphs without isolated nodes (obtained through simulations) and the probability that there are no isolated nodes $\mathbb{P}\{\text{no isolated nodes}\}$, obtained through (4.58), is provided for different values of $\sigma$. In both cases, simulation results are absolutely consistent to those obtained through the theoretical model. For comparison purposes, the rate of fully connected graphs, obtained through simulations without border effects, is also reported. It can be noted that, independently from $\sigma$, the rate of graphs without isolated nodes results to be an upper bound for the rate of fully connected graphs. The bound is achieved for high nodes density. For example, the difference between the two rates becomes negligible when $\mathbb{P}\{\text{no isolated nodes}\} < 0.9$ for $\sigma = 0$ and $\mathbb{P}\{\text{no isolated nodes}\} < 0.8$ for $\sigma = 5$. Summarizing, we can

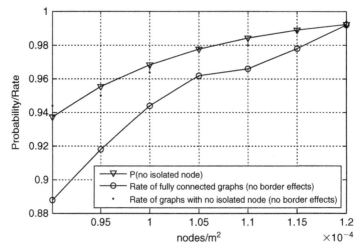

**Figure 4.25**  *Rate of graphs with no isolated nodes and probability that there are no isolated nodes in the absence of shadowing*

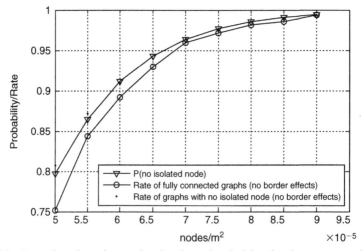

**Figure 4.26**  *Rate of graphs without isolated nodes and probability that there are no isolated nodes in the case σ = 5*

evaluate the minimum nodes density, $\rho$, required to have $\mathbb{P}\{$no isolated nodes$\} = 0.99$ ($\rho \approx 1.15 \cdot 10^{-4}$ [nodes/m$^2$] when $\sigma = 0$, $\rho \approx 8.50 \cdot 10^{-5}$ [nodes/m$^2$] when $\sigma = 5$) and use this value as a tight lower bound for the node density required to obtain $\mathbb{P}\{$con$\} = 0.99$ ($\rho \approx 1.2 \cdot 10^{-4}$ [nodes/m$^2$] for $\sigma = 0$, $\rho \approx 9 \cdot 10^{-5}$ [nodes/m$^2$] for $\sigma = 5$).

## 4.7.5  Probability of node isolation based on nodes degree

In this section, a different approach to evaluate the required CTR for achieving a certain $\mathbb{P}\{$no isolated nodes$\}$ is presented (Bettstetter & Zangl, 2002).

Let us consider the following scenario: $n$ nodes are independently placed at random, with uniform distribution, on an area of size $A$. The deterministic disk model (4.2) is considered where $r_0$ is the TR for each node. We are interested in finding the probability that the minimum node degree of the network, $d_{min}$, is greater than 0. If $d_{min} > 0$ then no isolated nodes will be present in the network. Because each node has a range $r_0$, it covers an area $A_0 = r_0^2 \pi$, where $A_0 \ll A$. The probability that a node has $i$ neighbours, i.e., the node degree $d$ is $i$, $\mathbb{P}\{d = i\} = (\mu^i/i!)e^{-\mu}$, where from (4.14) $\mu = \rho A_0 = nA_0/A = \rho r_0^2 \pi$ is the expected node degree $\mathbb{E}(d)$. In this derivation, we ignored border effects. Clearly, nodes close to the border cover an area smaller than $r_0^2 \pi$, and therefore the expected node degree in the network decreases, that is, $\mathbb{E}(d) < \rho r_0^2 \pi$. Thus, the above result is only applicable if border effects are not present or if methods to avoid border effects are applied. For example, no border effects occur if we regard an area $C$ which is a inner sub-area of $A$ and each borderline of $A$ is at least $r_0$ away from the borderline of $C$. The remaining area $B = A \backslash C$ is denoted as guard area. Only nodes in $C$ are considered for evaluation, but these nodes may have links to nodes in $B$.

However, ad hoc networks in real life also have natural borders and taking them into account leads to a more realistic result. Let us consider the area $A$ to be circular with radius $R$ and define the two disjoint sub-areas: a *central area* $C$ ($0 \leqslant r \leqslant R - r_0$), where all nodes do not suffer from border effects, and a *border area* $B$ ($R - r_0 < r \leqslant R$), where nodes do suffer from these effects.

The probability that a random chosen node in $C$ is isolated is $\mathbb{P}\{d = 0\} = e^{-\mu}$. The probability that none of the $n$ nodes in $C$ is isolated is $\mathbb{P}\{d_{min} \neq 0\} = (1 - e^{-\mu})^n$, since the probabilities $\mathbb{P}\{d = 0\}$ are 'almost independent' for $n \gg 1$ and $r_0/R \ll 1$. Nodes at the borders instead have higher isolation probability, since they cover a smaller area.

Our goal is to compute a lower bound for $\mathbb{P}\{d_{min} \neq 0\}$ accounting for both the sub-areas $B$ and $C$. In the following, the random variables $X(B)$ and $X(C)$ denote the number of nodes in $B$ and $C$, respectively; and $Y(A)$, $Y(B)$, and $Y(C)$ denote the number of isolated nodes in $A$, $B$, and $C$, where $Y(A) = Y(B) + Y(C)$. There are no isolated nodes in the network if and only if $Y(B) = Y(C) = 0$. We first derive an expression for $\mathbb{P}\{Y(C) = 0\}$, then consider $\mathbb{P}\{Y(B) = 0\}$, and finally give a lower bound for $\mathbb{P}\{Y(A) = 0\} = \mathbb{P}\{d_{min} \neq 0\}$.

- *Central area*   If we assume that $n_C$ nodes are located in $C$, we obtain as usual

$$\mathbb{P}\{Y(C) = 0 | X(C) = n_C\} = (1 - e^{-\mu})^{n_C}, \tag{4.59}$$

with $\mu = \rho A_0 = \rho r_0^2 \pi$. We do not know how many nodes $n_C$ are located in $C$, but we can write

$$\mathbb{P}\{Y(C) = 0\} = \sum_{n_C=0}^{\infty} \mathbb{P}\{Y(C) = 0 | X(C) = n_C\} \cdot \mathbb{P}\{X(C) = n_C\}$$

$$= \sum_{n_C=0}^{\infty} (1 - e^{-\rho A_0})^{n_C} \frac{(\rho C)^{n_C}}{n_C!} e^{-\rho C}. \tag{4.60}$$

By defining $\mu_* \triangleq (1 - e^{-\rho A_0})\rho C$ and considering that $\sum_{n_C=0}^{\infty} ((\mu_*)^{n_C}/n_C!)e^{-\mu_*} = 1$, we obtain

$$\mathbb{P}\{Y(C) = 0\} = \exp(-\rho C e^{-\rho A_0}). \tag{4.61}$$

This term computes, in general, the probability that there is no isolated nodes in an area that does not suffer from border effects.

- *Border area*   The range of nodes in $B$ covers an area that is smaller than $r_0^2\pi$, that is, $A_0(B) < r_0^2\pi$. In the worst case, a node is located at $r = R$ and covers

$$A_0(R) = r_0^2 \arccos \frac{r_0}{2R} + R^2 \arccos \frac{2R^2 - r_0^2}{2R^2} - \frac{r_0}{2}\sqrt{4R^2 - r_0^2}. \tag{4.62}$$

We denote this worst case coverage as $A_0(B)_{\min} = A_0(r = R)$ and can state

$$\mathbb{P}\{Y(B) = 0\} > \exp(-\rho B e^{-\rho A_0(B)_{\min}}). \tag{4.63}$$

For low values of $r_0/R$, one can simplify $A_0(R) \approx \frac{1}{2}r_0^2\pi$.

- *Complete system area*   The product of the right-hand side of (4.61) and (4.63) is a lower bound for $\mathbb{P}\{Y(A) = 0\}$, and the following theorem can be given.

**Theorem: a lower bound for $\mathbb{P}\{d_{\min} \neq 0\}$**   *In an ad hoc network with n $\gg$ 1 randomly uniformly distributed nodes on a disk of radius R, each node with transmission range $r_0 \ll$ R, we have*

$$\mathbb{P}\{\text{no isolated nodes}\} = \mathbb{P}\{d_{\min} \neq 0\} > \exp(-\rho(Ce^{-\rho A_0} + Be^{-\rho A_0(R)})). \tag{4.64}$$

with $C = (R - r_0)^2\pi$, $B = (2R \cdot r_0 - r_0^2)\pi$, $A_0 = r_0^2\pi$, and $A_0(R)$ according to (4.62). The value of (4.64) depends only on $n$ and the ratio of $r_0/R$.

The bound that we have obtained turns out to be tight only for values of $\mathbb{P}\{\text{no isolated nodes}\}$ close to 1. In fact we applied the worst case coverage area of a node located at the border $r = R$ for all nodes in $B$ and this leads to an underestimation of $\mathbb{P}\{\text{no isolated nodes}\}$. However, the approach can be generalized and applied in an iterative fashion in order to increase the accuracy as much as desired. The idea is to divide the area $B$ into $N - 1$ concentric rings $B_1, B_2, \ldots, B_{N-1}$, so that we have $N$ sub-areas in total. For each of these border areas $B_i$, $1 \leq i \leq N - 1$, we calculate the worst case coverage $A_0(B_i)_{\min}$, and can then write

$$\mathbb{P}\{Y(B_i) = 0\} > \exp(-\rho B_i e^{-\rho A_0(B_i)_{\min}}). \tag{4.65}$$

Clearly, $A_0(B_i)_{\min} \geq A_0(B)_{\min}$ hold $\forall i$, and thus $\prod_{i=1}^{N-1} \mathbb{P}\{Y(B_i) = 0\}$ is a tighter bound for $\mathbb{P}\{Y(B) = 0\}$ than the right-hand side of (4.63).

## 4.8   Network connectivity for WSANs

We consider now an actual WSN in a limited area, which is composed of some sensor nodes (with density $\rho_s$) and some sink nodes (with density $\rho$). A sensor is not isolated provided that it can directly transmit its data to a sink and, unlike the ad hoc case, the network is fully connected if no sensors are isolated.

Let us consider a squared region A: if $k$ is the number of sensors in A and $n$ the number of sensors that can hear at least one sink, we want to estimate the probability that $n = k$, denoted as $p_n(n = k)$.

Since a PPP has been considered, $k$ will be Poisson distributed and the probability $Z$ that the WSN is fully connected can be expressed as

$$Z = \mathbb{E}_k\{p_n(n = k)\} = \sum_{k=1}^{\infty} p_n(n = k) \cdot \frac{e^{-\rho_s A}}{k!}(\rho_s A)^k. \qquad (4.66)$$

The following section will focus on computing $p_n(n = k)$.

As explained in Orriss & Barton, 2003, let us define the random variable $N_{r_1, r}$ to be the number of sinks between the distances $r_1$ and $r$ from the sensor node with power loss $\leq L_{th}$, and $P_n(r)$ the probability that this number is exactly $n$. Then, for $n \geq 1$

$$P_n(r + \delta r) = P_n(r)[1 - 2\pi\rho r\delta r] + P_{n-1}(r)2\pi\rho r\delta r\Phi\left(\frac{L_{th} - k_0 - k_1 \ln r}{\sigma}\right)$$

$$+ P_n(r)2\pi\rho r\delta r\left[1 - \Phi\left(\frac{L_{th} - k_0 - k_1 \ln r}{\sigma}\right)\right] + o(\delta r). \qquad (4.67)$$

where we consider the log-normal link connectivity model (4.7) and $\Phi(x) = \int_{-\infty}^{x}(1/\sqrt{2\pi})e^{-u^2/2}\,du$.

The three products on the right correspond to the three possible cases:

1. exactly $n$ sinks between $r_1$ ed $r$, none between $r$ ed $r + \delta r$;
2. exactly $n - 1$ sinks between $r_1$ ed $r$, and one, whose loss does not exceed $L_{th}$, between $r$ and $r + \delta r$;
3. exactly $n$ sinks between $r_1$ ed $r$, and one, whose loss does exceed $L_{th}$, between $r$ and $r + \delta r$.

Writing $a_1 \triangleq (L_{th} - k_0)/\sigma$, $b_1 \triangleq k_1/\sigma$, (4.67) becomes, after rearrangement in terms of the incremental ratio of $P_n(r)$ and letting $\delta r \to 0$

$$P'_n(r) = 2\pi\rho r\Phi(a_1 - b_1 \ln r)[P_{n-1}(r) - P_n(r)], \qquad n \geq 1. \qquad (4.68)$$

Similarly,

$$P_0'(r) = -2\pi\rho r\Phi(a_1 - b_1 \ln r)P_0(r). \tag{4.69}$$

Define the probability generating function

$$\Pi(r, z) = \sum_{n=0}^{\infty} P_n(r)z^n. \tag{4.70}$$

From (4.68), (4.69), (4.70),

$$\frac{\partial}{\partial r}(\ln \Pi) = 2\pi(z - 1)\rho r\Phi(a_1 - b_1 \ln r). \tag{4.71}$$

Define

$$\Psi(a_1, b_1; r) = r^2\Phi(a_1 - b_1 \ln r) - e^{\frac{2a_1}{b_1} + \frac{2}{b_1^2}}\Phi(a_1 - b_1 \ln r + 2/b_1), \tag{4.72}$$

which is an indefinite integral of $2r\Phi(a_1 - b_1 \ln r)$. Hence

$$\ln \Pi = \pi(z - 1)\rho[\Psi(a_1, b_1; r) + G(z)], \tag{4.73}$$

where

$$G(z) \equiv -\Psi(a_1, b_1; r_1). \tag{4.74}$$

The solution of the equation is, therefore

$$\ln \Pi = \pi(z - 1)\rho[\Psi(a_1, b_1; r) - \Psi(a_1, b_1; r_1)], \tag{4.75}$$

so the distribution of 'audible' sinks within a range of distances from a generic sensor node is Poisson with mean

$$\mu_{r_1, r} = \pi\rho[\Psi(a_1, b_1; r) - \Psi(a_1, b_1; r_1)]. \tag{4.76}$$

The (4.76) can be adjusted to show that the number of audible sinks within a sector of an annulus having radii $r_1$ and $r$ and subtending an angle $2\theta$ to a generic sensor, is Poisson distributed with mean

$$\mu_{r_1, r; \theta} = \theta\rho[\Psi(a_1, b_1; r) - \Psi(a_1, b_1; r_1)], \quad 0 \le \theta \le \pi. \tag{4.77}$$

If the annulus extends from $r$ to $r + \delta r$, and $\theta = \theta(r)$, this mean value becomes

$$\mu_{r, r+\delta r; \theta(r)} = \theta(r)\rho\frac{\delta\Psi(a_1, b_1; r)}{\delta r}\delta r, \quad 0 \le \theta \le \pi. \tag{4.78}$$

As a consequence, if a region is located within the two radii $r_1$ and $r_2$ and its points at a distance $r$ from the sensor node are defined by a $\theta(r)$ law, then the number of sinks in this region are again Poisson distributed with mean

$$\mu_{r_1,r_2;\theta(r)} = \int_{r_1}^{r_2} \theta(r)\rho\,\frac{d\Psi(a_1,b_1;r)}{dr}\,dr, \tag{4.79}$$

that is, from (4.72),

$$\mu_{r_1,r_2;\theta(r)} = \int_{r_1}^{r_2} 2\theta(r)\rho r\Phi(a_1 - b_1\ln r)dr. \tag{4.80}$$

At this point, equation (4.80) is suitable for an arbitrary point $(x, y)$ in $A$, provided that such point is considered as a new origin and that the bounds of $A$ are expressed with respect to the new origin. This allows us to obtain a mean value $\mu(x, y)$ of sinks audible in the location $(x, y)$. The diagram in Figure 4.27 shows what happens for a point $(x, y)$. Without loss of generality, we can refer to a square of side 2 and take $0 \leqslant y \leqslant x \leqslant 1$, that is, the lower half of the first quadrant. The axes are transformed so that $(x, y)$ is the new origin. Then, the area inside the square is divided into 8 sub-regions by circles whose centres lie on the new origin. The radius, $r$, of each circle determines what portion of the circle lies inside the square, and also the value of $\theta(r)$. As an example, in the innermost region (region 1), the circle with radius $r$ lies completely in $A$, so $\theta(r) = \pi$; when $r$ grows, the portion of the

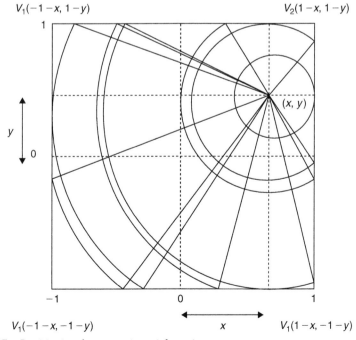

**Figure 4.27** *Partitioning the square into eight regions*

**Table 4.2**　*Boundary values of r in the eight regions*

| Region | Range |
|---|---|
| 1 | $0 \le r \le 1 - x$ |
| 2 | $1 - x \le r \le 1 - y$ |
| 3 | $1 - y \le r \le ((1-x)^2 + (1-y)^2)^{\frac{1}{2}}$ |
| 4 | $((1-x)^2 + (1-y)^2)^{\frac{1}{2}} \le r \le 1 + y$ |
| 5 | $1 + y \le r \le ((1-x)^2 + (1+y)^2)^{\frac{1}{2}}$ |
| 6 | $((1-x)^2 + (1+y)^2)^{\frac{1}{2}} \le r \le 1 + x$ |
| 7 | $1 + x \le r \le ((1+x)^2 + (1-y)^2)^{\frac{1}{2}}$ |
| 8 | $((1+x)^2 + (1-y)^2)^{\frac{1}{2}} \le r \le ((1+x)^2 + (1+y)^2)^{\frac{1}{2}}$ |

**Table 4.3**　*Boundary values to the angle $\theta$ as functions of r in the eight regions*

| Region | $\theta(r)$ |
|---|---|
| 1 | $\pi$ |
| 2 | $\dfrac{\pi}{2} + \arcsin \dfrac{1-x}{r}$ |
| 3 | $\dfrac{\pi}{2} + \arcsin \dfrac{1-x}{r} - \arccos \dfrac{1-y}{r}$ |
| 4 | $\dfrac{\pi}{2} + \dfrac{1}{2}\left( \arcsin \dfrac{1-x}{r} - \arccos \dfrac{1-y}{r} \right)$ |
| 5 | $\dfrac{\pi}{2} - \arccos \dfrac{1+y}{r} + \dfrac{1}{2}\left( \arcsin \dfrac{1-x}{r} - \arccos \dfrac{1-y}{r} \right)$ |
| 6 | $\dfrac{\pi}{2} - \dfrac{1}{2}\left( \arccos \dfrac{1+y}{r} + \arccos \dfrac{1-y}{r} \right)$ |
| 7 | $\dfrac{1}{2}\left( \arcsin \dfrac{1-y}{r} + \arcsin \dfrac{1+y}{r} \right) - \arccos \dfrac{1+x}{r}$ |
| 8 | $\dfrac{1}{2}\left( \arcsin \dfrac{1+y}{r} - \arccos \dfrac{1+x}{r} \right)$ |

circle that falls into $A$ gets proportionally smaller and also $\theta(r)$ decreases. The edges $V_1$, $V_2$, $V_3$, $V_4$ of the square have coordinates, with respect to the new origin: $(-1 - x, 1 - y)$, $(1 - x, 1 - y)$, $(1 - x, -1 - y)$, $(-1 - x, -1 - y)$. In Tables 4.2 and 4.3 the bounds of the 8 regions are reported for the radius $r$ (as a function of $x, y$) and for the angle $\theta$.

Now, by considering the values of $r$ and $\theta(r)$ reported in the tables, (4.80) leads to the average number of sinks located in $A$ which are 'audible' by a generic sensor placed in $(x, y)$:

$$\mu(x, y) = \sum_{i=1}^{8} \int_{r_i}^{r_{i+1}} 2\theta_i(r) \cdot \rho \cdot r \cdot \Phi(a_1 - b_1 \ln r)dr$$

The probability that the sensor will be connected to at least one sink in $A$, that is, will not be isolated, is thus $q(x, y) = 1 - e^{-\mu(x,y)}$. Noting that we have chosen a point belonging to the lower-half of the first quadrant and exploiting the symmetry of the integral at issue, the probability that a randomly chosen sensor in $A$ will not be isolated turns out to be

$$p = \frac{8}{A} \int_0^1 \int_0^x q(x, y) dy\, dx \tag{4.81}$$

Finally, recall that in a PPP the positions of nodes are independent and note that the expression of $p$ in (4.81) counts for any given sensor in $A$: since statistical independence holds, we can assume $p_n(k) = p^k$ and substitute it into the expression (4.66) of $Z$, which becomes

$$Z = \sum_{k=1}^{+\infty} p^k \cdot \frac{e^{-\rho_s A}}{k!} (\rho_s A)^k. \tag{4.82}$$

Let us now examine some numerical results. We considered $k_0$, $k_1$, $\sigma$, $L$, $\rho$, $\rho_s$ as variables and evaluated (4.81) and (4.82). Though, in the following graphs we fixed $k_0 = 40$ dB, $k_1 = 15$, $L = 1000$ m. In Figure 4.28 the densities are also fixed ($\rho = 100/L^2$ and $\rho_s = 1000/L^2$) and the probability that a randomly chosen sensor will not be isolated vs the loss threshold $L_{th}$[dB] is evaluated. We observed that when the sink density $\rho$ is too low, the probability $p$ never reaches 1, even though the TR approaches infinity. As an example, for $\rho = 10/L^2$ we find an asymptote for $p$ at 0.999954. This is due to the fact that in a PPP the probability of not having any sink in the square is not negligible at this

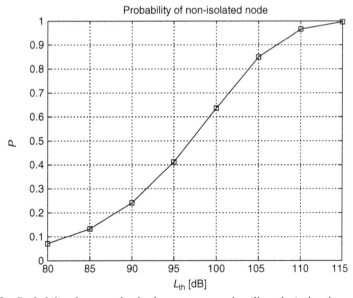

**Figure 4.28** *Probability that a randomly chosen sensor node will not be isolated*

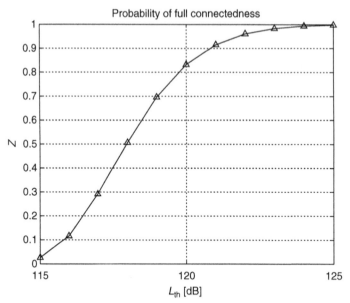

**Figure 4.29**   *Probability Z that the network is fully connected*

density value. Actually, this asymptote equals the probability that there is at least one sink in the area, which is $1 - e^{-\rho L^2} (1 - e^{-10} = 0.9999546$ when $\rho = 10/L^2$).

In Figure 4.29 we still refer to a ratio $\Delta = 10$ between the densities and we plot the probability $Z$ that the WSN is fully connected (i.e., no isolated sensor). The corresponding values of $p$ vary pretty much slowly in this range of losses (from 0.996353 to 0.999998). It turns out that the network is fully connected when $L_{th} = 125\,$dB, that is, in the absence

of fluctuations ($\sigma = 0$), we have $r_0 = e^{(L_{th} - k_0)/k_1} \simeq 289$. We also verified that $p$ and $Z$ do not change till the ratio $r_0/L$ remains constant: roughly speaking, the value $L_{th}^*$ that must be associated to the square side $L^*$ in order to obtain the same $p$ and $Z$ as with $(L_{th}, L)$, is $L_{th}^* = L_{th} - k_1 \ln(L/L^*)$.

In Figure 4.30, instead, we compare the probability of fully connected network $Z$ obtained by considering different densities of both sensor and sink nodes. It is clear that when sensor density grows and sink density remains constant, the probability of fully connectivity $Z$ decreases (with $L_{th}$ being fixed, central curve). If, instead, sensor density remains unchanged while sink density decreases, $Z$ gets even smaller (right-most curve). As an example, in order to have a probability $Z \approx 0.9$ of fully connectivity, we need to have $L_{th} = 120\,$dB if $\rho_s = 1000$ [nodes/m²] and $\rho = 100$ [nodes/m²], $L_{th} = 122$ dB if $\rho_s = 2000$ [nodes/m²] and $\rho = 100$ [nodes/m²], $L_{th} = 127\,$dB if $\rho_s = 1000$ [nodes/m²] and $\rho = 50$ [nodes/m²].

While in the previous graphs we fixed $\sigma = 4$, in Figure 4.31 we fix $L_{th}$ [dB] and $\sigma$ is variable: it appears evident how channel fluctuations positively affect the connectivity of the WSN.

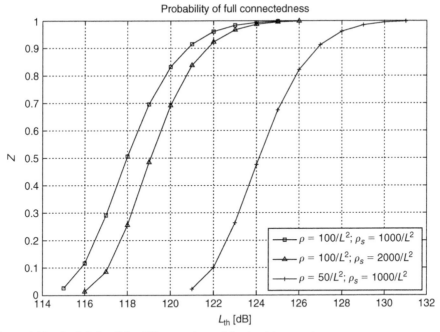

**Figure 4.30** *Probability Z for different values of the densities ρ and ρ_s*

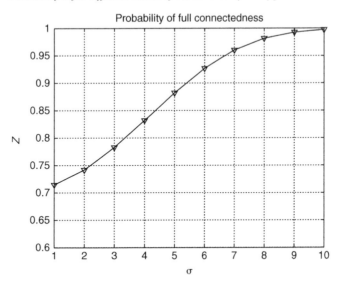

**Figure 4.31** *Probability Z vs the standard deviation σ of the channel fluctuations*

## 4.9 Alternate models for network connectivity

We have seen that the connectivity of a network may be analyzed from a variety of viewpoints, depending on both the scenario and the application considered. The aim of the present section is to give the most recent literature overview in order to find those research trends which are in some way different from the previous ones.

## 4.9.1   An information-theoretic view of connectivity

The analysis of connectivity based on the disk model (4.2) suffers from some weaknesses. First, it ignores the interference generated by a potentially large number of interferers. Second, it wrongly suggests that the packet reception probabilities decrease if all nodes scale their power by the same factor.

Moreover, the disk model leads to a *hard* estimation of connectivity, in that two nodes can be connected or not. This is misleading in practice, because two nodes that are very far apart might still be able to communicate reliably, even though at a lower data rate. That said, a *soft* definition of connectivity may also be given: *the network is connected at rate R(n) if, for a randomly picked source-destination pair, we can guarantee that the source can communicate with the destination at rate R(n), assuming all other nodes act as relay nodes* (Liu & Srikant, 2004).

As a consequence, we can talk about the *achievable rate of a wireless sensor network*: such data rate will be upper-bounded by the ratio of the total received power vs noise power. On the other hand, a specific interference cancellation transmission scheme can be used to achieve a lower bound. Anyway, this goes beyond the scope of the present book.

## 4.9.2   Reachability

In Section 4.7.3 we mentioned the 'clusterization' effect, which will now be further discussed. Let us consider an ad hoc network with randomly distributed nodes in a squared region. Taking channel fluctuations into account, it can be stated that the probability of connected graph $\mathbb{P}\{\text{con}\}$ increases as long as the power transmitted by nodes get larger. For a certain value of the transmitted power $P_t$ it will approach 1, since there will be a path connecting any pair of nodes with probability 1.

We have seen that as $P_t$ gets larger, more edges will be observed in the graph. However, simulations showed that the graph does not 'grow' uniformly, but several connected subgraphs (clusters) appear before the entire graph is connected. In particular, when $P_t$ is very close to the critical value, there will be the giant component (which comprehends almost all nodes) and a subset of few sparse nodes. In such a situation, $\mathbb{P}\{\text{con}\}$ will be very small but the network can still perform very well, since the isolated nodes only represent a small percentage of the totality.

Thus it appears that $\mathbb{P}\{\text{con}\}$ does not properly assess the extent of communication that a wireless multi-hop network can support, especially if we deal with a sparse one. An alternate connectivity measure, more suitable for sparse networks, is called *reachability* and is defined as the fraction of node pairs that are connected (Perur & Iyer, 2006):

$$\text{Rch} = \frac{\text{No. of connected node pairs}}{\binom{N}{2}} \tag{4.83}$$

The advantage of adopting this measure is illustrated through an example: $N$ nodes, each with a uniform radio transmission range $r_0$, are to be deployed in a $2000\,m \times 2000\,m$ area. The nodes are uniformly distributed in the area of operation. The growth curves for both connectivity and reachability are computed for increasing $N$ when $r_0 = 300\,m$. The connectivity curve remains at 0 till $N$ grows to almost 70. But even with less than 70 nodes, the network affords a significant degree of communication. For example, with 60 nodes, 45% of node pairs have a multi-hop path connecting them. As a consequence, the CTR can be chosen to be less than the one computed with respect to connectivity.

Another advantage of reachability arises from how it can be modelled: it has been observed from simulation data that it consistently obeys the *logistic growth model* (i.e., the law that models the growth of population over time) (Kingsland, 1982):

$$\mathrm{Rch}_{N,r} = \frac{1}{1 + e^{\alpha_N - \beta_N r}}. \tag{4.84}$$

The (4.84) is referred to a network with $N$ nodes and $r = R/L$ is the transmission range normalized to the side $L$ of the square area. The constants $\alpha_N$ and $\beta_N$ are to be determined through regression methods.

### 4.9.3 Global connectivity from local constraints

In Section 4.7 we focused on determining the optimum value for the node density $\rho$ such that the network would be fully connected with probability $\mathbb{P}\{con\}$, given a fixed uniform transmission range.

Now we look at the same problem from a different perspective. Let us consider a network of arbitrary, fixed topology: we want to find the optimum power level for each node that results in a fully connected network, assuming that any node can choose a transmission power level regardless of any other.

We want to emphasize the fact that power control is a link level algorithm, while connectivity refers to the overall network: we are trying to achieve a global specification by applying some local rules. The present approach also uses minimal overhead, even when locations of nodes and their linkages can change over time. It is based on purely local information, namely the angles between the neighbouring edges originating on each node $v$ of the graph.

The algorithm, called adaptive power topology control (APTC), works as follows (Souza, Moore, Galvin & Randall, 2006). Each initially isolated node begins by transmitting at low power, and then ramping up until its neighbourhood satisfies a local geometric constraint. As the node ramps up its power, it broadcasts connection requests and processes acknowledgments of these requests, thus establishing communication links with other nearby nodes.

The node will first establish a link with the most accessible node within its communication range, then with the next most accessible, etc. With each new connection made, the geometric information is assessed. In general, at each step, we consider the vectors drawn originating from a node and ending at its, say, $m$ neighbours. These vectors divide the area around the central node into $m$ disjoint sectors.

If the angle of each sector is less than some threshold $\theta$, the constraint is satisfied and the node sets its operating power at the current value. If any angle is greater than or equal to $\theta$, the construction continues. If a node reaches its maximum operating power before satisfying the constraint, it halts execution and lowers its power back down to the level where the last new connection to a neighbour was made (or to zero if it has no neighbours in its broadcast range).

As mentioned earlier, each node can locally determine where it succeeded in satisfying the $\theta$-constraint and communicate that fact to the rest of the network. If every node has sufficient power to satisfy the $\theta$-constraint, the resulting graph $G_\theta$ (which is some sub-graph of the maximum power graph $G_R$), turns out to be fully connected.

The issue still remains of determining the proper value of $\theta$. We first subdivide the nodes in *boundary nodes* (those belonging to the convex hull of the network) and *interior nodes*. Several theorems on how to choose $\theta$ are available, depending on the restrictions imposed on the boundary nodes and on the wireless '*footprint*', which is the shape of the broadcast region.

As an example, if boundary nodes are known to be connected, for an arbitrary footprint, the following holds: *if* G(V, E) *satisfies the* $\theta$-*constraint at every internal node with* $\theta < \pi$, *then* G(V, E) *is fully connected.*

If, instead, we relax the requirement on the boundary nodes, a few more definitions must be introduced. The wireless footprint is said to be *monotone* if, given three nodes $i, j, k$ and provided that $\vec{ij}$ is an edge of the graph, $\vec{ik}$ is also an edge, with $|\vec{ik}| < |\vec{ij}|$. In other words, if a certain node can be reached, a closer node will be necessarily reached (which is not true in general, due to channel fluctuations). A less restrictive concept is said to be *weak-monotonicity*: if $\vec{ij}$ is an edge, then $i$ has a link to all other vertices in the circle of diameter $d(i, j)$ centred at the midpoint of the edge $\vec{ij}$.

We consider now a variant on the APTC algorithm to produce $(\theta_I, \theta_B)$ graphs where internal nodes satisfy the $\theta_I$-constraint and boundary nodes satisfy the $\theta_B$-constraint. We call the output of the algorithm a $G_{\theta_I, \theta_B}$ graph. The following theorem holds:

*Let* $G_{\theta_I, \theta_B}$ *be the graph formed by APTC with the weak-monotone footprint model. If* $\theta_I < \pi$ *and* $\theta_B < 3\pi/2$ *then* $G_{\theta_I, \theta_B}$ *is fully connected.*

In general, the more restrictive the boundary constraints, the less restrictions need to be imposed on the wireless footprint to guarantee connectivity. It must also be noted that each node is supposed to be capable of carrying out geometric measurements and hence positioning devices, such as GPS receivers or directional antennas, are required.

## 4.10 Coverage vs energy efficiency

In this section the energy and SNR advantages (coverage) will be investigated in faded channels in order to show as relaying and multi-hop communications enable a gain that can be properly translated into energy saving or improvement of the received SNR.

As an example, let us consider the scenario in which a sensor reaches the destination through $n \in \{0,1,\ldots,N_H\}$ hops on $n$ relays nodes. Being $D$ the total distance between source and destination, each link has a length $D/(n+1)$ (sensors are here assumed in line to simplify the discussion but without loss of generality). The received power at a distance $d$ for the case with $n$ hops, $P_r^{(n)}(d)$ is given by

$$P_r^{(n)}(d) = P_t^{(n)} - L(d), \tag{4.85}$$

where $P_t^{(n)}$ is the transmitted power from each node in the case of $n$ hops and $L(d)$ is the power loss at distance $d$ both expressed in dB units. Since the TX–RX distance is $D$ without hop and $D/(n+1)$ for $n$ hops, the required transmitted power to receive the minimum admitted power level (i.e., the receiver sensitivity equal to $P_{r_{min}}$) is

$$P_t^{(n)} = P_{r_{min}} + L + 10\log_{10}\left(\frac{D}{n+1}\right). \tag{4.86}$$

Appropriate channel models for connectivity have been discussed previously. For our purpose let us consider here the log-normal model (4.4) (a power decaying with distance having path-loss exponent $\alpha$ and a normal shadowing (in dB) $S$ with standard deviation $\sigma$). The loss in dB at a normalized distance $y = d/D$ is

$$L(y) = -A - S + 10\alpha\log_{10}y, \tag{4.87}$$

where $A$ is a constant dependent on the operating frequency, antenna gains and distance $D$. Following Jakes, 1995, Chiani, Conti & Verdone, 2001, the coverage probability at a normalized distance $y$ is given by

$$P_{cov|y}^{(n)}(y) = \mathbb{P}\left\{P_r^{(n)}(y) \geq P_{r_{min}}\right\}, \tag{4.88}$$

and after some algebra

$$P_{cov|y}^{(n)}(y) = \frac{1}{2}\text{erfc}\left[\frac{-M^{(n)} + 10\alpha\log_{10}y}{\sqrt{2}\sigma}\right], \tag{4.89}$$

where $M^{(n)} = P_t^{(n)} - P_{r_{min}} + A$ is the link margin against shadowing. By considering sensors uniformly distributed in a circular area of radius $D$, the mean coverage probability becomes, following methodology in (Chiani et al., 2001).

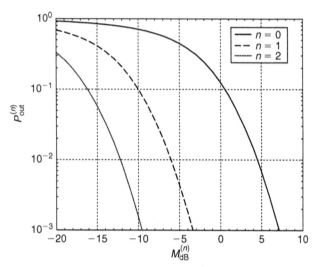

**Figure 4.32** *Outage probability vs the link margin for* n = 0, 1, 2 *hops,* α = 3.5 *and* σ = 3

$$P_{cov}^{(n)} = \int_0^{1/(n+1)} (n+1)^2 \, y \, \text{erfc} \left\{ \frac{-M^{(n)} + 10\alpha \log_{10} y}{\sqrt{2}\sigma} \right\} dy \qquad (4.90)$$

$$= \int_0^1 y \, \text{erfc} \left\{ \frac{-M^{(n)} - 10\alpha \log_{10}(n+1) + 10\alpha \log_{10} y}{\sqrt{2}\sigma} \right\} dy. \qquad (4.91)$$

From (4.90) it appears that for a fixed coverage probability, multi-hop communication allows a reduction of the margin, thus a reduction of the required transmitted power for receiving $P_{r_{min}}$, equal to $10\alpha \log_{10}(n+1)$ that depends on the number of relays and the propagation coefficient. This represents, for a fixed coverage probability, an energy advantage due to multi-hop. In Figure 4.32 the mean outage probability for the case of $n = 0, 1, 2$ hops, that is

$$P_{out}^{(n)} = 1 - P_{cov}^{(n)}, \qquad (4.92)$$

is shown versus the link margin $M^{(n)}$. Note that from (4.90) one can obtain the outage probability for a general number of hops $n$ by translating the curve for $n = 0$ of the quantity $10\alpha \log_{10}(n+1)$, as well as an increasing of the number of hops from $n-1$ to $n$ enables a reduction of the link margin in dB as given by

$$M^{(n)} = M^{(n-1)} - 10\alpha \log_{10} \left( \frac{n+1}{n} \right). \qquad (4.93)$$

Starting from the no hopping case depicted in Figure 1 of Chiani et al., 2001 and properly scaling the link margin to the case of $n$ hops, one can play with $M^{(n)}/\sigma$ and $\alpha/\sigma$ for a fixed coverage probability.

Let us now investigate the SNR advantage. Performance mostly depends on the SNR at the receiving sensor that is defined as

$$\mathrm{SNR}^{(n)} = P_\mathrm{r}^{(n)} + 10 \log_{10}\left(\frac{D}{n+1}\right) - P_\mathrm{noise}, \tag{4.94}$$

where $P_\mathrm{noise}$ is the noise power level and SNR is expressed in dB. Thus, gain in terms of SNR when multi-hop techniques are adopted instead of single link is then given by

$$\eta^{(n)} = \mathrm{SNR}^{(n)} - \mathrm{SNR}^{(0)} = P_\mathrm{t}^{(n)} - P_\mathrm{t}^{(0)} + 10\alpha \log_{10}(n+1) + \tilde{S}, \tag{4.95}$$

where $\tilde{S}$ is a Gaussian r.v. with zero mean and variance $2\sigma^2$. One can easily obtain the c.d.f. of $\eta^{(n)}$ and other statistics (the median value being $P_\mathrm{t}^{(n)} - P_\mathrm{t}^{(0)} + 10\alpha \log_{10}(n+1)$ in dB). If the transmitted power is the same independently on the number of hops, the SNR advantage results in an increasing function of the received SNR equal to $10\alpha \log_{10}(n+1)$. On the other hand, this would be obtained at the expense of the energy saving. Hence, from both (4.90) and (4.95) it is possible to see relaying techniques enable a total gain of $10\alpha \log_{10}(n+1)$ that, depending on WSN requirements, can be properly divided into energy and SNR advantages as a trade-off.

## 4.11 Further reading

Here we present a non-exhaustive list of important papers related to connectivity in WSNs, in addition to those already treated in this chapter.

The one-dimensional scenario plays an important role for example in automotive environments where sensors are mounted on cars with the purpose to improve drivers' safety and provide advanced services (Andrisano, Verdone & Nakagawa, 2000). Some further results can be found in Dousse, Thiran & Hasler, 2002; Foh & Lee, 2004. The two-dimensional scenario has received most of the attention. Some other studies related to the characterization of the critical range for network connectivity are Philips, Panwar & Tantawi, 1989; Gupta & Kumar, 1998; Santi & Blough, 2003. Basic results related to the capacity of WSNs can be found in the pioneer work of Gupta and Kumar (2000). The effect of channel randomness on network connectivity is addressed in Ferrari, Tonguz & Bhatt, 2004; Booth, Bruck, Cook & Franceschetti, 2003; Betstetter, 2005.

# 5

# Network lifetime

This short chapter deals with aspects related to WSAN lifetime. The operational duration of a WSAN after its deployment is a very relevant issue. Once nodes are distributed over the monitored area, it is desirable that they can operate with no need to change their batteries (if they are not plugged) as long as possible. On the other hand, as mentioned in the introduction, energy is consumed at the transceiver in all operational states (reception, transmission, sensing the channel, etc.). Therefore, the more the node is active, the shorter is its lifetime before energy fades away.

Depending on the type of application running in the WSAN, it is desirable that all nodes keep alive, or at least a given fraction of them. So, the expiration of a specific node might let the network be operational or not, depending on the specific role of that node and the running application. Network lifetime is thus a significant metric whose definition is not trivial, being related to the application and to the protocols used (Kumar, Arora & Lai, 2005; Chen & Zhao, 2005).

This chapter reports some definitions of network lifetime, discusses them, and shows some simulation results that are useful to discuss the meaning of the different definitions.

## 5.1 Definition of node lifetime

A node consumes energy when receiving and transmitting packets, and in any other transceiver state (Shnayder, et al., 2004). When energy falls below a given threshold, then the transceiver circuitry is no longer working properly and the node is isolated, with no possibility to communicate to others. In principle the energy consumed at a node might be calculated by summing all energies consumed in the various states, weighted by the amount of time spent in each of them. Then, starting from the initial battery charge, one can think of computing the time when the residual energy falls below the threshold and therefore the node lifetime.

However, a number of problems arise with this approach. The threshold is not deterministically set, as external conditions (such as temperature) may impact on it. Moreover, batteries are affected by dynamic effects (Chiasserini & Rao, 1999&2000): for instance, a battery can regenerate part of the energy consumed during inactivity periods. Finally, the energy consumed by a node cannot be computed as the sum of energies consumed in the different states, because the amount of energy needed to move from one state to another should also be considered and is sometimes relevant (Shnayder, et al., 2004).

Nevertheless, in most cases node lifetime is measured according to such a simple approach; in fact, proper modelling of dynamic effects makes node lifetime evaluation much more complex.

Computation of node lifetime requires knowledge of the time spent in the various states, (which depends on the communication protocols used and the activity in the network), and of the amount of energy per unit time consumed by the circuitry in the different states. The latter data is usually available in the data sheet of the transceiver. The former is sometimes difficult to compute, because of the random behaviour of the radio access procedures, the unknown number of retransmissions needed to overcome problems due to packet collisions, etc.

However, in principle node lifetime can be estimated if protocol characteristics are known.

## 5.2   Definitions of network lifetime

Probably the simplest definition of network lifetime is as follows (Chen & Zhao, 2005; Xue & Ganz, 2004).

**Definition 1**   *Network lifetime can be defined as the interval of time, starting with the first transmission in the wireless network and ending when the percentage of nodes that have not terminated their residual energy falls below a specific threshold, which is set according to the type of application (it can be either 100% or less).*

This definition is related to that of node lifetime given before. It does not take into account the specific role of the nodes which expire in the network. If the percentage of desired alive nodes is set to 100%, then once the first node expires the network is considered expired as well, regardless of whether the node is a router forwarding information gathered by many others or a leaf in a tree topology providing one single bit of information. Also, after deployment a node might be disconnected from the rest of the network because its transmission range is smaller than the minimum distance from every other node; if such a node expires, the network does not change its performance. This definition does not take into account such an issue.

Other definitions might be used, that are more closely related to the degree of connectivity and the topology of the network.

A second option is to define network lifetime as in the following (Kumar, Lai & Balogh, 2004; Esseghir, Bouabdullah & Pujolle, 2005; Yen, et al., 2007).

**Definition 2**   *The interval of time, starting with the first transmission in the wireless network and ending when the percentage of nodes still reachable in the network by any* *sinks falls below a specific threshold, which is set according to the type of application (it can be either 100% or less).*

This definition also considers connectivity issues. In fact, if a node which previously belonged to a set of disconnected nodes expires, its expiration does not provide any contributions to the shortage of network lifetime. On the other hand, if a node expires which was the only one able to forward information gathered by a cluster to a sink, then the entire cluster remains disconnected and such a single expiration plays a very significant role.

Moving further in this direction, we can introduce another definition (Hu & Li, 2004).

**Definition 3**   *Network lifetime can be defined as the interval of time, starting with the first transmission in the wireless network and ending when the percentage of reports from nodes falls below a specific threshold, which is set according to the type of application (it can be either 100% or less).*

This definition also considers QoS (i.e., MAC and network) issues. During network life, nodes expire. It might happen that some nodes need to deliver their report to the sink(s) through a path which increases loss probability. Loss probability on the average gets larger as network evolves. In general, according to this definition network lifetime is shorter than the one obtained through the previous definition, since expiration of a single node can make many reports follow new paths.

Finally, we introduce a definition which is meaningful in the case of applications of type process estimation.

**Definition 4**   *Network lifetime can be defined as the interval of time, starting with the first transmission in the wireless network and ending when the spatial random process estimation error falls below a specific threshold, which is set according to the type of application.*

Such error depends on the density of nodes and the uniformity of their spatial distribution. If nodes expire in a non-uniform way, the estimation error becomes larger with respect to a situation where the same number of nodes is uniformly distributed in the monitored area. Thus, definition 4, which is the closest to the application requirements, also takes account of the possible uneven spatial distribution of energy consumption in the network.

On the other hand, this definition is also very complex to be used for a priori (pre-deployment) evaluation of network performance. In fact, it requires a precise relationship between node locations and the spatial process estimation error, which is very difficult to achieve.

Other definitions could be provided. However, those provided here are sufficient to allow discussion of the role played by the different definitions.

## 5.3 Communication protocols and network lifetime: How to choose

From the energy consumption viewpoint, the best protocol set is the one providing the longest network lifetime. However, this depends on the application running in the network. To clarify, let us consider as a reference definition 1 (same considerations might be done using other definitions). As an example, let us assume Definition 1 with 100% nodes needed to measure network lifetime, and that two different sets of protocols are tested, namely, A and B. If set A provides the first node expiration after 1 year, and set B after 2 years, then the latter option clearly outperforms set A. On the other hand, the set B might cause simultaneous expiration of all nodes after two years, whereas option A might let the nodes expire in different times, with 50% nodes still alive after three years. In such a situation, if the application allows sufficient quality of service even if only 50% of nodes are alive, the set A outperforms B, since in the former case network lifetime equals three years, whereas for set B it is two years.

In general, one can say that the set of protocols to be used depends on the application type, and no protocol set fits all applications. This is a very relevant feature of WSANs. As shown in Chapter 2, many types of applications can run over WSANs. Therefore, the best protocol set is the one that can work satisfactorily with many different types of applications, unless the nodes are designed for a specific type of application.

## 5.4 Some numerical examples

Let us now discuss some simulation results where network lifetime is measured according to definitions 1, 2 or 3.

Let us consider a square, with side equal to 50 m. One hundred nodes are randomly and uniformly deployed in the square. They have an ideal transmission range set to 54 m. However, the power loss is computed assuming random channel fluctuations which are Gaussian distributed in logarithmic scale with zero mean and standard deviation set at 3.5. The MAC protocol used is IEEE 802.15.4 (with no use of CFP) and routing of information follows a LEACH protocol, where cluster heads are randomly selected. One single sink is present in the scenario, in the centre of the square. Every time the sink sends a beacon to sample the environment, cluster heads are elected and the topology is formed, and all nodes try to send their samples to the sink. The time between two subsequent beacons is denoted as round. Given the size of the square, compared to the transmission range, the network is fully connected with probability tending to one.

The next figures refer to an initial battery charge set to 1 joule to make simulations shorter. Figures 5.1, 5.2 and 5.3 report network lifetime according to definitions 1, 2 and 3, respectively. The application requirement is set at 75% (horizontal line).

Figures 5.1 and 5.2 show similar curves. This is because the network is fully connected with high probability and no difference is found in the performance measured with the two definitions. The network lifetime is equal to about 100 rounds.

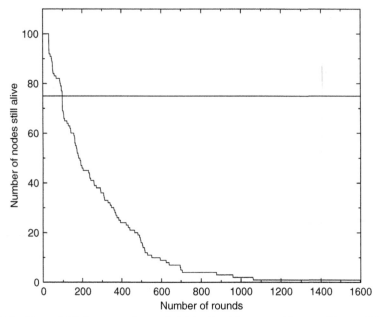

**Figure 5.1**   *Network lifetime according to definition 1, with square side set to 50 m and 100 nodes*

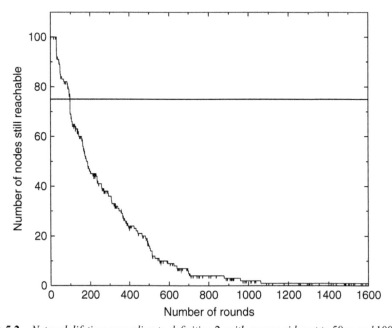

**Figure 5.2**   *Network lifetime according to definition 2, with square side set to 50 m and 100 nodes*

Figure 5.3 shows a different behaviour, with deep peaks. This is due to the random multiple access mechanism used, which makes the number of reports from sensor nodes unpredictable. In particular, large deviations (i.e., many collisions) are found when all nodes are alive, competing for the channel. If one uses definition 3, then the network lifetime

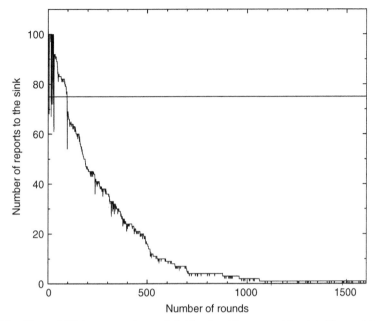

**Figure 5.3** *Network lifetime according to definition 3, with square side set to 50m and 100 nodes*

can be much less than 100 rounds, as in one of the very first rounds the number of reports is significantly below 75. On the other hand, after that event, the network was able to report 100 samples several times. Clearly, such random behaviour must be kept under control. Therefore, a new definition of network lifetime can be introduced.

**Definition 3/enhanced** *Network lifetime can be defined as the interval of time, starting with the first transmission in the wireless network and ending when the percentage of reports from nodes, averaged over a time window, falls below a specific threshold, which is set according to the type of application (it can be either 100% or less).*

In Figure 5.4 the time window for the average is set to 50 rounds. Once more, network lifetime is about 100 rounds.

Let us now change one parameter in the scenario. The number of nodes deployed is set to 500. All other parameters, and the protocols, are unchanged.

Figures 5.5, 5.6, 5.7 and 5.8 report network lifetime, measured with respect to 75% of nodes, with definitions 1, 2, 3 and 3/enhanced, respectively. Considerations similar to those reported above apply. However, network lifetime is close to 200 rounds. A larger number of nodes brings to a longer network lifetime. This is basically because, with the same percentage of cluster heads, a larger number of nodes provides a better spatial distribution of the cluster heads, and a more balanced topology which is more energy efficient. This shows as the node density can have a significant impact on network lifetime, given the type of protocol implemented.

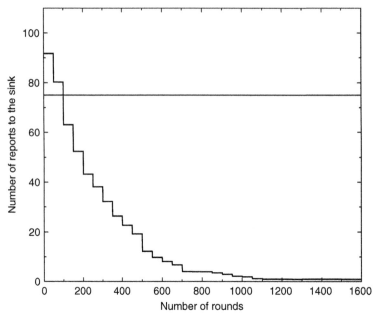

**Figure 5.4**   *Network lifetime according to definition 3/enhanced, with square side set to 50 m and 100 nodes*

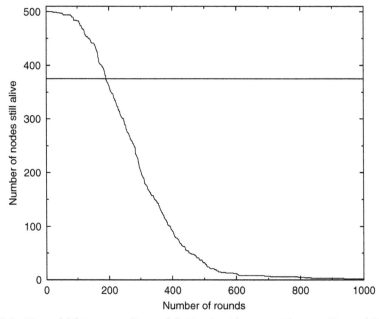

**Figure 5.5**   *Network lifetime according to definition 1, with square side set to 50 m and 500 nodes*

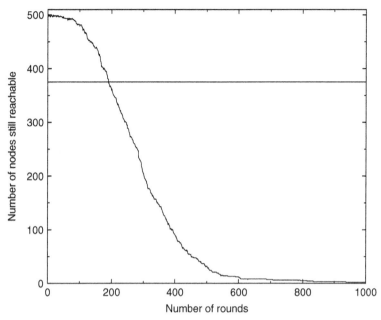

**Figure 5.6**  *Network lifetime according to definition 2, with square side set to 50 m and 500 nodes*

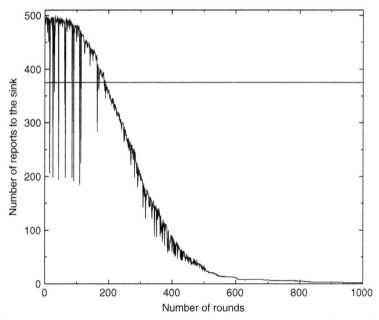

**Figure 5.7**  *Network lifetime according to definition 3, with square side set to 50 m and 500 nodes*

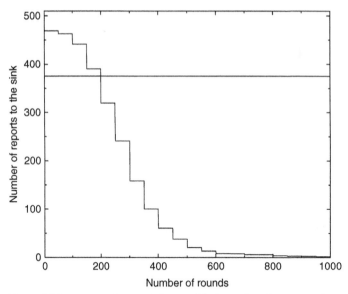

**Figure 5.8**   *Network lifetime according to definition 3/enhanced, with square side set to 50 m and 500 nodes*

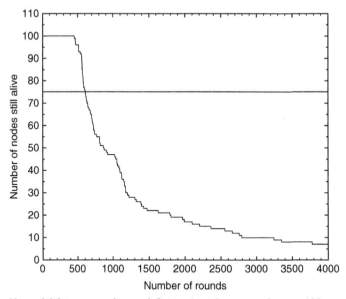

**Figure 5.9**   *Network lifetime according to definition 1, with square side set to 100 m and 100 nodes*

Finally, let us consider once more 100 nodes deployed in the square, having a larger side set to 100 metres. In this case, owing to the larger area, some nodes can be isolated.

Figures 5.9, 5.10, 5.11 and 5.12 report network lifetime, measured with respect to 75% of nodes, with definitions 1, 2, 3 and 3/enhanced, respectively. Figure 5.9 shows a network

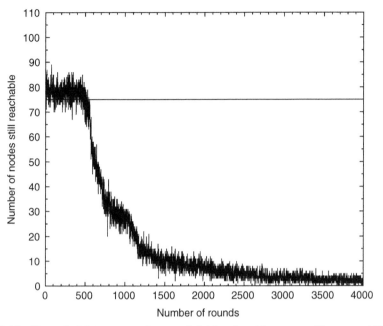

**Figure 5.10** *Network lifetime according to definition 2, with square side set to* 100 m *and* 100 *nodes*

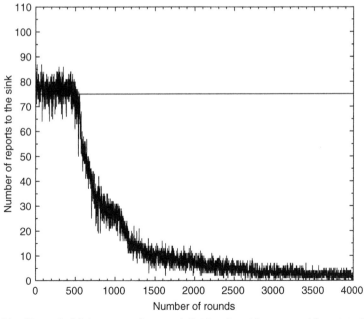

**Figure 5.11** *Network lifetime according to definition 3, with square side set to* 100 m *and* 100 *nodes*

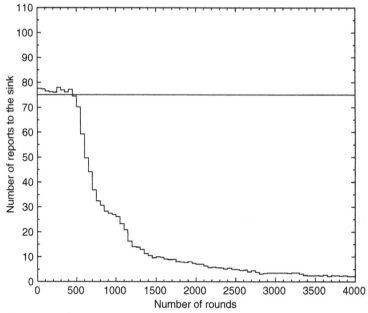

**Figure 5.12** *Network lifetime according to definition 3/enhanced, with square side set to* 100 *m and* 100 *nodes*

lifetime equal to about 600 rounds, much larger than in the case of a square side of 50 m. The reason for this is in the fact that the nodes that are isolated do not send any message and as a result they consume much less energy per round. Since at each round the role of cluster head rotates, the set of isolated nodes changes and all nodes are on average isolated the same fraction of time. So, network lifetime is larger. However, this is at the expense of the average number of nodes that can be reached (and the average number of reports collected) by the sink. This is clearly shown by Figures 5.10 and 5.11, where the maximum ordinate is significantly lower than 100. With definition 2, network lifetime would be close to zero. On the other hand, with definition 3/enhanced (see Figure 5.12), network lifetime is about 500, that is, significantly shorter than what is measured by definition 1.

If the threshold on the percentage of nodes were set to 90%, in the latter case network lifetime would be equal to zero.

These examples clearly show the relevance of the definition used to measure network lifetime.

# 6

# Technologies for WSANs

While no standard explicitly and specifically devoted to WSANs exists, the enabling technologies whose standardization process has already provided or is currently producing steady releases are discussed; namely, ZigBee, with two options as regards the PHY layer (IEEE 802.15.4 and UWB IEEE 802.15.4a), and Bluetooth (IEEE 802.15.1). The main characteristics of those standards, at PHY and MAC layers, are described, with emphasis on the aspects related to the fulfilment of the requirements anticipated in Chapter 2. The description of the various aspects discussed by the relevant SIGs (e.g., Zigbee Alliance) with reference to higher layers will also be included.

## 6.1   ZigBee technology

ZigBee wireless technology is a short-range communication system for applications with relaxed throughput and latency requirements in wireless personal area networks. The key features of ZigBee wireless technology are low complexity, low cost, low power consumption, low data rate transmissions, supported by cheap fixed or moving devices. The main field of application of this technology is the implementation of WSNs.

The IEEE 802.15.4 Working Group[1] focuses on the standardization of the bottom two layers of the ISO/OSI protocol stack. The other layers are normally specified by industrial consortia such as the ZigBee Alliance.[2]

In the following the main specifications related to both the physical layer and the MAC sublayer as defined in the IEEE 802.15.4 and IEEE 802.15.4a standards will be reported. Moreover, some characteristics related to higher layers will be presented, such as possible network topologies supported, since they are useful to understand some MAC functionalities.

---

[1] See also the IEEE 802.15.4 websites: http://www.ieee802.org/15/pub/TG4.html
[2] See also the ZigBee Alliance websites: http://www.zigbee.org/en/index.asp

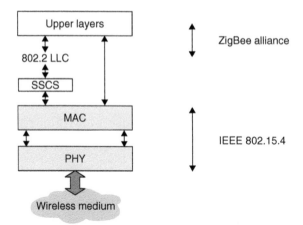

**Figure 6.1**  *ZigBee protocol stack*

## 6.1.1   Introduction to ZigBee Characteristics: IEEE 802.15.4

The ZigBee core system consists of an RF transceiver and the protocol stack, which is depicted in Figure 6.1. The system offers low rate services that enable the connection of possibly mobile low-complexity devices based on the carrier sensing multiple access with collision avoidance (CSMA/CA) channel access technique.

### ZigBee physical layer
The ZigBee physical layer operates in three different unlicensed bands with different modalities according to the geographical area where the system is deployed. However, direct sequence spread spectrum (DS-SS) is wherever mandatory to reduce the interference level in shared unlicensed bands.

The PHY layer provides the interface with the wireless medium. It is in charge of radio transceiver activation and deactivation, energy detection, link quality, clear channel assessment, channel selection, and transmission and reception of the message packets. Moreover, it is responsible for the establishment of the RF link between two devices, bit modulation and demodulation, synchronization between the transmitter and the receiver, and, finally, for packet level synchronization.

The IEEE 802.15.4 working group specifies a total of 27 half-duplex channels across the three frequency bands, whose channelization is depicted in Figure 6.2, organized as follows:

- The 868 MHz band mode, ranging between 868.0 and 868.6 MHz and used in the European area, adopts a raised-cosine-shaped binary phase shift keying (BPSK) modulation format, with DS-SS at chip-rate 300 kchip/s. Thus a pseudo-random sequence of 15 chips is transmitted in a 25 µs symbol period. Only a single channel with data rate 20 kbps is available and, with a minimum of −92 dBm RF sensitivity required, the ideal transmission range (i.e., without considering wave reflection, diffraction and scattering) is approximatively 1 km.

IEEE 802.15.4 Channelization at 868/915 MHz

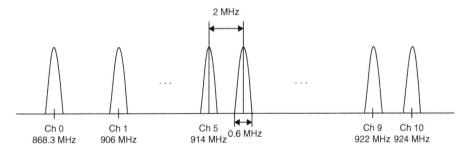

IEEE 802.15.4 Channelization at 2.4 GHz

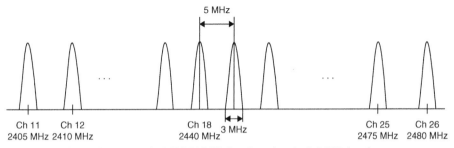

**Figure 6.2** *Channelization at the 868/915 MHz bands and at the 2.4 GHz band*

- The 915 MHz band mode, ranging between 902 and 928 MHz and used in the North American and Pacific area, adopts a raised-cosine-shaped BPSK modulation format, with DS-SS at chip-rate 600 kchip/s. Thus, a pseudo-random sequence of 15 chips is transmitted in a 50 µs symbol period. Ten channels with data rate 40 kbps are available and, with a minimum −92 dBm RF sensitivity required, the ideal transmission range is approximatively 1 km.
- The 2.4 GHz industrial scientific medical (ISM) band mode, which extends from 2400 to 2483.5 MHz and is used worldwide, adopts a half-sine-shaped offset quadrature shift keying (O-QPSK) modulation format, with DS-SS at 2 Mchip/s. Thus, a pseudo-random sequence of 32 chips is transmitted in a 16 µs symbol period. Sixteen channels with data rate 250 kbps are available and, with a minimum −85 dBm RF sensitivity required, the ideal transmission range is approximatively 220 m.

The ideal transmission range is computed considering that, although any legally acceptable power is permitted, IEEE 802.15.4-compliant devices should be capable of transmitting at −3 dBm. Since the 2.4 GHz band is shared with many other services, the other two available bands can be used as an alternative.

Power consumption is a primary concern. Therefore, to achieve long battery life the energy must be drained continuously at an extremely low rate, or in small amounts at a low power duty cycle: this means that IEEE 802.15.4-compliant devices are active only during a short fraction of time. The standard allows some devices to operate with both

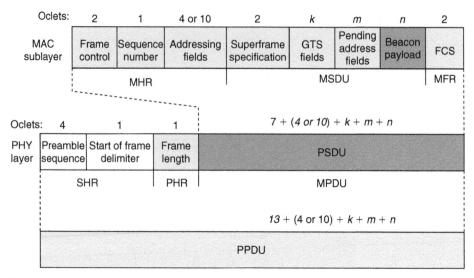

**Figure 6.3**  *Beacon frame structure*

the transmitter and the receiver inactive for over 99% of time. So, the instantaneous link data rates supported (i.e., 20 kbps, 40 kbps, and 250 kbps) are high with respect to the data throughput in order to minimize device duty cycle.

According to the IEEE 802.15.4 standard, transmission is organized in frames, which can differ according to the relevant purpose. In particular, there are four frame structures, each designated as a physical protocol data unit (PPDU): beacon frame, data frame, acknowledgement frame and MAC command frame. They are all structured with a synchronization header (SHR), a physical header (PHR), and a physical service data unit (PSDU), which is composed of a MAC payload data unit (MPDU). This is in turn constructed with a MAC header (MHR), a MAC footer (MFR), and a MAC service data unit (MSDU), excepting the acknowledgement frame, which does not contain an MSDU. The structure of each possible frame is depicted in Figures 6.3–6.6. To detect that a message has been correctly received, cyclic redundancy check (CRC) is used. The meaning of the different frame structures will be clarified after introducing the possible network topologies and MAC channel access strategies.

### ZigBee network topologies and operational modes

To overcome the limited transmission range, multihop self-organizing network topologies are required. They can be realized taking into account that IEEE 802.15.4 defines two type of devices: the full function device (FFD) and the reduced function device (RFD). The FFD contains the complete set of MAC services and can operate as either a network coordinator (from this point in time also denoted as 'personal area network (PAN) coordinator') or a simple network device. The RFD contains a reduced set of MAC services and can operate only as a network device.

Two basic topologies are allowed, though not completely described by the standard since the definition of higher layers functionalities is out of the scope of IEEE 802.15.4. The

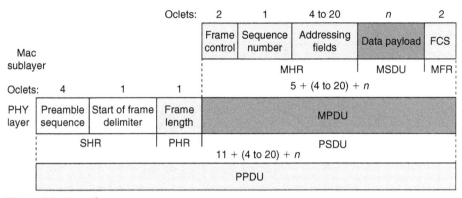

**Figure 6.4**  *Data frame structure*

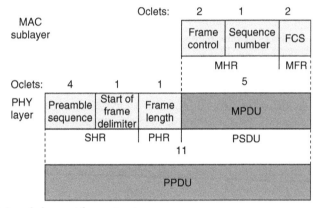

**Figure 6.5**  *Acknowledgement frame structure*

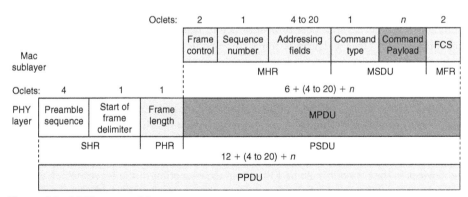

**Figure 6.6**  *MAC command frame structure*

star topology is formed around an FFD acting as a PAN coordinator, which is the only node allowed to form links with more than one device, and the peer-to-peer topology, where each device is able to form multiple direct links to other devices so that redundant paths are available. An example of both the IEEE 802.15.4-compliant network topologies is shown in Figure 6.7.

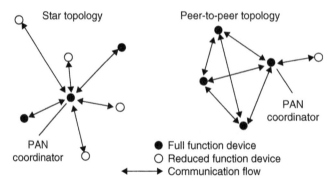

**Figure 6.7**   *The two IEEE 802.15.4-compliant network topologies: star and peer-to-peer topology*

Star topology is preferable in case the area covered is small and low latency is required by the application. Communication is controlled by the PAN coordinator that acts as network master sending packets, named beacons, for synchronization and device association. Network devices are allowed to communicate only with the PAN coordinator and any FFD may establish its own network by becoming a PAN coordinator according to a predefined policy. A network device wishing to join a star network listens for a beacon message and, after receiving it, sends an association request back to the PAN coordinator, which allows the association or not. Star networks support also a non-beacon-enabled mode. In this case, beacons are used for association purpose only, whereas synchronization is achieved by polling the PAN coordinator for data on a periodic basis. Star networks operate independently from their neighbouring networks.

Peer-to-peer topology is preferable in case a large area should be covered and latency is not a critical issue. This topology allows the formation of more complex networks and permits any FFD to communicate with any other FFD behind its transmission range via multihop. Each device in a peer-to-peer structure needs to proactively search for other network devices. Once a device is found, the two devices can exchange parameters to recognize the type of services and features supported. The drawback of this topology is that the introduction of multihop requires additional device memory for routing tables.

IEEE 802.15.4 can also support other network topologies, such as cluster, mesh, and tree. These last network topology options are not part of the IEEE 802.15.4 standard, but the tree topology is described in the ZigBee Alliance specifications. This topology, which is depicted in Figure 6.8, can be interpreted as a hierarchical tree of network devices. All the devices in the network must be FFDs with the exception of the leaves which may be either FFDs or RFDs, since they must do no message relaying. Specifically, one device in the network assumes the special role of the PAN coordinator. Surrounding each coordinator, a hierarchical tree may be formed in a typical parent–child relationship, but only one single device in the entire network functions as the PAN coordinator.

Regardless of the type of topology, each device belonging to a particular network uses its unique IEEE 64-bit address and a short 16-bit address is allocated by the PAN coordinator to univocally identify the network.

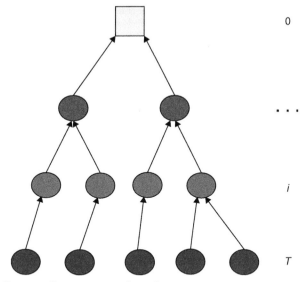

**Figure 6.8**   *ZigBee-compliant tree network topology*

Finally, the PAN coordinator election can be performed in different ways according to the application. In particular, for the applications where only one device can be the coordinator (e.g., a gateway), it is preferable to have a dedicated PAN coordinator, whereas in other applications it could be significant to have several eligible FFDs and an event-determined PAN coordinator. In those applications where it is not relevant which particular device is the PAN coordinator, it can be self-determined. The PAN coordinator may be selected because of its special computation capability, the bridging capability to other network protocols, or simply because it was among the first participants in the formation of the network.

### ZigBee MAC sublayer

The MAC sublayer, with the logical link control (LLC) sublayer, comprises the data link layer in the ISO/OSI model. The MAC layer provides access control to a shared channel and reliable data delivery.

IEEE 802.15.4 uses a fully acknowledged protocol based on the CSMA/CA algorithm, which requires listening to the channel before transmitting to reduce the probability of collision with other ongoing transmissions. The main functions performed by the MAC sublayer are: generation of acknowledgement frames, association and disassociation, security control, optional star network topology functions (such as beacon generation and guaranteed time slot management), and finally provision of application support for the two possible network topologies described in the standard.

By default IEEE 802.15.4 does not support isochronous communications and multiple class of services within a single PAN. However, the standard introduces the implementation of an optional superframe structure managed by the PAN coordinator possible in case of star and tree topology only. As shown in Figure 6.9, the superframe is bounded

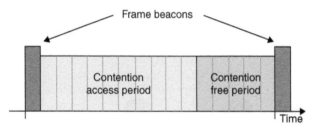

**Figure 6.9**  *Superframe structure*

by two successive beacon messages and is sent by the PAN coordinator at regular intervals. So, each beacon provides synchronization and contains information such as the network identifier, beacon periodicity, and superframe structure. The superframe is divided
into two parts: a contention access period (CAP), during which network devices wishing to communicate with the PAN coordinator can attempt by using CSMA/CA, and
a contention free period (CFP), organized in up to seven contiguous time slots, named
'granted time slots' (GTSs), managed by the PAN for synchronous low-latency communications. Granted time slots (GTSs) make higher level protocol service possible.
Therefore, for those networks not implementing higher levels of service, the employ-
ment of GTS is optional. The use of beacon-enabled or non-beacon-enabled communications depends on the applications: beacon transmissions are disadvantageous when no
messages are expected from the coordinator to the network and only light traffic from
the network device to the coordinator is envisaged.

As a consequence of the different type of topologies and the possibility of implementing the beacon-enable mode, three different MAC data transfer protocols are defined by
IEEE 802.15.4:

- In the case of beacon-enabled star topology, a network device wishing to send data to
  the PAN coordinator needs to listen for a beacon. If the device does not have a GTS
  assigned, it transmits its data frame in the contention access period with CSMA/CA.
  If the device has a GTS assigned, it waits for the appropriate one to transmit its data
  frame. Afterwards, the PAN coordinator sends back an acknowledgement to the network
  device, as shown in Figure 6.10. When the PAN coordinator has data for a network
  device, it sets a special flag in its beacon. Once the appropriate network device detects
  that the PAN coordinator has pending data for it, it sends back a 'data request' message. The PAN coordinator responds with an acknowledgment followed by the data
  frame, and, finally, an acknowledgement is sent from the network device, as depicted
  in Figure 6.11.
- In the case of non-beacon-enabled star topology, a network device wishing to transfer
  data sends a data frame to the PAN coordinator using CSMA/CA. The PAN coordinator responds to the network device, sending an acknowledgement message, as shown
  in Figure 6.12. When a PAN coordinator requires a data transfer to a network device,
  it will keep the data until the network device sends a data request message. Then,
  the PAN coordinator sends an acknowledgement immediately followed by the data

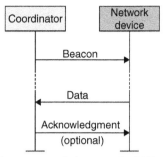

**Figure 6.10** *Communication from a network device to the PAN in a beacon-enabled network*

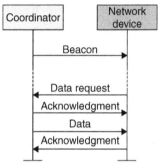

**Figure 6.11** *Communication from the PAN to a network device in a beacon-enabled network*

frame. Finally, the network device acknowledges reception of the data frame, as depicted in Figure 6.13.

- In the case of peer-to-peer topology, the strategy is governed by the specific network layer managing the wireless network. A given network device may stay in reception mode scanning the radio channel for ongoing communications or can send periodic 'hello' messages to achieve synchronization with other potential listening devices.

In the case of tree topology, as specified by the ZigBee Alliance, beacon scheduling is necessary to prevent the beacon frames of one device from colliding with either the beacon frames or data transmissions of its neighbouring devices. Communication in a tree network will be accomplished using the parent–child links to route along the tree. Since every child tracks the beacon of its parent, transmissions from a parent to its child will be completed using the indirect transmission technique. Transmissions from a child to its parent will be completed during the CAP of the parent.

Having in mind that power consumption should be as low as possible, to support low-duty cycles in the case of beacon-enabled networks, the IEEE 802.15.4 beacon packet can be as short as 544 µs in the 2.4 GHz band, while the superframe period may be extended from 15.36 ms to over four minutes, whereas the non-beacon mode enables the slaves in a master-slave star network to remain in standby mode indefinitely.

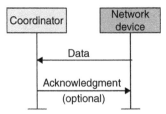

**Figure 6.12**  *Communication from a network device to the PAN in a non-beacon-enabled network*

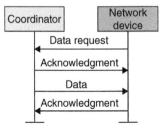

**Figure 6.13**  *Communication from the PAN to a network device in a non-beacon-enabled network*

Finally, as far as the MAC frame structure is concerned, a MAC frame consists of three parts: header, variable length payload, and footer. The MAC header contains a frame control field and an addressing field. The MAC payload contains information specific to the type of transaction being handled by the MAC. The MAC footer consists of a 16-bit CRC algorithm. When the three components of the MAC frame are assembled into the PHY packet, it is called the MPDU. Four types of MAC frame are defined: beacon, data, acknowledgment, and MAC command.

### ZigBee higher levels overview

The purpose of the ZigBee Alliance is to univocally describe the ZigBee protocol standard in such a way that interoperability is guaranteed also among devices produced by different companies, provided that each device implements the ZigBee protocol stack.

The ZigBee stack architecture is composed of a set of blocks called layers, as depicted in Figure 6.14. Each layer performs a specific set of services for the layer above.

Given the IEEE 802.15.4 specifications on PHY and MAC layer, the ZigBee Alliance defines the network layer and the framework for the application layer. The responsibilities of the ZigBee network layer include: mechanisms to join and leave a network, frame security, routing, path discovery, one-hop neighbours discovery, neighbour information storage. The ZigBee application layer consists of the application support sublayer, the application framework, the ZigBee device objects, and the manufacturer-defined application objects. The responsibilities of the application support sublayer include: maintaining tables for binding (defined as the ability to match two devices together based on their services and their needs) and forwarding messages between bound devices. The responsibilities of the ZigBee device objects include: defining the role of the device within the network (e.g., PAN coordinator or end device), initiating and/or responding to binding requests, establishing secure relationships between network devices, discovering devices in the network, and determining which application services they provide.

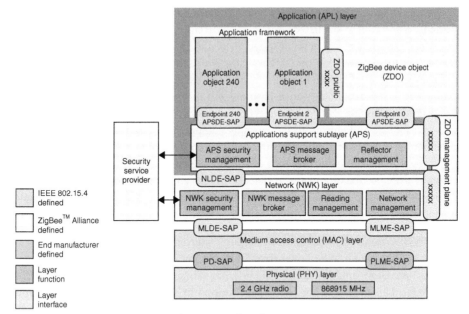

**Figure 6.14** *A detailed overview of ZigBee stack architecture*

## 6.2 Ultrawide bandwidth technology

### 6.2.1 Introduction

Ultrawide bandwidth radio is a fast emerging technology with uniquely attractive features that has attracted a great deal of interest from academia, industry, and global standardization bodies. The UWB technology has been around since 1960, when it was mainly used for radar and military applications, whereas nowadays it is a very promising technology for advances in wireless communications, networking, radar, imaging, positioning systems and, in particular, WSNs.

As already mentioned in Section 3.4, the most widely accepted definition of a UWB signal is a signal with instantaneous spectral occupancy in excess of 500 MHz or a fractional bandwidth of more than 20%.

In 2002, the US Federal Communications Commission (FCC) issued the First Report and Order (R&O), which permitted unlicensed UWB operation and commercial deployment of UWB devices. There are three classes of devices defined in the R&O document: (1) imaging systems (e.g., ground-penetrating radar systems, wall-imaging systems, through-wall imaging systems, surveillance systems, and medical systems), (2) vehicular radar systems, and (3) communications and measurement systems. The FCC allocated a block of unlicensed radio spectrum from 3.1 to 10.6 GHz at the noise floor of −43 dBm/MHz for the above applications where each category was allocated a specific spectral mask (as described in Federal Communications Commission adopted 14 February, 2002, released 22 April, 2002) and UWB radios overlaying coexistent RF systems can operate.

In Figure 6.15 an example of FCC a spectral mask for indoor commercial systems is reported. With similar regulatory processes currently under way in many countries worldwide, government agencies responded to this FCC ruling. Regarding Europe, it is important to mention that on 21 February, 2007, the Commission of the European Communities released a decision on allowing the use of the radio spectrum for equipments using UWB technology in a harmonized manner in the European Community (European Commission, 2007). The decision concerns the use of the radio spectrum on a non-interference and non-protected basis by equipments using UWB technology, with the definition of maximum allowed effective isotropic radiated power (EIRP) densities both in the absence and in the presence of appropriate interference mitigation techniques. In Figure 6.15 the maximum EIRP density in the absence of appropriate mitigation techniques is reported. These limits can be relaxed if interference mitigation techniques are adopted.

Such UWB systems can be realized through conventional modulation schemes by stressing the bandwidth to be larger than 500 MHz, for example by adopting orthogonal frequency division multiplexing (OFDM) signaling leading to multiband UWB (MB-UWB). This approach has been first followed by the IEEE 802.15.3a Task Group (now officially disbanded by the IEEE Standards Association) then included in the WiMedia alliance standard for high-speed applications which involve imaging and multimedia (WiMedia Alliance, 2005; ECMA368 ecma-international, 2006).

Another very promising UWB technique, especially for WSN applications, is named impulse radio-UWB (IR-UWB). The IR-UWB technique relies on ultra-short (nanosecond scale) waveforms that can be free of sine-wave carriers and do not require intermediated frequency (IF) processing because they can operate at baseband. As information-bearing

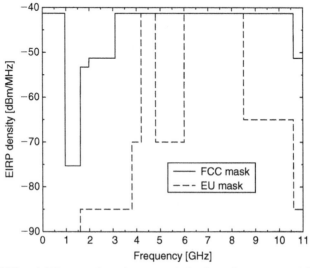

**Figure 6.15** *FCC and EU spectral masks, respectively, for indoor commercial systems in the absence of appropriate mitigation techniques*

pulses with ultra-short duration have UWB spectral occupancy, UWB radios come with unique advantages that have long been appreciated by the radar and communications communities, such as the enhanced capability to penetrate through obstacles, the ultra-high precision ranging at the centimetre level, the very high data rates in harsh multipath environments along with a commensurate increase in user capacity, and potentially small size and processing power. The IR-UWB technique has been selected as the PHY layer of the IEEE 802.15.4a Task Group for wireless personal area network (WPAN) low rate alternative PHY layer (IEEE 802.15.4a Standard 2006).

Readers who are interested in UWB IR and its coexistence with narrowband system can refer to (Win, Scholtz & Fullerton, 1996; Win & Scholtz, 1998, 2000; Giorgetti, Chiani & Win, 2005). For the performance evaluation and the design of UWB systems it is important to properly model the channel. Channel measurement and modeling for UWB are addressed in (Win & Scholtz, 1998a,b, 2002; Cramer, Scholtz & Win, 2002; Cassioli, Win & Molisch, 2002; Cassioli et al., 2007; Molisch et al., 2006) while UWB spectral analysis is given in (Win, 2002a,b; Ridolfi & Win, 2006). For receiver design and analysis a reader can refer to (Win & Kostic, 1999a,b; Win, Chrisikos & Sollenberger, 2000a,b), while transmitted reference systems are described in (Quek & Win, 2004a,b, 2005a,b; Gifford & Win, 2004; Quek, Win & Dardari, 2005a,b, 2007a,b). There are several topics related to UWB systems and networks such as rapid acquisition for synchronization (Suwansantisuk & Win, 2005a,b,c,d, 2006a,b, 2007; Suwansantisuk, Win & Shepp, 2005a,b), localization (Dardari, Chong & Win, 2006, 2008; Dardari & Win, 2006; Jourdan, Dardari & Win, 2006, 2007; Jourdan et al., 2005; Falsi et al., 2006; Shen & Win, 2007; Shen, Wymeersch & Win, 2007; Shen et al., 2007), and modeling of aggregate interference (Pinto & Win, 2006a,b,c, 2007; Win et al., 2006; Pinto et al., 2006). The impact of power dispersion profile on the performance is given in (Win & Wen, 2005; Win, 2003).

### 6.2.2 Impulse radio UWB

As previously remarked, IR-UWB technique is particularly suitable for WSN applications. In IR-UWB the information is encoded using impulses. Typically, the adopted pulse $p(t)$, of duration $T_p$, is derived by the Gaussian pulse and its derivatives due its smallest possible time-bandwidth product which maximizes range-rate resolution and are readily available from antenna pattern. The $n$th derivative of the basic Gaussian pulse, $p_0(t) = \exp(-2\pi(t^2/\tau_p^2))$, is usually adopted. In this case, for $n > 0$, we have

$$p(t) = p_0^{(n)}(t) \sqrt{\frac{(n-1)!}{(2n-1)! \pi^n \tau_p^{(1-2n)}}}, \tag{6.1}$$

where $\tau_p$ is strictly related to the pulse width, $p_0^{(n)}(t)$ denotes the $n$th-order derivative of $p_0(t)$ with respect $t$, and the last fraction has been introduced to normalize the pulse energy to one. In Figure 6.16 and Figure 6.17, the 6th derivative of the Gaussian pulse with $\tau_p = 0.192$ ns is shown in the time and frequency domain, respectively. It can be

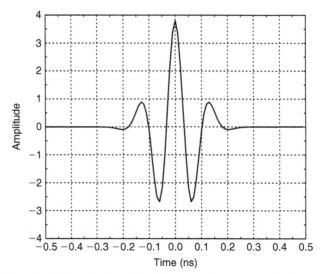

**Figure 6.16**   *Example of 6th derivative of the Gaussian pulse with $\tau_p = 0.192$ ns*

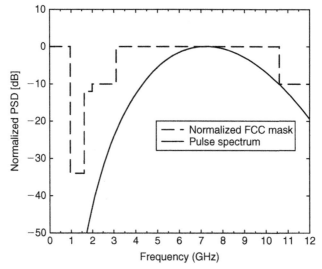

**Figure 6.17**   *Example of 6th derivative of the Gaussian pulse spectrum with $\tau_p = 0.192$ ns*

noted that this pulse is compliant with the FCC specifications in terms of power spectral density (PSD). In general, due to the short pulse duration (typically less than 1 ns), the bandwidth of the transmitted signal can be on the order of one or more GHz.

These impulses can be modulated either using pulse position modulation (PPM) or pulse amplitude modulation (PAM). The transmitter feeds these impulses to a very large bandwidth non-resonating antenna, or sometimes the antenna itself shapes the impulses to the required frequency of operation.

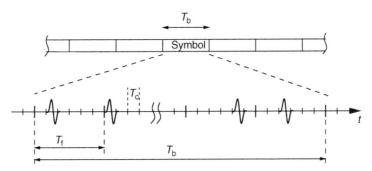

**Figure 6.18**   *Example of UWB time-hopping frame structure*

In a typical UWB system, a symbol of duration $T_b$ is divided in time intervals $T_f$ called frames (see Figure 6.18), which are further decomposed into smaller time slots $T_c$ called chips. To allow for multi-user access the pulse $p(t)$, with duration $T_p < T_c$, is transmitted in each frame in a chip position specified by a user-specific pseudo-random time-hopping (TH) code $\{c_k\}$ having period $N_s$, where $N_s$ is the number of frames per symbol (Win & Scholtz, 1998; Yang & Giannakis, 2004). The frame time $T_f$ is usually chosen to be greater than the maximum multipath delay to avoid intersymbol interference. By adopting the PPM scheme, the transmitted signal from the $u$th user can be written as

$$s_{tr}^{(u)}(t) = \sqrt{\frac{E_b^{(u)}}{N_s}} \sum_{n=-\infty}^{\infty} p\left(t - c_n^{(u)} T_c - nT_f - \delta d_{\lfloor n/N_s \rfloor}^{(u)}\right), \qquad (6.2)$$

where $\delta$ is the modulation index, $E_b^{(u)}$ is the transmitted energy per bit, and $d_j^{(u)} \in \{-1, 1\}$ is the data sequence of the $u$th user.

In the case where $U$ similar UWB interferers are present and assuming the user 1 as the useful one, the received signal is in multipath plus AWGN channel is given by

$$r(t) = \sum_{n=-\infty}^{\infty} w^{(1)}\left(t - c_n^{(1)} T_c - nT_f - \delta d_{\lfloor n/N_s \rfloor}^{(1)}\right) + d(t) + n(t), \qquad (6.3)$$

with

$$d(t) = \sum_{u=2}^{U} \sum_{n=-\infty}^{\infty} w^{(u)}\left(t - c_n^{(u)} T_c - nT_f - \delta d_{\lfloor n/N_s \rfloor}^{(u)}\right) \qquad (6.4)$$

and

$$w^{(u)}(t) = \sqrt{\frac{E_b^{(u)}}{N_s}} \sum_{l=1}^{L} a_l^{(u)} p(t - \tau_l^{(u)}), \qquad (6.5)$$

where $n(t)$ is additive white Gaussian noise (AWGN) with zero mean and two-sided power spectral density $N_0/2$, $L$ is the maximum number of multipath components, whereas

$\{\tau_1^{(u)}, \tau_2^{(u)}, \ldots, \tau_L^{(u)}\}$ and $\{a_1^{(u)}, a_2^{(u)}, \ldots, a_L^{(u)}\}$ are sets of parameters composed of the path amplitudes $a_l^{(u)}$'s and delays $\tau_l^{(u)}$'s related to the $u$th user. The component $d(t)$ represents multi-user interference (MUI) which can often be assumed a Gaussian random process when $U$ is large. Assuming the following normalization $\sum_l \mathbb{E}\{(a_l^{(u)})^2\} = 1$ where the average is taken over all fading realizations the average energy per received bit is given by $E_b^{(u)}$.

The most commonly used UWB receiver is a correlation receiver (Win & Scholtz, 1998), where the received signal is correlated with a local replica of the transmitted pulse (matched filter). In a single-user AWGN scenario, the bit error probability is simply

$$P_b = \frac{1}{2}\,\text{erfc}\sqrt{\frac{E_b(1-\rho)}{2N_0}}, \tag{6.6}$$

where

$$\rho = \int_{-\infty}^{\infty} p(t)\,p(t-\delta)dt \tag{6.7}$$

is the correlation coefficient and $\text{erfc}(\cdot)$ is the complementary error function. For example, using the 6th Gaussian derivative pulse the optimum value of $\delta$ that maximizes the performance is $\delta_{\text{opt}} = 0.07$ ns, which gives $\rho = -0.81$. In Figure 6.19 the performance of a matched filter (MF) receiver in AWGN is reported for this parameter choice and compared to the orthogonal and antipodal pulse cases.

The performance limits of a communication system are determined by the channel in which it operates. UWB propagation channels show fundamental differences from

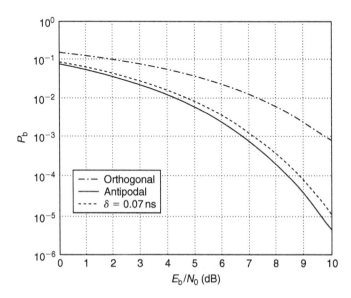

**Figure 6.19**   *Performance of a UWB received based on matched filter in AWGN for different pulses*

conventional narrowband propagation in many respects. The transmission of ultra-short pulses can potentially resolve extremely large numbers of paths experienced by the received signal, especially in indoor environments, thus eliminating significant multipath fading (Cassioli et al., 2002). This may considerably reduce fading margins in link budgets and may allow low transmission power operation. In addition, rich multipath diversity can be collected through the adoption of rake receivers (Win & Scholtz, 2002a). In Section 3.4 a brief description of UWB propagation characteristics and models is presented.

### 6.2.3 Low complexity receivers

To avoid the high sampling rate required to digitally represent a UWB signal, and hence expensive analog-to-digital converters, and computational complexity associated with the channel estimation in a rake receiver, non-coherent receivers are gaining a particular interest. One simple solution is to detect the position/presence of the transmitted pulse in PPM/on-off keying (OOK) signaling schemes through an ED (Sahin, Guvenc & Arslan, 2005; Arias-de Reyna, D'Amico & Mengali, 2006; Nemati, Mitra & Scholtz, 2006).

An interesting alternative is offered by the adoption of the transmitted reference signaling in conjunction with UWB. Transmitted reference signaling involves the transmission of a reference and data signal pulse pair separated in time by $T_r$ seconds. In order for both signals to experience the same channel, the time separation must be less than the channel coherence time. Due to the simplicity of TR signaling, there is renewed interest in its use for UWB systems (Choi & Stark, 2002; Quek & Win, 2005; Chao & Scholtz, 2005; Quek, Win & Dardari, 2007), which can exploit multipath diversity inherent in the environment without the need for channel estimation and stringent acquisition. A simple autocorrelation receiver (AcR) is sufficient as shown in Figure 6.20, where the incoming signal, passed through a band-pass zonal filter (BPZF), is correlated with a delayed version of the reference signal, thus collecting the received signal energy. The integration interval $T < T_f$ determines the number of multipath components (or equivalently, the amount of energy) captured by the receiver, as well as the amount of noise accumulation.

The drawback of this scheme is that the noisy reference signal in the AcR and the additional energy spent to transmit the reference pulse lead to some performance degradation. To avoid noise collection from time regions within the integration interval that do not contain useful energy contributions, a stop-and-go strategy based on energy detection in the AcR to selectively collect portions of effective signal is proposed in Dardari, Giorgetti, Chiani & Win, 2006.

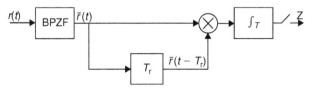

**Figure 6.20** *Transmitted reference receiver scheme*

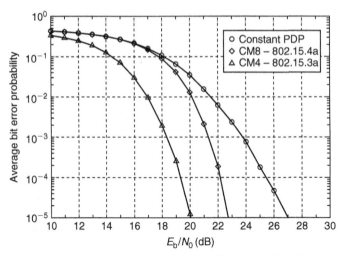

**Figure 6.21** *Average bit error probability of the AcR for three channel models*

In Figure 6.21 the average bit error probability of the AcR is plotted for three different channel models: the CM4 type proposed by the 802.15.3a Task Group; the CM8 type proposed by the 802.15.4a Task Group; and a constant PDP channel model with $L = 120$ independent paths spaced $\Delta = T_p$ apart, each with Nakagami-$m$ distributed amplitudes with parameter $m = 3$ and a probability $a = 0.3$ of having a path at a given time slot $\Delta$ (Dardari, Giorgetti, Chiani & Win, 2006). As can be noted, the channel behaviour has a strong impact on system performance.

## 6.2.4 UWB standards for WSNs: the IEEE 802.15.4a PHY

The Task Group TG4a was formed in March 2004 with the objective of providing an amendment to IEEE 802.15.4 (IEEE 802.15.4 Standard, 2006) for an alternative PHY layer (IEEE 802.15.4a Standard, 2006). The aim is to offer both communications and high precision ranging capability (i.e., 1 m accuracy and better). In January 2005, 26 companies turned in proposals to be considered in the standard and in March 2005 all proposals were merged and the baseline specification was selected without enacting the down-selection procedures. The baseline is based on two optional PHYs consisting of a UWB impulse radio (operating in unlicensed UWB spectrum) and a chirp spread spectrum (CSS) (operating in unlicensed 2.4 GHz spectrum), where the former will be able to deliver communications and high precision ranging. The group finalized the standard on August 2007 (IEEE 802.15.4a Standard, 2007). In particular, the UWB PHY supports an over-the-air mandatory data rate of 851 Kbit/s, with optional data rates of 110 Kbit/s, 6.81 Mbit/s, and 27.24 Mbit/s. The typical range is of 10–50 metres. The choice of the PHY depends on the local regulations, application, and user preferences. Table 6.1 reports the frequency bands foreseen by the standard (some of them are optional). The modulation used combines both BPSK and PPM signaling so that both coherent and low

**Table 6.1** *IEEE 802.15.4a PHY layer frequency bands*

| PHY mode | Frequency band (MHz) |
| --- | --- |
| UWB sub-GHz | 250–750 |
| 2450 CSS | 2400–2483.5 |
| UWB low-band | 3244–4742 |
| UWB high-band | 5944–10,234 |

complexity non-coherent receivers can be used to demodulate the signal. The key protocol adopted for ranging is a two-way frame exchange which will be explained in Chapter 8. Ranging measurements can be used by the application level to derive the node position information. Some techniques suitable to this scope will be described in Chapter 8.

## 6.3 Bluetooth technology

Bluetooth wireless technology is a short-range communication system intended to replace the cables in WPAN. The key features of Bluetooth wireless technology are robustness, low power, and low cost, and many features of the core specification are optional, allowing product differentiation.[3]

The IEEE project 802.15.1 has derived a WPAN standard based on the Bluetooth v1.1 foundation specifications.

In the following we will report some technical details useful for the usage in WSANs and we describe some frameworks able to capture main effects affecting the physical layer performance in fading channel and in the presence of interference with single or multiple antenna reception.

### 6.3.1 Introduction to Bluetooth characteristics

The Bluetooth core system consists of an RF transceiver, baseband, and protocol stack. The system offers services that enable the connection of devices and the exchange of a variety of data classes between these devices. Bluetooth specifications for more details are in Bluetooth™, 2004.

---

[3]The name Bluetooth is from the 10th century Danish King Harald Blatand (Harold Bluetooth in English) who was instrumental in uniting warring factions in parts of what is now Norway, Sweden, and Denmark; just as Bluetooth technology is designed to allow collaboration between differing industries such as the computing, mobile phone, and automotive markets.

*Bluetooth physical layer*

The Bluetooth RF (physical layer) operates in the unlicensed ISM band, for the majority of countries around 2.4 GHz in [2400, 2483.5]MHz. The system employs a frequency hopping (FH) transceiver (the nominal hop rate is 1600 hops/s) to combat interference and fading, and provides many FH-spread spectrum (SS) carriers. RF operation uses a Gaussian-shaped binary frequency shift keying (GFSK) modulation to minimize transceiver complexity, and a forward error correction (FEC) coding technique. The symbol rate is 1 Msps supporting the bit rate of 1 Mbps or, with enhanced data rate, a gross air bit rate of 2 or 3 Mbps. These modes are known as basic rate and enhanced data rate, respectively.

The equipment is classified into three power classes (given as transmitted power levels at the antenna connector of the equipment; if the equipment does not have a connector, a reference antenna with 0 dBi gain is assumed): class 1 with maximum output power of 20 dBm, class 2 with maximum output power of 4 dBm, and class 3 with maximum output power of 0 dBm. A power control is required for power class 1 equipment. The power control is used for limiting the transmitted power over 0 dBm. Power control capability under 0 dBm is optional and could be used for optimizing the power consumption and overall interference level. The power steps will form a monotonic sequence, with a maximum step size of 8 dB and a minimum step size of 2 dB. A class 1 equipment with a maximum transmit power of 20 dBm must be able to control its transmit power down to 4 dBm or less.

During typical operation, a physical radio channel is shared by a group of devices that are synchronized to a common clock and frequency hopping pattern. One device provides the synchronization reference and is known as the master. All other devices are known as slaves. A group of devices synchronized in this fashion form a piconet. This is the fundamental form of communication for Bluetooth wireless technology.

Devices in a piconet use a specific frequency hopping pattern which is algorithmically determined by certain fields in the Bluetooth specific address and clock of the master. The basic hopping pattern is a pseudo-random ordering of the 79 frequencies[4] with channel spacing of 1 MHz in the ISM band (e.g., $f = 2402 + k$MHz, with $k = 0, \ldots, 78$). To comply with out-of-band regulations in each country, a guard band is used at the lower and upper band edge, respectively of 2 MHz and 3.5 MHz. The hopping pattern may be adapted to exclude a portion of the frequencies that are used by interfering devices. The adaptive hopping technique improves Bluetooth technology coexistence with static (non-hopping) ISM systems when these are co-located.

The physical channel is subdivided into time units known as slots with duration 625 us. Data is transmitted between Bluetooth-enabled devices in packets that are positioned in these slots. When circumstances permit, a number of consecutive slots may be allocated to a single packet. Frequency hopping takes place between the transmission or reception

---

[4] In some countries, like France, the number of frequencies is 23.

of packets. Bluetooth technology provides the effect of full duplex transmission through the use of a time division duplexing (TDD) scheme.

## Layering

Above the physical channel there is a layering of links and channels and associated control protocols. The hierarchy of channels and links from the physical channel upwards is: physical channel, physical link, logical transport, logical link, and L2CAP channel.

Between master and slave(s), different types of links can be established. Two link types have been defined: (a) synchronous connection-oriented (SCO) link and (b) asynchronous connection-less (ACL) link. The SCO link is a point-to-point link between a master and a single slave in the piconet. The master maintains the SCO link by using reserved slots at regular intervals. The ACL link is a point-to-multipoint link between the master and all the slaves participating on the piconet. In the slots not reserved for the SCO link(s), the master can establish an ACL link on a per-slot basis to any slave, including the slave(s) already engaged in an SCO link. FEC schemes adopted for the payload are different for SCO and ACL links.

The L2CAP layer provides a channel-based abstraction to applications and services. It carries out segmentation and reassembly of application data and multiplexing and demultiplexing of multiple channels over a shared logical link. L2CAP has a protocol control channel that is carried over the default ACL logical transport. Application data submitted to the L2CAP protocol may be carried on any logical link that supports the L2CAP protocol.

## Piconet

Any time a Bluetooth wireless link is formed, it is within the context of a piconet. A piconet consists of two or more devices that occupy the same physical channel (which means that they are synchronized to a common clock and hopping sequence). The common (piconet) clock is identical to the Bluetooth clock of one of the devices in the piconet, known as the master of the piconet, and the hopping sequence is derived from the master's clock and the master's Bluetooth device address. All other synchronized devices are referred to as slaves in the piconet. The terms master and slave are only used when describing these roles in a piconet.

Within a common location a number of independent piconets may exist. Each piconet has a different physical channel (i.e., a different master device and an independent piconet clock and hopping sequence).

A Bluetooth-enabled device may participate concurrently in two or more piconets. It does this on a time-division multiplexing basis. A Bluetooth-enabled device can never be a master of more than one piconet.[5] A Bluetooth-enabled device may be a slave in many independent piconets.

---

[5] Since the piconet is defined by synchronization to the master's Bluetooth clock it is impossible to be the master of two or more piconets.

A Bluetooth-enabled device that is a member of two or more piconets is said to be involved in a *scatternet*. Involvement in a scatternet does not necessarily imply any network routing capability or function in the Bluetooth-enabled device. The Bluetooth core protocols do not, and are not intended to offer such functionality, which is the responsibility of higher-level protocols and is outside the scope of the Bluetooth core specification.

Logical transports, logical links, and L2CAP channels are used to provide capabilities for the transport of data.

### Operational procedures and modes
The typical operational mode of a Bluetooth-enabled device is to be connected to other Bluetooth-enabled devices (in a piconet) and exchanging data with that Bluetooth-enabled device. As Bluetooth wireless technology is an ad-hoc wireless communications technology there are also a number of operational procedures that enable piconets to be formed so that the subsequent communications can take place. Procedures and modes are applied at different layers in the architecture and therefore a device may be engaged in a number of these procedures and modes concurrently, such as: Inquiry (Discovering), Paging (Connecting), Hold, Sniff, Parked State.

### Enhanced data rate
Enhanced data rate (EDR) is a method of extending the capacity and types of Bluetooth packets for the purposes of increasing the maximum throughput, providing better support for multiple connections, and lowering power consumption, while the remainder of the architecture is unchanged.

EDR may be selected as a mode that operates independently on each logical transport. Once enabled, the packet type bits in the packet header are interpreted differently from their meaning in basic rate mode. This different interpretation is clarified in conjunction with the logical transport address field in the header. The result of this interpretation allows the packet payload header and payload to be received and demodulated according to the packet type.

## 6.3.2   Bluetooth physical layer performance

Bluetooth devices use the ISM band at 2.4 GHz and are expected to work in the presence of interference caused by other devices (e.g., IEEE802.11 WLAN) that could be present in the proximity with a consequent performance degradation. Most of solutions proposed in the literature to allow the coexistence of Bluetooth and WLAN devices are based on power control adjustments and scheduling policies, while others involves antenna diversity (see, e.g., Premkumar & Srinivasan, 2005; Golmie, Chevrollier & Rebala, 2003; Bektas, Vondra, Veith, Faltin, Pohl & Scholtz, 2003). The performance of Bluetooth at PHY layer in fading channels with and without multiple antennas reception and different combining techniques, as well as in the presence or absence of interference, are respectively investigated in (Masini, Conti, Dardari & Pasolini, 2006; Masini, Dardari, Conti & Pasolini, 2006a; Masini, Dardari, Conti & Pasolini, 2006b; B. M. Masini, A. Conti, G. Pasolini &

D. Dardari, 2008). Here a common framework is reported, enabling the analytical PHY layer performance evaluation which results to be useful to avoid time-consuming bit-level simulations in upper layers simulators still considering all aspects affecting the performance at PHY layer.

The impact of diversity reception on the performance of Bluetooth packet transmission in wireless channels with small-scale fading (also called fast fading) and large-scale fading (shadowing) and in the presence of IEEE 802.11g interference is investigated. In particular, for low-cost non-coherent demodulation the selection diversity (SD) scheme is considered, whereas in the case of coherent demodulation, maximal ratio combining (MRC) is also studied (for diversity technique see, e.g., Simon & Alouini, 2004); the result is that diversity techniques allow coverage extension.

Diversity techniques, in particular, have been shown to effectively counteract the effect of fading in wireless communications; thus, we consider the adoption of multiple antennas at the receiver side of the Bluetooth link in order to assess the possible performance improvement also in the presence of interference. Here in particular, we consider a simple SD combiner scheme, that is, the simplest form of diversity reception, whereby the received signal is selected as one (typically the one with strongest power) among the $N$ available diversity branches. It is worth noting that, owing to its easy implementation, the SD scheme is fully compliant with the low-cost nature of Bluetooth devices; moreover, the adoption of multiple antennas (placed, for instance, in the back of a Bluetooth-enabled laptop screen) at the receiver side changes neither the spectral shape nor the modulation format, thus being perfectly consistent with Bluetooth specifications, even with reference to the EDR.

A tight parametric exponential approximation for the instantaneous bit error probability (BEP) in additive white Gaussian noise is firstly derived; then, from this expression, the mean block error probability (BLEP), the mean packet error probability (PEP), and the coverage probability ($P_{cov}$) in fading channels can be obtained. In particular, the impact of the diversity order on the BLEP, PEP and $P_{cov}$ is shown in the presence and in the absence of IEEE 802.11g interference.

The contribution lies in the derivation of an analytical framework able to express the Bluetooth performance when multiple antennas are adopted at the receiver side and the communication is performed in the presence of AWGN, fast fading, log-normal shadowing as well as IEEE 802.11g interference,[6] taking into account also the modulation format, the presence of FEC coding and frequency hopping. By considering real propagation environments, both small-scale and large-scale effects (due to fast fading and shadowing, respectively) are properly taken into account (Andrisano, Tralli & Verdone, 1998).

Note also that, when real-time applications are considered, figures of merit averaged over fast fading, such as the mean BEP or the mean PEP derived in the following, are not sufficient to suitably characterize the system performance, hence the coverage probability

---

[6] As far as the IEEE 802.11g WLAN technology is concerned, for our purposes it is sufficient to recall that it is based on the OFDM modulation scheme and operates in the 2.4 GHz ISM band with a bandwidth of about 22 MHz.

$P_{\text{cov}}$, that is, the probability the PEP does not exceed a tolerable given threshold is also derived here as an important index of the system performance over large-scale effects (see, e.g., Conti, Win, Chiani & Winters, 2003).

*Bit and block error probability evaluation*
Through fitting with bit-level simulations, we approximate the BEP with the following parametric exponential expression:

$$P_b(\gamma) \simeq a \cdot e^{-b\cdot\gamma}, \tag{6.8}$$

where $\gamma$ is the instantaneous SNR in terms of symbol energy over one-side thermal noise power spectral density, and parameters $a$ and $b$ have to be properly chosen depending on the normalized maximum frequency deviation $f_d T$, being $f_d$ the maximum frequency deviation and $T$ the symbol time.

The accuracy of (6.8) has been validated through simulations showing a good agreement, as shown in Figure 6.22(a) which is referred to a non-coherent demodulation and different modulation parameters (i.e., different $f_d T$ values).[7] For instance, in the typical case $f_d T = 0.16$, compliant with the BT specifications (i.e., $f_d T$ within the interval [0.14, 0.175]), we found that a tight approximation can be obtained when $a = 0.08$ and $b = 0.13$.[8] By passing through, note that with proper choice of parameters $a$ and $b$, we verified that the accuracy of (6.8) results were satisfactory also for coherent schemes for which $a = 0.47$ and $b = 0.52$ when $f_d T = 0.165$ (Masini, Conti, Dardari & Pasolini, 2006); in this case a good agreement with simulation results (Soltanian & Van Dyck, 2001) is also verified (see Figure 6.22).

The relevance of (6.8) is that it allows the mathematical derivation of overall performance figures (such as the packet error probability) thus avoiding time-consuming bit-level simulations.

Assuming independent errors on a block of $N_{\text{BL}}$ bits and by means of (6.8), the instantaneous BLEP, that is, the probability to have at least an error in a block of bits, can be written as

$$P_{\text{BL}}(\gamma) = 1 - (1 - P_b(\gamma))^{N_{\text{BL}}}$$
$$= \sum_{k=1}^{N_{\text{BL}}} \binom{N_{\text{BL}}}{k} (-1)^{k+1} a^k \, e^{-kb\gamma}. \tag{6.9}$$

We assume the small-scale fading to be constant over a block, with mean value of the SNR equal to $\bar{\gamma}$, and independent identically distributed (i.i.d.), over antennas.

---

[7] The parameters $a$ and $b$ have been estimated by fitting simulative results with through the minimum mean square error technique.
[8] At the receiver side we considered a band-pass butterworth filter with 10 poles and a 3 dB bandwidth set to the optimum value for each $f_d T$ and a non-coherent demodulation given by limiter-discriminator detection.

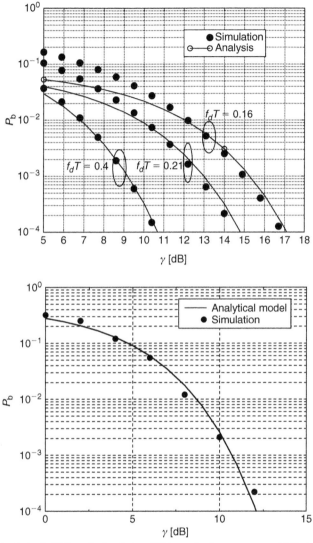

**Figure 6.22** *Analytical BEP and simulative BER vs the instantaneous SNR (a) Non-coherent schemes with different values of* $f_dT$ *(b) Coherent scheme with* $f_dT = 0.165$

Hence, by averaging the instantaneous BLEP over fast fading statistics, we obtain the following expression for the mean BLEP:

$$\overline{P_{\text{BL}}}(\overline{\gamma}) = \sum_{k=1}^{N_{\text{BL}}} \binom{N_{\text{BL}}}{k} (-1)^{k+1} a^k \mathbb{E}_{\gamma} \{e^{-bk\gamma}\}. \tag{6.10}$$

Recalling the definition of the moment generating function (MGF) of $\gamma$ (see, e.g., Simon & Alouini, 2004; Win & Winters, 2001; Alouini & Simon, 2000; Ma & Chai, 2000;

Win, Mallik & Chrisikos, 2003), that is, $\Phi_\gamma(s) \triangleq \mathbb{E}_\gamma\{e^{s\gamma}\}$, (6.10) takes the following form

$$\overline{P_{\mathrm{BL}}}(\bar\gamma) = \sum_{k=1}^{N_{\mathrm{BL}}} \binom{N_{\mathrm{BL}}}{k}(-1)^{k+1}a^k\Phi_\gamma(-bk). \tag{6.11}$$

Note that expression (6.11) is general with respect to the diversity reception technique and combining scheme as well as the fast fading statistics, once the proper MGF is known (as example results, some particular cases for the signal knowledge at the combiner and small-scale fading statistics will be considered).

*Block error probability evaluation in the presence of interference*
Now we extend the analysis to the case of BT transmission interfered by an IEEE 802.11g signal. We assume the small-scale fading level for both the useful and the interfering links as i.i.d.

In the scenario considered, an IEEE 802.11g and a BT link are simultaneously active in an indoor environment. We assume that both the BT link and the IEEE 802.11g link are affected by small-scale fading (please note that the framework is general with respect to fast fading statistics, that will be specified only for obtaining numerical results) and shadowing. We denote with $C$ and $I$ the short-term useful and interfering powers, received by the BT device, given by:

$$C = \frac{P_{\mathrm{U}}}{L_0(d_{\mathrm{U}})}\alpha_{\mathrm{U}}^2 S_{\mathrm{U}}^2 \tag{6.12}$$

$$I = \frac{P_{\mathrm{I}}}{L_0(d_{\mathrm{I}})}\alpha_{\mathrm{I}}^2 S_{\mathrm{I}}^2, \tag{6.13}$$

where $P_{\mathrm{U}}$ and $P_{\mathrm{I}}$ are the useful (BT) and interfering (IEEE 802.11g) transmitted power levels, respectively, $L_0(d)$ is the path-loss law (here in linear scale) as a function of the distance $d$ between the transmitter and the receiver, $d_{\mathrm{U}}$ is the distance between the useful transmitter and the receiver and $d_{\mathrm{I}}$ is the distance between the BT receiver and the IEEE 802.11g interfering transmitter. The parameters $\alpha_{\mathrm{U}}^2$ and $\alpha_{\mathrm{I}}^2$ represent the small-scale fading power gains, while $S_{\mathrm{U}}^2$ and $S_{\mathrm{I}}^2$ denote the shadowing affecting the useful and interfering links, respectively. Hence, the mean powers averaged over the fast fading statistics become (Andrisano, Conti, Dardari & Pasolini, 2003):

$$\bar{C} = \mathbb{E}_{\alpha_{\mathrm{U}}^2}\{C\} = \frac{P_{\mathrm{U}}}{L_0(d_{\mathrm{U}})}S_{\mathrm{U}}^2 \tag{6.14}$$

$$\bar{I} = \mathbb{E}_{\alpha_{\mathrm{I}}^2}\{I\} = \frac{P_{\mathrm{I}}}{L_0(d_{\mathrm{I}})}S_{\mathrm{I}}^2. \tag{6.15}$$

In Masini, Dardari, Conti & Pasolini, 2006a; Masini, Dardari, Conti & Pasolini, 2006b it is verified that at the input of the decision device the sampled interference in the useful bandwidth can be modelled as Gaussian distributed. Thus, by means of this result and following the procedure in Andrisano et al., 2003, the BLEP averaged over fading statistic in the presence of interference is still given by (6.11) simply replacing $\bar{\gamma}$ with the mean signal-to-noise-plus-interference ratio (SNIR), $\bar{\gamma}^{(I)}$, and using proper MGFs. We obtain in the following $P_{BL}$ averaged over fading realizations:[9]

$$\overline{P_{BL}}^{(I)}(\bar{\gamma}, \bar{I}, N_{BL}, a, b) = \overline{P_{BL}}(\bar{\gamma}^{(I)}, N_{BL}, a, b) \tag{6.16}$$

with mean SNIR given by

$$\bar{\gamma}^{(I)} = \cfrac{1}{\cfrac{1}{\bar{\gamma}} + \cfrac{F\Gamma_{\Delta f}}{\overline{C/I}}}. \tag{6.17}$$

In (6.17), the parameter $F$ is a coefficient depending on the interfering signal pulse shaping and the receiving filter, whereas $\Gamma_{\Delta f}$ is a coefficient depending on the frequency offset $\Delta f$ between the useful and interfering carriers (Andrisano et al., 2003). The parameter $\Gamma_{\Delta f}$ is normalized such that $\Gamma_1 = 1$ when $\Delta f = 1$ MHz, and $F$ is tuned as a consequence. The coefficient $\Gamma_{\Delta f}$ can be obtained by calculating the interfering power at the output of the BT receiver filter, normalized to the reference case of $\Delta f = 1$ MHz. Note that, when a IEEE 802.11g signal is involved, the evaluation of $\Gamma_{\Delta f}$ is made following the OFDM nature of the considered signal. In fact, the power spectral density of IEEE 802.11g is significant in 22 MHz band and it can be considered constant within this interval. Hence, it can be found that $\Gamma_{\Delta f} = 1$ for each value of $f$ in $[f_0 - 11\text{ MHz}, f_0 + 11\text{ MHz}]$, whereas, when $|\Delta f| > 11$ MHz the effect of the interference can be neglected (hence, $\Gamma_{\Delta f} = 0$ in this case).

### Packet error probability and coverage evaluation

We now derive the mean PEP of the Bluetooth link taking both thermal noise and interference generated by IEEE 802.11g into account, as well as the channel coding techniques adopted by Bluetooth. In fact, Bluetooth data and voice packets are composed by three fields (access code, header, and payload) that are protected against errors through different FEC block codes (Bluetooth™, 2004).

Thus, to evaluate the instantaneous PEP, the error correction capabilities of the FEC codes adopted in the different packet fields have to be considered. However, as shown in Andrisano et al., 2003, since in a BT data packet the payload is the longest and least protected field, the PEP almost coincides with the payload error probability $PE_{pl}$ after decoding:

$$PEP(\gamma) \simeq PE_{pl}(\gamma). \tag{6.18}$$

After some algebra it is possible to derive the following expression:

$$\begin{aligned} PE_{pl}(\gamma) &= 1 - (1 - P_e^{(cw)})^{N_c} \\ &= \sum_{k=1}^{N_c} \binom{N_c}{k} (-1)^{k+1} (P_e^{(cw)})^k, \end{aligned} \tag{6.19}$$

---

[9] Superscript $(I)$ stands for 'in the presence of interference'.

where $P_e^{(cw)}$ is the codeword error probability (CEP) and $N_c$ is the number of codewords in the payload. For codewords of $N_b$ bits and FEC codes able to correct up to $t$ errors per codeword with hard decision, the CEP results in:

$$P_e^{(cw)}(\gamma, t, N_b) = \sum_{n=t+1}^{N_b} \binom{N_b}{n} P_b(\gamma)^n (1 - P_b(\gamma))^{N_b-n}$$

$$= \sum_{n=t+1}^{N_b} \binom{N_b}{n} P_b(\gamma)^n \sum_{k=0}^{N_b-n} \binom{N_b-n}{k} (-1)^k P_b(\gamma)$$

$$= \sum_{n=t+1}^{N_b} \sum_{k=0}^{N_b-n} \binom{N_b}{n} \binom{N_b-n}{k} (-1)^k P_b(\gamma)^{n+k}. \tag{6.20}$$

Hence, by averaging over fast fading and using (6.18), the mean PEP can be approximated as follows:

$$\overline{PEP}(\bar{\gamma}) \simeq \sum_{k=1}^{N_c} \binom{N_c}{k} (-1)^{k+1} \, \mathbb{E}_{\gamma}\{(P_e^{(cw)})^k\}. \tag{6.21}$$

For the coding techniques and BEP of interest, we carefully checked that the CEP can be approximated by considering as erroneous decoding only the case of $t + 1$ errors (see, e.g., Proakis, 2001) giving

$$P_e^{(cw)}(\gamma, t, N_b) \simeq \binom{N_b}{t+1} P_b(\gamma)^{t+1}. \tag{6.22}$$

Thus, we obtain

$$\overline{PEP}(\bar{\gamma}) \simeq \sum_{k=1}^{N_c} \binom{N_b}{t+1}^k \binom{N_c}{k} (-1)^{k+1} \, a^{k(t+1)} \, \mathbb{E}_{\gamma}\{e^{-bk(t+1)\gamma}\}, \tag{6.23}$$

that can be rewritten, through (6.11), as

$$\overline{PEP}(\bar{\gamma}) \simeq \overline{P_{BL}}(\bar{\gamma}, N_c, A, B), \tag{6.24}$$

with $A = \binom{N_b}{t+1} a^{t+1}$ and $B = b(t + 1)$.

The impact of interference effects can be investigated by replacing the variable $\bar{\gamma}$ with $\bar{\gamma}_{\Delta f}^{(I)}$, leading to the following mean PEP expression:

$$\overline{PEP}^{(I)}(\bar{\gamma}_{\Delta f}^{(I)}) \simeq \overline{P_{BL}}(\bar{\gamma}_{\Delta f}^{(I)}, N_c, A, B). \tag{6.25}$$

Now, by denoting with $L_{wl}$ the probability of time collision between the transmitted Bluetooth and interference packet ($L_{wl}$ being an indicator of the IEEE 802.11g activity factor), and averaging $\overline{PEP}^{(I)}(\bar{\gamma}_{\Delta f}^{(I)})$ over all possible $\Delta f$, we obtain

$$\overline{PEP}_m^{(I)} = L_{wl} \sum_{\Delta f=-N_f}^{N_f} \mathbb{P}\{\Delta f\}\overline{PEP}^{(I)}(\bar{\gamma}_{\Delta f}^{(I)}), \qquad (6.26)$$

where $\mathbb{P}\{\Delta f\}$ is the probability to have a Bluetooth transmission in a particular frequency hop admitted by specifications (Bluetooth™, 2004). We assume that $\Delta f$ is uniformly distributed among all the allowed 79 hops, thus $\mathbb{P}\{\Delta f\} = 1/79$ and $N_f = 39$, while the activity factor $L_{wl}$ is a function of the traffic offered by the interfering IEEE 802.11g transmitter. Note that (6.26) represents a meaningful indicator of the system performance taking into account noise, interference, propagation, FEC codes, frequency hopping technique, traffic, and the diversity reception technique.

For different settings of system parameters (i.e., the interferer activity factor $L_{wl}$, the useful and interference transmitted power levels $P_U$ and $P_I$, and the diversity order $N$), from (6.14), (6.15), (6.26) and the path loss law, it is possible to derive the required mean SNIR to obtain a given level of $\overline{PEP}_m^{(I)}$ and thus relate it to the distances ratio $d_I/d_U$.

In addition to the mean PEP, another relevant figure of merit for BT interfered by IEEE 802.11g is given by the coverage probability, $P_{cov}$, that is, the probability that the $\overline{PEP}_m^{(I)}$ is below a maximum tolerable level $\overline{PEP}_m^{(I)*}$. Thus,

$$P_{cov} = \mathbb{P}\{\overline{PEP}_m^{(I)} \le \overline{PEP}_m^{(I)*}\}$$

$$= \mathbb{P}\left\{\frac{\bar{C}}{\bar{I}} \ge \left(\frac{\bar{C}}{\bar{I}}\right)^*\right\}, \qquad (6.27)$$

being $(\bar{C}/\bar{I})^*$ the value of $\bar{C}/\bar{I}$ giving $\overline{PEP}_m^{(I)}$ equal to $\overline{PEP}_m^{(I)*}$ for a fixed value of the mean SNR, $\bar{\gamma}$. By recalling that

$$\frac{\bar{C}}{\bar{I}} = \frac{P_U}{P_I} \frac{L_0(d_I)}{L_0(d_U)} \frac{S_U^2}{S_I^2}, \qquad (6.28)$$

then in the case of log-normal shadowing, that is, $S_U^2$ and $S_I^2$ in dB are Gaussian distributed with mean zero and variances $\sigma^2_{U\,dB}$ and $\sigma^2_{I\,dB}$, respectively, we have

$$S_{dB}^2 = 10 \log_{10} \frac{S_U^2}{S_I^2} \sim \mathcal{N}(0, \sigma_{dB}^2 = \sigma_{U\,dB}^2 + \sigma_{I\,dB}^2). \qquad (6.29)$$

Then, we can rewrite the coverage probability as follows:

$$P_{cov} = \mathbb{P}\left\{10 \log_{10}\left(\frac{P_U}{P_I} \frac{L_0(d_I)}{L_0(d_U)}\right) + S_{dB}^2 > \left(\frac{\bar{C}}{\bar{I}}\right)^*_{dB}\right\}$$

$$= \mathbb{P}\{S_{dB}^2 > \xi_{dB}\} = \frac{1}{2} \operatorname{erfc}\left(\frac{\xi_{dB}}{\sqrt{2}\sigma_{dB}}\right), \qquad (6.30)$$

where

$$\xi_{dB} = \left(\frac{\bar{C}}{\bar{I}}\right)^*_{dB} - 10\log_{10}\left(\frac{P_U}{P_I}\frac{L_0(d_I)}{L_0(d_U)}\right). \tag{6.31}$$

Now, by fixing a target value for the performance in terms of $\overline{PEP}_m^{(I)}$ and $P_{cov}$, the fast fading statistics, the diversity reception technique and the path loss law, we can obtain the tolerable value for $\bar{C}/\bar{I}$ and the relation between distances $d_I$ and $d_U$ satisfying the requirements.

### Some example results

In Masini, Dardari, Conti & Pasolini, 2006a; Masini, Dardari, Conti & Pasolini, 2006b, some numerical results was derived for the proposed framework in the case of non-coherent demodulation and SD combining (i.e., the low-cost solution feasible for WSANs), as well as some for coherent demodulation with SD and MRC reception (more results for the case of coherent demodulation when also MRC technique is adopted are reported in Masini, Conti, Dardari & Pasolini, 2006. At this point some parameter setting and assumptions have to be specified as a practical example for the general methodology.

We adopt the path loss model as a function of the distance, $L_0(d)$ as given in (3.5) where dB scale and LOS propagation for the first 8 metres are assumed. As far as the BEP model for Bluetooth is concerned, we will consider the couple $(a,b)$ corresponding to $f_d T = 0.16$ and non-coherent demodulation, whereas $F$ has been found to be 0.11.

For what concern the combining technique, we assume that the SD combiner has the knowledge of the SNIR per branch and selects the signal from the branch with strongest SNIR (The MGF related to other kind of knowledge, e.g., useful signal power or addition of useful signal power with total disturbance power, can be found in Yang & Alouini, 2006; Fu & Kam, 2005). Thus, from Win & Winters, 2001; Annamalai, Deora & Tellambura, 2002; Simon & Alouini, 2004, and assuming statistical i.i.d. fading levels among branches, the MGF for a $N$-branches SD receiver in Ricean fading channel (typical of an indoor environment) is given by

$$\Phi_\gamma(s) = N\int_0^\infty e^{s\xi} f_\gamma(\xi)\,[F_\gamma(\xi)]^{N-1}\,d\xi, \tag{6.32}$$

where $f_\gamma(\xi)$ is the Ricean probability density function with Ricean factor $K$, given by:

$$f_\gamma(\xi) = \begin{cases} \dfrac{1+K}{\bar{\gamma}}e^{-K-\frac{(1+K)\xi}{\bar{\gamma}}}\,I_0\left[2\sqrt{\dfrac{K(1+K)\xi}{\bar{\gamma}}}\right] & \text{if } \xi > 0 \\ 0 \end{cases}$$

otherwise, $I_0(\cdot)$ denotes the modified Bessel function of the first kind and $F_\gamma(\xi)$ is the Ricean cumulative density function. It can be found that (Annamalai et al., 2002):

$$F_\gamma(\xi) = 1 - Q\left(\sqrt{2K}, \sqrt{\frac{2(K+1)\xi}{\bar{\gamma}}}\right), \quad \text{for } \xi > 0 \tag{6.33}$$

where $Q(\sqrt{2c}, \sqrt{2d}) = \int_d^\infty e^{-t-c} I_0(2\sqrt{ct}) dt$ is the first order Marcum Q-function.

Finally, substituting (6.32) in (6.11), we derive the mean BLEP for a $N$-branches SD receiver and Ricean fading channel.

For the better performing but more complex case of $N$-branches MRC coherent receiver in Ricean fading we have:

$$\Phi_\gamma(s) = N \int_0^\infty e^{s\xi} f_\gamma(\xi) [\Phi_1(s, \xi)]^{N-1} d\xi, \tag{6.34}$$

where $\Phi_1(s, x) = \int_x^\infty e^{-st} f_\gamma(t) dt$ is the marginal MGF of $\gamma$ of a single diversity branch. Hence, one can obtain also the mean BLEP expression in case of MRC receiver by simply substituting (6.34) in (6.11).

While the framework is general, in the following example results are presented for DM1 and DH1 BT packet types (Bluetooth™, 2004) in Rayleigh fading. For DM1 packets we assume, in particular, $N_c = 15$, $N_b = 15$ and $t = 1$, whereas $N_c = 120$, $N_b = 1$ and $t = 0$ for DH1 packets.

In Figure 6.23, the asymptotical (large SNR, i.e., the system is interference limited) $\overline{PEP}^{(I)}$ conditioned to $\Delta f$ (with $|\Delta f| < 11\,\text{MHz}$) as a function of $\bar{C}/\bar{I}$ is plotted for different diversity orders, $N$, in both cases of DH1 and DM1 packets.

Note that the performance is strongly affected by the diversity order: for example, by fixing $\overline{PEP}^{(I)} = 10^{-2}$ for DM1 packets, a $\bar{C}/\bar{I}$ about 21 dB is required for a single antenna system, while about 11 dB are required with $N = 2$ antennas.

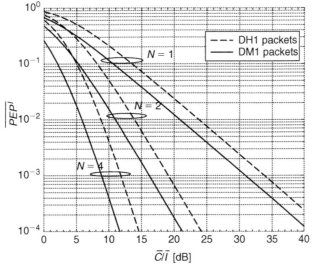

**Figure 6.23** *Asymptotical mean PEP vs $\bar{C}/\bar{I}$ in Rayleigh fading channel adopting a SD receiver varying the number of branches N for DM1 and DH1 BT packets type*

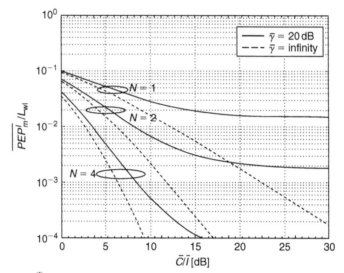

**Figure 6.24**  $\overline{PEP}_m^{(I)}/L_{wl}$ *vs*  $\overline{C/I}$ *for DM1 packets varying the number of antennas and* $\overline{\gamma}$; *SD receiver*

In Figure 6.24, we reported the $\overline{PEP}_m^{(I)}$ normalized to the inteference activity factor $L_{wl}$ for DM1 packets as a function of $\overline{C/I}$ for $\overline{\gamma} = 20$ dB and approaching infinity, by varying the number of antennas, $N$ with SD.

It is possible to jointly observe the effect of thermal noise (for $\overline{\gamma} = 20$ dB) and interference, thus regions in which the system is noise-limited or interference-limited. Note also the significant performance improvement just using two-branches SD instead of single antenna reception.

Figure 6.25 reports the coverage probability as a function of the desired $\overline{PEP}_m^{(I)*}/L_{wl}$ for interference-limited BT systems. Here, $d_U = 8$ m and $d_I = 4$ m and 8 m. The SD receiver is considered. As can be observed, the coverage probability increases, as expected, for larger values of the $\overline{PEP}_m^{(I)*}/L_{wl}$.

Figure 6.26 shows the coverage probability giving a $\overline{PEP}_m^{(I)*}/L_{wl} = 0.01$ for high SNRs as a function of $d_I$ when the useful distance $d_U$ is 2 m and for the two extreme cases of $P_U = 20$ and $0$ dBm, respectively, and SD receiver.

It follows that to obtain a 90% of coverage with $P_U = 0$ dBm the interferer has to be at a distance $d_I$ greater than 30 metres with a single branch receiver, while 21.5 metres are sufficient with two antennas. By increasing $P_U$ a distance $d_I$ less than before is required.

Please note the presence of slope variation when distances assume the limit value of 8 m that represents a discontinuity for the assumed path loss law in (3.5).

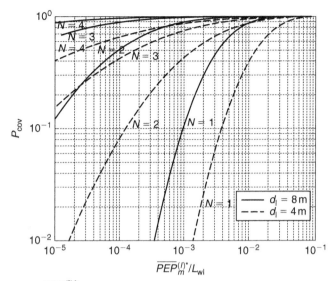

**Figure 6.25** $P_{cov}$ vs $\overline{PEP}_m^{(I)*}$ when $\bar{\gamma} = \infty$ for $d_U = 8\,m$, $L_{wl} = 1$ and $P_U = 20\,dBm$ varying the diversity order; SD receiver

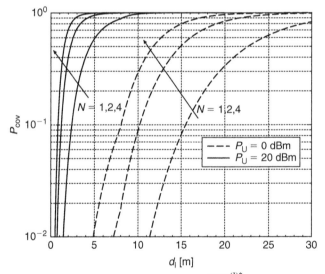

**Figure 6.26** $P_{cov}$ giving asymptotical (i.e., high SNR) $\overline{PEP}_m^{(I)*}/L_{wl} = 10^{-2}$ vs $d_I$ varying the diversity order and the transmitting power level when $d_U = 2\,m$ for a SD receiver in Rayleigh fading; SD receiver

In Figure 6.27, the joint effect on $P_{cov}$ of noise ($\bar{\gamma} = 20$ dB) and interference can be observed as a function of $d_I/d_U$ when $d_U = 2\,m$ and target $\overline{PEP}_m^{(I)*}/L_{wl}$ equal to $10^{-2}$. The strong impact of noise can be observed, in particular, for a single branch receiver, where a coverage of 90% can be hardly reached with a close interferer especially with

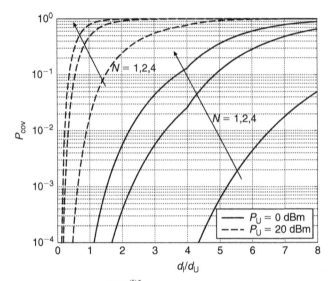

**Figure 6.27**   $P_{cov}$ *vs* $d_I/d_U$ *giving* $\overline{PEP_m}^{(I)^*} = 10^{-2}$ *when* $\overline{\gamma} = 20\,dB$ *and* $L_{wl} = 0.5$ *varying the diversity order and the useful power level when* $d_U = 2\,m$; *SD receiver*

$P_U = 0\,$dBm. Instead, the adoption of multiple antennas drastically reduce the effect of disturbances.

Some other results on the outage for different fading statistics and also considering the multiwall path loss are reported in Masini, Conti, Dardari & Pasolini, 2006. Here we would like to remark that the WAF path loss model (see Chapter 3) can be taken into account by adding to the path loss in dB, $L_0$, the term $nA_{wall}$ (dB) with $n$ equal to the number of walls and $A_{wall}$ being the attenuation per wall (typical value of about 6 dBs at 2.4 GHz).

## 6.4   Comparison among technologies

Different technologies, comprising Bluetooth, Zigbee and UWB among others, are compared in the following.

### ZigBee

The nine promoter companies of the ZigBee Alliance include Philips, Honeywell, Mitsubishi Electric, Motorola, Samsung, BM Group, Chipcon, Freescale and Ember; then are more than 70 members.

- Capacity of 250 Kbit/s at 2.4 GHz, 40 Kbit/s at 915 Mhz, and 20 Kbit/s at 868 Mhz with a range of 10–100 metres.
- Its purpose is to become a wireless standard for remote control in the industrial field.

- The ZigBee technology is targeting the control applications industry, which does not require high data rates, but must have low power, low cost and ease of use (remote controls, home automation, etc.).
- The specification was formally adopted in December 2004.
- Security was not considered in the initial development of the specification. Currently there are three levels of security.
- It operates in the 3.2–10.2 GHz band.
- ZigBee chips are low cost.

### Ultra-wideband

- UWB is a revolutionary wireless technology for transmitting digital data over a wide spectrum of frequency bands with very low power. It can transmit data at very high rates (for WPAN applications).
- Ideally, it will have low power consumption, low price, high speed, use a wide swath of radio spectrum, carry signals through obstacles (doors, etc.) and apply to a wide range of applications (defense, industry, home, etc.).
- Currently, there are two competing UWB standards for WPAN applications: the UWB Forum is promoting one standard based on direct sequence (DS-UWB), while the WiMedia Alliance is promoting another standard based on multiband OFDM.
- Each standard allows for data rates from approximately 0 to 500 Mbps at a range of 2 metres and a data rate of approximately 110 Mbps at a range of up to 10 metres.
- The Bluetooth SIG announced in May 2005 its intentions to work with both groups behind UWB to develop a high rate Bluetooth specification on the UWB radio.
- For application to WSANs the standard proposed, which can be used as the PHY layer of the ZigBee protocol stack, is the IEEE 802.15.4a based on IR-UWB. It allows for typical data rate of 850 Kbit/s (up to 27 Mbit/s optional) with a range of 10–50 metres and location capability.

### Bluetooth wireless technology

- Bluetooth wireless technology is geared towards voice and data applications.
- It operates in the ISM unlicensed 2.4 GHz spectrum.
- Typically, it can operate over a distance of few metres (up to 10 metres, depending on the device class also up to 100 metres). The peak data rate with EDR is 3 Mbps.
- It is omni-directional and in some conditions can work also in NLOS.
- The Bluetooth specification allows for three modes of security.
- The cost of Bluetooth chips is actually under 3USD.

### Infrared (IrDA)

- IrDA is used to provide wireless connectivity for devices that would normally use cables to connect. IrDA is a point-to-point, narrow angle (30 cone), ad hoc data transmission standard designed to operate over a distance of 0 to 1 metre and at speeds of 9600 bps to 16 Mbps.
- IrDA is not able to penetrate solid objects and has limited data exchange applications compared to other wireless technologies.

- IrDA is mainly used in payment systems, in remote control scenarios or when synchronizing two PDAs with each other.

### Radio frequency identification (RFID)
- There are over 140 different ISO standards for RFID for a broad range of applications.
- With RFID, a passive or unpowered tag can be powered at a distance by a reader device. The receiver, which must be within a few feet, pulls information off the tag, and then looks up more information from a database. Alternatively, some tags are self-powered, active tags that can be read from a greater distance.
- RFID can operate in low frequency (less than 100 MHz), high frequency (more than 100 MHz), and UHF (868 to 954 MHz).
- Uses include tracking inventory both in shipment and on retail shelves.

# Part 2

# Communication protocols, localization and signal processing techniques for WSANs

This part covers important issues for the design of WSANs such as those related to routing, localization and time synchronization and signal processing. Several protocols and algorithms presented in the literature are reported and described, and some of them are evaluated through the use of suitable frameworks. Not only the scientific literature is reviewed, but based on theoretical analyses some protocols and algorithms are also numerically evaluated.

In particular, both MAC and routing protocols for WSANs are reviewed in Chapter 7; some of them are also considered in the example of WSN design and case studies given in the following chapters.

Localization and time synchronization issues are investigated in Chapter 8. There, time estimation and distance estimation in realistic scenarios are reported and related to position tracking. Theoretical performance limits of positioning in single-hop and multihop WSN are also given.

Signal processing affects several aspects of WSANs design under a cross-layer view. In Chapter 9 we review the literature and propose frameworks and performance analysis for cooperative distributed detection techniques as well as for distributed scalar field estimation. Also some compression techniques for WSNs are reported.

# 7

# Communication protocols
# for WSANs

## 7.1 Introduction

Classic MAC and routing protocols studied for wireless voice and data traditional networks are not suitable for wireless sensor networks. Wireless sensor networks in fact are
characterized by many specific aspects that differentiate them from the traditional communication networks. First of all, nodes are based on battery supply and a typical applications environment does not allow to simply change batteries. Second, often nodes are
deployed in an ad hoc way rather than with careful pre-planning; they must then organize themselves to create a communication network. Third, sensor networks are generally
composed by a high number of nodes; furthermore, node density can vary in different
places, having both sparse areas and other areas with nodes with many neighbours, and
it can also vary during time. Finally, most traffic in the network is generated by sensing
events, and it can be extremely bursty. MAC and routing protocols for wireless sensor
networks have therefore to face each of the mentioned issues. In particular, energy efficiency is one of the most important goals to achieve in designing protocols for wireless
sensor networks. To do that, the protocols should address the major sources of energy
waste on MAC and routing levels.

## 7.2 MAC protocols

More specifically, the causes of energy waste at the MAC layer of a wireless network are
the following: retransmissions of the corrupted packets during a collision; overhearing,
that is, receiving packets destined for other nodes; control packet overhead; idle listening to receive possible traffic.

The basic approach adopted in order to decrease energy consumption is the possibility
to put nodes into a sleep state as much as possible; this implies the need to address the

compatibility of aggressive sleep modes with connectivity of the whole network, time synchronization, and routing issues.

According to the mechanism for collision avoidance, MAC protocols can be divided into two groups: scheduled based and contention based.

## 7.2.1 Scheduled protocols

Many MAC layers in WSNs adopt scheduled protocols because of their energy efficiency; since channels are pre-allocated to individual nodes, there is no energy wasted on collisions due to channel contention.

Among protocols belonging to this group, time division multiple access (TDMA) seems particulary suitable for WSNs, for many reasons: for example, it can support low-duty-cycle operation; it can avoid overhearing by turning off the radio during the slots of other nodes.

Generally, TDMA is based on a cluster structure: nodes form clusters; one of the nodes within the cluster is selected as the cluster head, and it acts as a base station; peer-to-peer communications are not supported, so each node within a cluster has to communicate through its cluster head; communications between different clusters and interference between them is managed thanks to other approaches, such as frequency division multiple access (FDMA) or code division multiple access (CDMA).

In general, scheduled protocols can provide good energy efficiency, but they also present some disadvantages. For example, TDMA protocols are not very scalable; when new nodes join or old nodes leave a cluster, the base station must adjust the slot allocation. Frame length and static slot allocation can limit the throughput. Slotted structure needs a precise synchronization.

Protocols presented in the following introduce some variations into classic scheduled protocols, in order to meet specific wireless network requirements and to reduce the mentioned disadvantages. For example, the base station may dynamically allocate slot assignments on a frame-by-frame basis; the role of the cluster head may be rotated among the nodes of a cluster in order to balance energy consumption.

## 7.2.2 LEACH protocol

LEACH (low-energy adaptive clustering hierarchy) (Heinzelman, Chandrakasan & Balakrishnan, 2000) is an example of utilizing TDMA in WSNs (Figure 7.1).

LEACH is a clustering based protocol that utilizes randomized rotation of local cluster base stations in order to fairly balance the energy consumption among nodes in the network. The channel access within each cluster is regulated by the TDMA protocol.

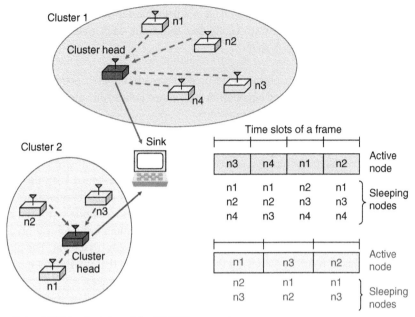

**Figure 7.1** *TDMA schedules of the LEACH protocol*

To reduce interference between different clusters, each cluster communicates using different CDMA codes.

In LEACH, the nodes organize themselves into local clusters; in each cluster there is one node called cluster head, acting as the local base station. Adopting the typical selection of the cluster heads followed in general by classic clustering algorithms, that is, chosen a priori and fixed for all the duration of the network operation, the sensors chosen to be cluster heads would quickly consume all their battery power, dramatically decreasing the network lifetime. Thus, LEACH includes randomized rotation of the cluster-head role among all nodes in the cluster, in order to not drain the battery of a single sensor in each cluster.

Election of the cluster heads happens in the following way. At each round of the network life, each node has a certain probability to elect itself cluster head; the decision about whether or not to become a cluster head for the current round is based on some parameters: the suggested percentage $P$ of cluster heads that should be present in the network in order to achieve the best performance and the number of times the node has been a cluster head so far. The node $n$ chooses a random number between 0 and 1 and it elects itself cluster head for the current round if the number is less than a threshold $T(n)$. The threshold is calculated as

$$\frac{P}{1 - P \cdot \left(r \bmod \frac{1}{P}\right)} \tag{7.1}$$

if $n$ belongs to the groups of nodes that have not been cluster heads in the last $\frac{1}{p}$ rounds while it is equal to zero otherwise. Using this threshold, each node of a cluster will become a cluster head within $\frac{1}{p}$ rounds.

The decision of becoming a cluster head or not is made by each node independently without the need of any type of negotiation with other nodes; this avoids therefore any energy consumption dedicated to negotiations.

After the election, a cluster head broadcasts its status to the other sensors in the network, by using an advertisement message sent with a CSMA protocol; all cluster heads send the advertising message with the same energy. The other nodes must keep their receivers on during this phase of setup to hear the advertisements of all cluster-heads nodes. Then, each sensor node determines to which cluster it wants to belong to for the current round, by choosing the cluster head that requires the minimum communication energy. This decision is based on the received signal strength of the advertisement. The node has to inform the cluster head, by following the CSMA protocol, that it will be a member of its cluster. During this phase, all cluster-head nodes must be awake.

Each cluster head defines a TDMA schedule for its cluster and it broadcasts it to the nodes in the cluster. This allows each non-cluster-head node to switch off its radio except during its dedicated transmit time slot, thus minimizing the dissipation of energy.

The cluster head aggregates all the data it has received from the nodes inside its cluster and then it transmits the compressed data to the base station; this reduces the amount of information that must be transmitted to the base station.

The authors find that there exists an optimal percentage of nodes $N$ that should be cluster heads and that it depends on several parameters, such as the network topology and the costs of computation versus communication.

### 7.2.3   Guo protocol

Guo, Zhong & Rabaey, 2001 propose the design of a multichannel MAC with low-power wakeup radio.

The available band is divided in multiple channels and each node is assigned a locally unique channel. In order to avoid interferences, all neighbours of each node have to own a channel different from the channel of the node. This channel assignment is equivalent to the two hops colouring problem in graph theory, that requires to find how to colour the nodes of a graph such that any pair of nodes two hops away are not assigned the same colour.

The authors propose a heuristic distributed solution to the NP problem of the colour assignment (Figure 7.2). All nodes can listen to a common control channel, called CCC.

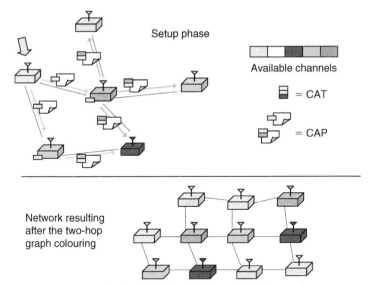

**Figure 7.2** *The two-hop graph colouring solution proposed by Guo et al.*

Periodically, each node sends over this channel a packet, called CAP packet, to its neighbours; CAP contains the current colour assignment. Each node stores the information it has received from the CAP packet inside a channel assignment table (CAT): in this way it can learn the channel usage of its one-hop and two-hop neighbours and it can verify that it is using a channel different from the ones used by the neighbours.

The protocol begins with a setup phase, that consists of two steps. During the first step, every node monitors CCC for a random period within a predefined contention window; it randomly chooses a channel from the ones available, and broadcasts to its neighbours at one hop the CAP with only its own channel information. The second step allows to spread the information of the chosen channel to the two-hop neighbours: the one hop neighbours get this CAP packet and add it to their CAT; they send a CAP with their own channels as well as all the one-hop neighbour information. After this setup mechanism all nodes make channel selection decisions based on their current CAT; in case a node finds a conflict it has to switch to another channel.

By using a separate low-power radio as a wake-up radio to monitor the channel, the normal data radio can be turned off as far as the node is neither transmitting nor receiving any packet. If the node has to transmit a packet, it simply wakes up by itself; if a neighbour node is sending a packet to this node, it will send a short wake-up beacon using the wake-up radio channel in order to allow the node to switch on its radio and receive the packet.

## 7.2.4 TRAMA protocol

Traffic adaptive MAC protocol (TRAMA) (Rajendran, Obraczka & Garcia-Luna-Aceves, 2003) is a TDMA-based algorithm proposed to increase the energy efficiency of a classic

TDMA scheme. TRAMA exploits information about the traffic generated by each node in order to choose which node to assign a time slot; in this way it avoids the assignment of time slots to nodes with no traffic to send, allowing it to determine when a node can stop listening to the channel and going into an idle state.

TRAMA employs a time-slot assignment that is traffic adaptive, distributed and that selects receivers based on schedules announced by transmitters. Transmission schedules specify, in chronological order, the intended receivers' nodes of their traffic; this information is exchanged by nodes among their two-nodes neighbours; thanks to this information transmitters select the nodes that should transmit and receive during each time slot.

TRAMA consists of three components: the neighbour protocol (NP) and the schedule exchange protocol (SEP), which allow nodes to exchange two-hop neighbour information and their schedules; the adaptive election algorithm (AEA), which uses neighbourhood and schedule information to select the transmitters and receivers for the current time slot, leaving all other nodes going in a low power idle state.

TRAMA assumes a single, time-slotted channel for both data and signalling transmissions. Time is divided into random-access and scheduled-access periods. Random-access period is used to establish two-hop topology information where channel access is contention based. Transmission slots are used for collision-free data exchange and also for schedule propagation.

### 7.2.5 Contention-based protocols

Contention protocols present some advantages compared to scheduled protocols. They are more flexible with respect to changes in traffic load, network topology, and node density, since resources allocation occurs in an on-demand fashion. There is no need of clusters creation since they support peer-to-peer communications. Finally, they do not require fine-grained time synchronizations as in TDMA protocols. Nevertheless, contention-based protocols often do not achieve the same energy efficiency as the scheduled schemes; this is due especially because nodes waste energy listening to the channel during the contention for the media and by retransmitting the packets that have experienced collisions on the media.

### 7.2.6 Zhong protocol

Zhong, Shah, Guo & Rabaey, 2001 present the MAC studied for the Pico-radio project. The proposed protocol exploits multiple channels and random-access technique (Figure 7.3); therefore the overhead of scheduling and reservation is eliminated and synchronization is not needed. The protocol relies on a distributed algorithm, which does not require the existence of a central base station: each node can operate independently to other nodes in the network.

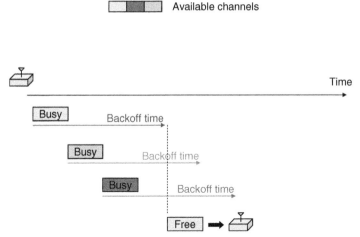

**Figure 7.3** *The node randomly selects a channel. If all channels are busy, the node sets a random timer for each of them and backs off. It will use the channel whose timer expires first and clear the timers for all other channels*

It extends CSMA into multichannel CSMA. A channel can be code, time slot, sub-carrier/ frequency or space. Before transmission, a node randomly selects a channel to see if it is busy. If the channel is busy, the node randomly selects another channel from the remaining channels. This process will go on until the node finds a channel that is idle. If all channels are busy, the node will set a random timer for each of them and back off. It will use the channel whose timer expires first and clear the timers for all other channels.

Low power consumption is being achieved by saving power in many aspects: the algorithm used to choose a communication channel is very simple and it does not imply any handshaking procedure; the presence of a sleep mode allows to switch off the radio when is not needed; thanks to the use of multiple channels, collisions and retransmission are reduced.

To support sleep mode, a wake-up radio is used in the physical layer. It wakes up the main radio and informs it of the channel it should tune to. Since the destination ID is modulated into the wake-up signal, only the destination node will be woken up.

## 7.2.7 DMAC protocol

Typically, sensor network applications collect information from multiple sources, the sensor nodes, and send the data to a common destination, the sink; the data forwarding paths often form a tree structure, called a data gathering tree. DMAC (Lu, Krishnamachari & Raghavendra, 2004) is an energy-efficient and low-latency MAC optimized for such type of data gathering trees. DMAC could be summarized as an improved slotted aloha algorithm where slots are assigned to the sets of nodes based on a data gathering tree.

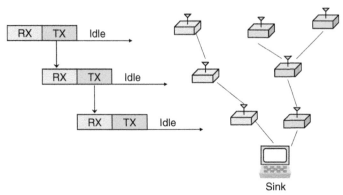

**Figure 7.4**   *Wake-up schedule proposed in DMAC*

DMAC is designed to solve a common problem that happens with MAC protocols based on active/sleep modes: nodes out of the range of the actual sender node do not hear its transmission and they switch off their radio; this introduces a latency for the next packet forwarding step, that experiences a queue interval waiting for the radio turning on again. DMAC tries to solve this inconvenience and allow continuous packet forwarding by distributing the activity schedules to the nodes in order to sequentially wake up the nodes in the tree.

The wake-up schedule proposed in DMAC gives the sleep schedule of a node an offset that depends upon its depth on the tree (Figure 7.4). Each interval is divided into three periods, respectively: receiving, sending, and sleep periods. The receiving and sending periods have the same length $\mu$, which is enough for one packet transmission and reception. Depending on its depth $d$ in the data gathering tree, a node skews its wake-up scheme $d\mu$ ahead from the schedule of the sink. Therefore, the data transferring follows an unidirectional path from the multiple sources to the sink that is in the tree root. Each source begins sending its data to its next hop. These nodes begin with the receiving period; when a node is in the receiving state it can receive a packet; when the packet has been successfully received the node replies with an ACK back to the sender. The forwarding process continues with these nodes in the sending state, during which each node tries to send a packet to its next hop and waits for an ACK confirming the transmission success. In sleep state, nodes turn off the radio to save energy.

Since the nodes' wake up is sequential the sleep delay is considerably reduced. Furthermore, since active periods are separated, collisions due to the medium contention are also reduced.

### 7.2.8   PAMAS protocol

Power aware multi-access protocol with signalling for ad hoc networks (PAMAS) (Singh & Raghavendra, 1998) is a MAC protocol that exploits nodes' sleeping states, by

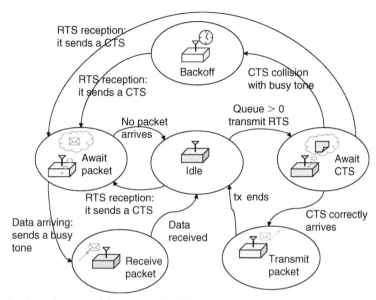

**Figure 7.5** *State diagram of the protocol PAMAS*

powering off the radio of the nodes that are not actively transmitting or receiving data. It is based upon an RTS/CTS message exchange performed over a dedicated signalling channel. The use of a separate signalling channel allows nodes to simply determine when and for how long they can power off themselves.

A node may be in one of six possible states: idle, await CTS, binary exponential backoff, await packet, receive packet, and transmit packet (Figure 7.5).

When a node is not receiving or transmitting a packet or it has packet to send but it cannot proceed because a neighbour is receiving packets it is in the idle state.

When a node has a packet to send it transmits an RTS and it goes into the await CTS state; if the CTS does not arrive or if it collides with the busy tone of a neighbour, it enters the binary exponential backoff state; if the CTS correctly arrives, it enters the transmit state and it begins transmitting the data. If it has gone into the binary exponential backoff state and it receives an RTS from a neighbour, it leaves the backoff state, it sends a CTS and it enters the await packet state.

The intended receiver can reply with a CTS if no neighbour is in the transmit packet state or in await CTS state. In case it is possible, the receiver sends the CTS and it enters the await packet state. If the packet does not arrive within a predefined interval of time the node returns into the idle state; if the packet begins arriving, it transmits a busy tone over the signalling channel and enters the receive packet state. When a node that is receiving a packet hears an RTS from a neighbour, it sends a busy tone, in order to ensure that the neighbour will not receive any CTS and that therefore neighbour transmission is blocked.

Ideally, a node remains powered off as it has no packet to send or if at least one of its neighbours is transmitting or receiving data; however, possible collisions may make the definition of the length of a transmission a difficult task for a node. Therefore, the PAMAS protocol has defined also a set of rules in order to estimate the power off duration for a node, that uses an exchange of probe packets over the signalling channel.

## 7.2.9  SMAC protocol

Ye et al. (C. S. Raghavendra, 2004) propose the self-organizing MAC (S-MAC) protocol for sensor networks, built on a contention-based scheme. It introduces a power-saving policy based on scheduling sleep/listen cycles, called frames, between the neighbouring nodes.

Each frame is divided into two parts: the first one is a listen period, followed by a sleep period. The listen period is needed to allow coordination among nodes that have data to send. During the sleep period, nodes that have no data to send can turn off while nodes with data to send remain asleep to communicate.

Each node independently chooses its own listen/sleep schedule and shares it with its neighbours: it periodically broadcasts its schedule in a SYNC packet, which provides simple clock synchronization; these operations require a period of time called synchronization period. In this way communication between nodes is possible: a node schedules

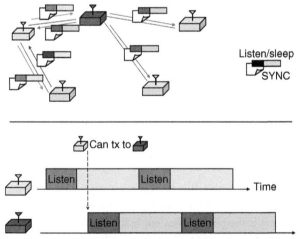

**Figure 7.6**  *In S-MAC each node independently chooses its own listen/sleep schedule periodically and broadcasts it in a SYNC packet. If a node wants to send a packet to a node that follows a different schedule, it just waits until the other node is listening*

its transmission during the listen time of its intended destination. For example, if a node wants to send a packet to a node that follows a different schedule, it just waits until the other node is listening. Contention only happens at a receiver's listen interval. Unicast packets are sent with CSMA combined with an RTS-CTS-DATA-ACK exchange; a broadcast packet uses only CSMA procedure. Each data packet contains also a field indicating the duration of the transmission: this allows neighbours to know how long they need to remain sleeping; ideally the node goes to sleep after receiving an RTS or CTS packet destined to other nodes, and it avoids overhearing subsequent data and ACK packets.

To reduce control overhead, S-MAC encourages neighbouring nodes to choose the same schedules. During the first configuration, a node listens for a synchronization period and adopts the first schedule it hears. In addition, nodes periodically perform neighbour discovery, listening for an entire frame, allowing them to discover nodes on different schedules that may have moved within range.

S-MAC allows the possibility to send multiple fragments from a message in a burst: in this modality only one RTS and one CTS are used to reserve the medium for the time needed to transmit all fragments. Besides RTS and CTS, each fragment or ACK also includes the duration of the remaining transmission, allowing neighbour nodes that wake up in the middle of the transmission to return to sleep.

## 7.3  Routing protocols

Routing in WSNs is a very challenging task due to the inherent characteristics that distinguish these networks from other wireless networks like cellular or mobile ad hoc networks. Traditional IP-based protocols may not be applied to WSN, due to the large number of sensor nodes and because getting the data is often more important than knowing the specific identity of the source sending it. Furthermore, almost all applications of sensor networks require the flow of sensed data from multiple sources to a particular base station, sink. Sensor nodes are constrained in terms of energy, processing, and storage capacities, thus, they require careful resource management. Sensor networks are strictly dependent from their application, and the design requirements of a sensor network change with the application. Furthermore, position awareness of sensor nodes is important since data collection is normally based on their location. Finally, since data collected by many sensors in WSN are typically based on common phenomena, they are often very correlated and contain a lot of redundancy. Such redundancy needs to be exploited by the routing protocols to improve energy and bandwidth utilization.

Many routing and dissemination protocols have been studied specifically for WSNs, following all the mentioned essential design issues, in particular the most important one concerning energy awareness and efficiency. The key point of energy-efficient network operation is the possibility to put nodes into sleep as much as possible. Since sleeping modes are usually defined at the MAC layer, dealing with routing in wireless sensor networks cannot be done independently of the lower layers; this implies addressing the compatibility of aggressive sleep modes with connectivity of the whole network and routing design.

Routing protocols can be classified according to the network structure as flat, hierarchical, or location based. In flat-based routing, all nodes are typically assigned equal roles or functionality. In hierarchical-based routing, nodes will play different roles in the network; for example, hierarchical protocols aim at clustering the nodes so that cluster heads can do some aggregation and reduction of data in order to save energy. Location-based routing exploits sensor nodes' positions to route data in the network.

## 7.3.1  Flat routing

In flat networks, sensor nodes typically play the same role and collaborate together to perform the sensing task.

The lack of a global identification due to the large number of nodes present in the network and their random placement, typical of many specific wireless sensor network applications, makes it hard to select a specific set of sensors to be queried. Often this could cause redundant transmission of data from every sensor node with consequent inefficient energy consumption. A useful solution is the definition of routing protocols able to select appropriate sets of forwarding sensor nodes and to utilize data aggregation during the relaying of data. This routing policy is known as data-centric routing, which is different from traditional address-based routing where routes are created between addressable nodes. In data-centric routing, the sink sends queries to certain regions and waits for data from the sensors located in the selected regions. Attribute-based naming is necessary to specify the properties of data requested in the queries. SPIN and directed diffusion are the two first data-centric protocols proposed and they have inspired a lot of other data-centric schemes.

## 7.3.2  Flooding and gossiping

Flooding and gossiping (Hedetniemi & Liestman, 1988) are two classical data-relay mechanisms that do not require any routing algorithms and topology maintenance. In flooding, each sensor receiving a data packet broadcasts it to all of its neighbours, regardless of whether or not a neighbour has already received the data from another node; this process continues until the packet arrives at the destination.

Gossiping is a slightly enhanced version of flooding: when a node receives a packet it randomly chooses one of its neighbours to which it forwards the packet. The mechanism of packet forwarding for both the two protocols is shown in Figure 7.7.

Flooding is very easy to implement, but it has several drawbacks: implosion, caused by duplicated messages sent to same node; overlap, when sensor nodes cover overlapping

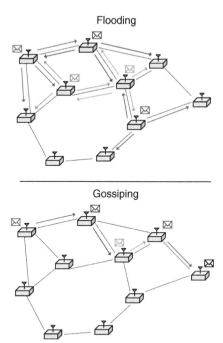

**Figure 7.7** *Flooding and gossiping. The black lines represent the wireless links of the nodes in reciprocal coverage radio, while the coloured lines show the data packets' forwarding process*

geographic areas and therefore they collect overlapping pieces of data; resource blindness, since in classic flooding nodes do not modify their activities based on the amount of energy available to them at a given time.

Gossiping does not solve the problem of overlap but it can avoid implosion by just selecting a random next node rather than broadcasting; however, this cause delays in propagation of data.

## 7.3.3 SPIN protocol

Sensor protocol for information via negotiation (SPIN) (Kulik, Heinzelman & Balakrishnan, 2002) adopts two innovative aspects that overcome the deficiencies of the classic flooding approach: negotiation and resource adaptation.

To overcome the problem of implosion and overlapping, SPIN proposes that nodes have to negotiate each other before transmitting their data, therefore only useful information will be delivered in the network. This allows a considerable energy saving. To successfully negotiate, nodes must be able to describe the data they collected. The idea proposed by SPIN is to name the data using high-level descriptors, called meta-data.

Meta-data have to completely resume and describe the true data collected by a sensor. There is no standard meta-data format since the way to describe data is application

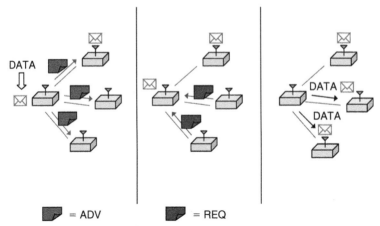

**Figure 7.8**   *The negotiation procedure of the SPIN protocol*

dependent. The size of the meta-data descriptor has to be much shorter than the size of the true data, in order to implement a beneficial SPIN algorithm.

Nodes in the network perform a meta-data negotiation before any data is transmitted to other nodes (Figure 7.8). Before transmission, meta-data are exchanged among sensors via a data advertisement mechanism. SPIN adopts three types of messages: ADV, REQ and DATA. ADV is used to advertise new data, REQ to request data, and DATA is the actual message itself. Each time a node receives new data, it advertises its neighbours by broadcasting an ADV message containing meta-data: in this way the neighbour that does not still have this data can obtain it by sending a request message REQ to the node that had advertised it before; DATA is sent to this neighbour node. The neighbour sensor node then repeats this process with its neighbours. As a result, the entire sensor area will receive a copy of the data.

This protocol works in a time-driven fashion and disseminates the information assuming that all nodes in the network are potential base stations; this enables a user to query any node and get the required information immediately.

SPIN does not specify any energy policy, rather it provides the application level with an interface to query the available resources. Before data is transmitted, nodes poll their resources to find out how much energy is available. Each node has a resource manager to keep track of resource consumption and calculate cost of performing computations and sending and receiving data.

## 7.3.4   Directed diffusion protocol

The structure of the directed diffusion (Intanagonwiwat, Govindan & Estrin, 2000) protocol directly comes from the intrinsic working model of wireless sensor networks, that can

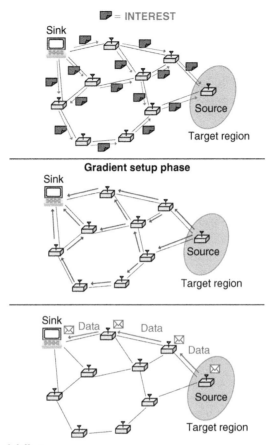

**Figure 7.9** *Directed diffusion*

be simplified as follows: a human operator, through an opportune sink node, makes a query to the network about some particular aspects of a target region; the nodes in that region collect the data necessary to reply to the task; once the data have been gathered, individual nodes collaborate with their neighbours and report the result back to the sink.

Directed diffusion is a data-centric and application-aware paradigm that aims at diffusing data through sensor nodes by using a naming scheme for the data gathered by the sensors. Basically, its functioning follows these steps (Figure 7.9): data is named using attribute-value pairs; a sensing task generated by a sink is spread over the network in the form of an interest for named data; this dissemination sets up gradients within the network that are used to route the events (data matching the interest) back to the sink; events are then forwarded down towards the sink, by following multiple paths; the network works by reinforcing one or a set of these paths; intermediate nodes can cache or aggregate data. Interests and data propagation and aggregation are performed by localized interactions, in terms of message exchanges between neighbour nodes.

A sensing task is described by a list of attribute-value pairs; for example, this list can comprise the type of the requested data (temperature), the interval of time at which the sink would like to receive information about the requested task, the duration of the entire task, the geographic area to monitor, etc. . . . Therefore, the task description specifies an interest for data matching the described attribute. Data collected by the sensors monitoring the area and sent in response to that interest use the same naming scheme, with a set of attribute-value pairs.

Each node has a cache to maintain the interests. Each interest entry has different fields: a timestamp field, that specifies the instant of the last received matching interest; some gradient fields, up to one per neighbour (each gradient contains a data rate field, that specifies the data rate requested by the neighbour and derived by the interval attribute of the interest arrived from that neighbour); a duration field, that indicates the lifetime of the interest. When a node receives an interest, it checks in its cache: if no matching entry exists, it creates an interest entry with the parameters specified in the received interest. This entry has a single gradient towards the neighbour from which the interest has been received. After receiving an interest, a node may decide to broadcast the interest to all its neighbours, or it may send the interest to a subset of its neighbours or it may suppress it if it recently resent a matching interest. The forwarded interest appears to the neighbours as an interest originated from the sending node, although it might have come from a distant sink; in this manner, interests diffuse throughout the network through local interactions.

A sensor node that is within the target region processes the collected data and it generates data at a rate equal to the highest rate indicated in its outgoing gradients. When a node receives a data message from one of its neighbours it searches for a matching interest entry in its cache. If no match exists, the data message is dropped. If a match exists, the node checks the associated data cache; this cache keeps track of recently received data items, as a loop prevention function. If a received data message has a matching data cache entry, the data message is dropped. Otherwise, the received message is added to the data cache and it is sent to the node's neighbours. Before sending a received data message, a node needs to examine the matching interest entry's gradient list. If all gradients have a data rate that is greater than or equal to the rate of incoming events, the node may send the received data message to the appropriate neighbours. If some gradients have a lower data rate than others it may convert down to the appropriate gradient.

The interest is periodically refreshed by the sink; the refresh rate is a protocol design parameter chosen as a trade-off between robustness and interest loss. The sink initially sends its interest at a low rate; once sources detect a matching target, they send low rate events, possibly along multiple paths, towards the sink. After the sink starts receiving these low data rate events, it reinforces one particular neighbour in order to receive higher data rate events from that particular path or from a smaller set of paths.

## 7.3.5 *Rumour routing*

Rumour routing (Braginsky & Estrin, 2002) is a variant of directed diffusion and is devoted to applications where geographic routing is not feasible and where the application requires little amount of data from sensors or simply needs to order the target area to initiate more intense sensing.

In general, directed diffusion uses flooding to disseminate the query to the entire network when there is no geographic information available to forward a task. If the amount of returning data is significant, it makes sense to invest in discovering short paths from the source to the sink, but this approach is not convenient, on the contrary.

Rumour routing implements a trade-off between flooding queries and flooding event notifications; it follows an alternative strategy: it floods the events and sets gradients towards them if the number of events is small and the number of queries is large and it routes the queries to the nodes that have observed a particular event. Nodes can form gradients towards the event, based on the number of hops to sink. After the cost field is set up, queries can be routed to the event along the shortest path. When a node generates a query, instead of flooding it, it can directly route the query to the event, in case a path to the event has been discovered; otherwise it sends a query on a random walk until it finds the event path and then the query can be routed directly to the event; in case a path cannot be found it can flood the query (Figure 7.10).

The algorithm used to set up the paths to the events uses specific long-lived packets, called agents. When a node detects an event, it adds it to its local event table and it generates

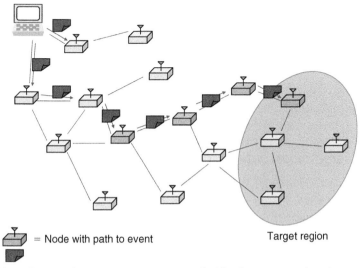

**Figure 7.10**  *When a node generates a query, instead of flooding it, it can directly route the query to the event, in case a path to the event has been discovered*

an agent. Agents propagate information about local events over the entire network: they live until a certain number of hops, $H$, the parameter chosen by the algorithm. Whenever an agent crosses a path leading to an event it has not yet seen, it creates a path state that leads to multiple events. An agent can update the nodes routing table to the more efficient path that it has discovered. By checking its event table each node that knows a route can reply to a received query.

Communications costs are reduced since there is no need to flood the whole network. On the other hand, rumour routing maintains only one path between source and destination while with directed diffusion data can be routed through multiple paths at low rates. Simulation results showed that rumour routing can achieve significant energy savings when the number of events is small. For a large number of events, the cost of maintaining agents and event tables in each node becomes infeasible if there is not enough interest in these events from the sink.

### 7.3.6   Gradient-based routing

Gradient-based routing (GBR) (Schurgers & Srivastava, 2001) is another variant of directed diffusion (Figure 7.11).

It stores the number of hops while the interest is diffused through the network. Each node can discover its height, defined as the minimum number of hops from that node to the sink. The difference between a node's height and that of its neighbour is considered the gradient on that link. A packet is forwarded on a link with the largest gradient. When there are two or more hops with the same gradient, the node chooses one of them at random. When a node's energy drops below a certain threshold, the node increases its height so that other sensors are discouraged from sending data to that node; this allows it to increment energy efficiency and network lifetime. Nodes acting as a relay for multiple paths perform data aggregation before forwarding data they have received from multiple neighbours.

Simulations have shown that GBR outperforms directed diffusion in terms of communication energy efficiency.

### 7.3.7   Hierarchical routing

In a hierarchical architecture, higher energy nodes can be used to process and send the information while low-energy nodes can concentrate each other in monitoring the interested area and gathering data. This means the creation of clusters with the assigning of special tasks to cluster heads, such as data fusion and data forwarding, in order to achieve system scalability, network lifetime increment, and energy efficiency. Hierarchical routing generally is composed of two steps: the first one for cluster heads selection and the other devoted to routing the messages.

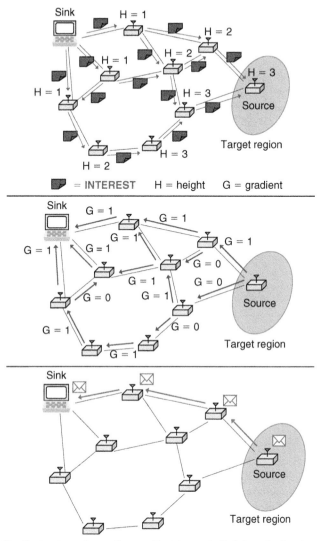

**Figure 7.11** *Gradient setup phase: the gradient in each link is calculated as the difference between the heights of the nodes in that link*

## 7.3.8  LEACH protocol

This protocol (Heinzelman et al., 2000) has already been presented as an example of TDMA MAC implementation in WSNs. As far as the routing is concerned, it can be classified as a hierarchical algorithm, due to its inherent creation of clusters.

The LEACH operation is composed by two phases: a setup phase and a steady-state phase. The setup phase is needed in order to create the clusters inside the network and elect the cluster heads in each cluster. Creation of clusters is described in the section dedicated to the MAC layer.

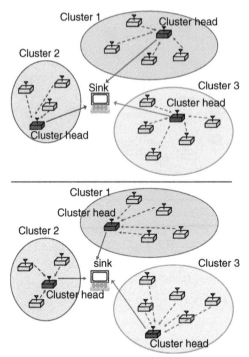

**Figure 7.12**   *LEACH routing protocol: the two examples evidence the rotation of cluster heads among nodes in each cluster*

During the steady-state phase the nodes inside each cluster sense data and transmit data to their cluster head. The cluster head collects all the data sent by the the nodes in its cluster, it aggregates it and sends it to the sink (Figure 7.12). Aggregation is useful if the data collected in a cluster are correlated. LEACH protocol assumes that all cluster heads can directly communicate with the central base station of the network; therefore it is not applicable in large regions. Periodically, the network goes back to the setup phase, to allow the selection of new cluster heads.

### 7.3.9   PEGASIS protocol

Power-efficient gathering in sensor information systems (PEGASIS) (Lindsey & Raghavendra, 2002) is an improvement of the LEACH protocol.

Rather than forming multiple clusters, PEGASIS forms chains of sensor nodes, so each node communicates only with its closest neighbours. Only one node of the chain, the leader, is then selected to transmit to the sink. The protocol assumes that all nodes are able to communicate directly with the sink, so the role of the leader is rotated among all the nodes forming the chain, in order to balance the energy consumption.

For gathering data in each round, each node receives data from one neighbour, fuses it with its own data, and transmits to the other neighbour on the chain. Gathered data

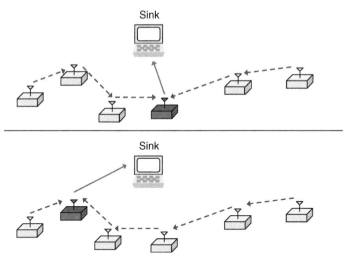

**Figure 7.13**  *PEGASIS: the two examples evidence the rotation of leaders among the nodes composing the chain*

moves from node to node, until it reaches the leader; the leader then sends data to the sink. The chain in PEGASIS is made up by those nodes that are closest to each other and form a path to the base station. To locate the closest neighbour node in PEGASIS, each node uses the signal strength to measure the distance to all neighbouring nodes and then adjust the signal strength so that only one node can be heard. The chain construction starts from the furthest node from the sink and continues in a greedy way. Whenever a node dies, the chain will be reconstructed.

Simulations have shown that the greedy chain construction performs well with different size networks and random node deployments. During the construction of a chain some nodes may become relatively distant from their closest neighbour and therefore they dissipate more energy. Performance of the protocol can be improved by avoiding these nodes to become leaders; this can be achieved by the introduction of a threshold in the leader election process; the threshold can be chosen adaptive to the remaining energy levels in nodes and to the distance between neighbours.

PEGASIS has been shown to outperform LEACH by about 100% to 300% in terms of network lifetime, for different network sizes and topologies. Such performance gain is due to various aspects. First, thanks to the elimination of the overhead caused by dynamic cluster formation. Second, PEGASIS uses multihop routing by forming chains and selecting only one node to transmit to the base station in each round of the communication instead of using multiple cluster heads. Finally, the amount of data for the PEGASIS leader to receive is at most two messages instead of all the messages that the cluster head of LEACH has to receive from all nodes inside its cluster. However, PEGASIS introduces excessive delay with distant nodes in the chain.

Hierarchical-PEGASIS is an extension to PEGASIS, which aims at decreasing the delay incurred for packets during transmission to the sink and proposes a solution to the data gathering problem by considering energy $\times$ delay metric. It introduces simultaneous transmission of data messages. To avoid collisions and possible signal interference among sensors, two approaches have been investigated: CDMA or allowing transmissions only to spatially separated nodes at the same time.

### 7.3.10   TEEN protocol

*T*hreshold sensitive *e*nergy *e*fficient sensor *n*etwork protocol (TEEN) (Manjeshwar & Agarwal, 2001) is a hierarchical protocol, useful for time-critical applications in which the network operates in a reactive way. Closer nodes form clusters and elect a cluster head. Each cluster head is responsible for directly sending the data to the sink. After the clusters are formed, the cluster head broadcasts two thresholds to the nodes. These are hard and soft thresholds for sensed attributes. Hard threshold is the minimum possible value of an attribute to trigger a sensor node to switch on its transmitter and transmit to the cluster head (Figure 7.14). Thus, the hard threshold allows the nodes to transmit only when the sensed attribute is in the range of interest, thus reducing the number of transmissions significantly. Once a node senses a value at or beyond the hard threshold, it transmits data only when the value of that attribute changes by an amount equal to or greater than the soft threshold. As a consequence, soft threshold will further reduce the number of transmissions if there is little or no change in the value of sensed attribute. One can adjust both hard and soft threshold values in order to control the number of packet transmissions. However, TEEN is not good for applications where periodic reports are needed since the user may not get any data at all if the thresholds are not reached.

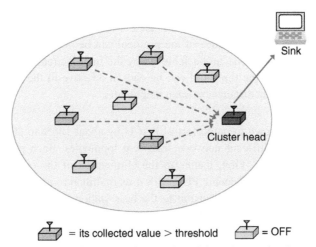

**Figure 7.14**   *Sensors in each cluster switch on and send their data to the cluster head only when collected values are greater than a predefined threshold*

## 7.3.11  MECN protocol

Minimum energy communication network (MECN) (Rodoplu & Meng, 1999) identifies a relay region for every node. The relay region consists of nodes in a surrounding area where transmitting through those nodes is more energy efficient than direct transmission (Figure 7.15). The enclosure of a node is then created by taking the union of all relay regions that node can reach. The main idea of MECN is to find a sub-network, which will have fewer nodes and require less power for transmission between any two particular nodes. In this way, global minimum power paths are found without considering all nodes in the network. This is performed using a localized search for each node considering its relay region. The protocol has two phases. During the first one it takes the position of a two-dimensional plane and constructs a sparse graph, which consists of all the enclosures of each transmit node in the graph; this construction requires local computations in the nodes; the enclosure graph contains globally optimal links in terms of energy consumption. During the second phase it finds optimal links on the enclosure graph with distributed Belmann-Ford shortest path algorithm, with power consumption as the cost metric.

## 7.3.12  SPAN protocol

SPAN (Chen, Jamieson, Balakrishnan & Morris, 2001) selects some nodes as coordinators; coordinators form a network backbone that is used to forward messages (Figure 7.16); coordinators are selected based on their positions. For these reasons it can be classified both as a hierarchical and local-based algorithm.

Coordinators stay awake continuously and forward packets over the network; the other nodes can remain in a sleep state, saving energy; in fact when a region has a sufficient high density of nodes, only a small number of them are needed at any time in order to assure network connectivity. SPAN ensures that enough coordinators are selected, so every

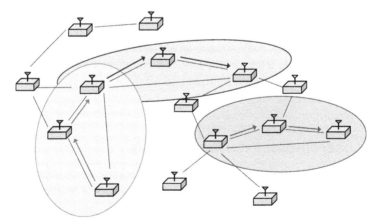

**Figure 7.15**  *The relay region of the MECN protocol consists of nodes in a surrounding area where transmitting through those nodes is more energy efficient than direct transmission*

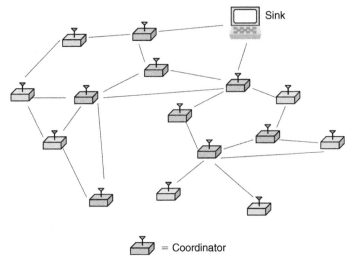

**Figure 7.16**   *Coordinators forming the backbone in SPAN*

node is in radio range of at least one coordinator, but at the same time it tries to maintain the number of coordinators as low as possible, in order to save nodes' resources and
without too much increment latency. The coordinators forming the backbone rotate with
time, in order to assure fairness in the energy consumption of the whole network. SPAN
is proactive: each node periodically broadcasts hello packets containing the node's status
(coordinator or not), its current coordinators' and its neighbours.

SPAN is a distributed randomized algorithm where nodes make local decisions on
whether to sleep or to join a forwarding backbone as coordinator. This decision is based
on the estimate of how many neighbours of that node will benefit if it became a coordinator and of the amount of energy available to it. Periodically, each node determines if
it should become a coordinator, by following this rule: a node should become a coordinator if, from the information gathered from hello packets of its neighbours, it sees that
two of its neighbours cannot communicate directly or through one or two coordinators.
When more than one node discovers the lack of a coordinator at the same time, SPAN
solves the contention by using a randomized backoff timer associated to the coordinator
hello announcement.

## 7.3.13   *Location-based routing protocols*

This kind of protocol exploits information about the location of the sensors in order to
forward data among the network in an energy-efficient way. The location of nodes may
be available directly from a GPS system or by implementing some localization protocol
specifically studied for wireless sensor networks; many solutions for nodes localization
have been proposed in literature, based on many different techniques.

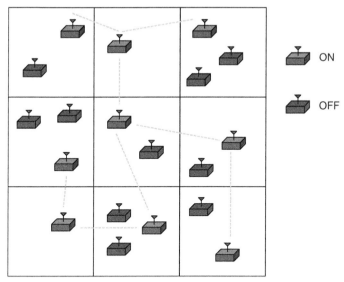

**Figure 7.17** *GAF: nodes inside the same square zone are equivalent for routing metrics; only one node at a time needs to be awake in each square zone, all the others can be put in a sleep state*

## 7.3.14  GAF protocol

Geographic adaptive fidelity (GAF) (Xu, Heidemann & Estrin, 2001) is an energy-aware location-based routing algorithm born for mobile ad hoc networks, but may be applicable to sensor networks as well.

The network area is divided into fixed zones to form a virtual grid (Figure 7.17); GAF uses equal and square zones, whose size is dependent on the required transmitting power and the communication direction. Each node associates itself to a zone in the virtual grid by using GPS information. Nodes associated with the same zone of the grid are considered equivalent in terms of the cost of packet routing. The virtual grid is constructed in a way such that for two adjacent zones $G_1$ and $G_2$, all nodes in $G_1$ can comunicate with all nodes in $G_2$; therefore, all nodes in a zone are equivalent for routing. Assume that the radio range of the nodes is equal to $R$ and the side of a square zone of the grid is equal to $l$; as the definition of virtual grid states, the distance between two possible farthest nodes in any two adjacent zones must be shorter than $R$, therefore we have: $l^2 + (2l)^2 \leq R^2$ or $l \leq (R/\sqrt{5})$.

GAF exploits the equivalence of all nodes inside the same zone by keeping at least one node per zone awake for a certain period of time and turning all the others in that zone into sleep state during that time; nodes in the same zone can coordinate to decide who will sleep and for how long; this choice does not affect the network connectivity and the routing fidelity, while it improves its energy efficiency by turning off nodes in proximity of the selected node. The awake node is representative of the data collected in its zone and it is responsible for monitoring and forwarding data to the base station or to other awake nodes.

GAF can be classified as both a location-based protocol and a hierarchical protocol, since the zones of the grid can be seen as clusters based on geographic location. The leader node, however, does not do any aggregation or fusion.

In order to balance the load and the energy consumption among the nodes in a zone, nodes change states from sleeping to active in turn. This change is regulated by three states: discovery, active, and sleep. Discovery state is used by a node to determine its neighbours in the zone; when in discovery state, a node turns on its radio and exchange discovery messages to find other nodes in the same zone; the discovery message contains the node identifier, the zone identifier, the node remaining energy and the node state. Active state is used for routing data in the network; when a node enters active state it sets a timeout $T_a$ to define how long it will remain in this state and after this time it goes back to the discovery state. A node in discovery or active state can go to sleep state when it detects some other equivalent node that will handle the routing; nodes negotiate which node will handle routing through an application-dependent ranking procedure; during the sleep state the radio of the node is turned off; the duration of the sleeping period depends on the specific application and the related parameters are tuned accordingly during the routing process. The ranking procedure is needed to select the new node to handle routing and it consists in ordering the nodes that should be active. This ordering is defined through several rules based on load balancing and remaining energy information, in order to maximize network lifetime.

GAF can also support mobility: each node in a zone estimates its leaving time of the zone and sends this to its neighbours; the sleeping neighbours adjust their sleeping time accordingly in order to keep the routing fidelity; before the leaving time of the active node expires, sleeping nodes wake up and one of them becomes active.

GAF is independent of the specific ad hoc routing protocols used to route packets among the awake nodes; the authors simulate GAF over AODV and DSR. Simulations showed that GAF does not degrade performances of classic ad hoc routing protocols in terms of of latency and packet loss while it can consume around half of the energy consumed by unmodified ad hoc routing protocols; because of the mechanism of maintaining only one node per zone as active at a time, the network lifetime with GAF increases proportionally to the node density.

### 7.3.15  GEAR protocol

Since data queries often include geographic attributes, the protocol geographic and energy aware routing (GEAR) (Yu, Estrin & Govindan, 2001) exploits geographic information while propagating queries only to appropriate regions. It can be classified as a data-centric algorithm with geographic information knowledge. Many location-aware systems use this to disseminate information to a geographic region, for example a sensor net application may be interested in what is the average temperature in a certain region *R* during a specified period of time $\Delta t$. GEAR restricts the number of interests in directed diffusion by routing the query directly to the interested region rather than sending the interests to the whole network. By doing this, GEAR can conserve more energy than directed diffusion. The GEAR algorithm uses energy-aware neighbour selection to route a packet towards the target region and recursive geographic forwarding or restricted flooding algorithm to disseminate the packet inside the destination region.

Consider the following assumptions: each query packet has a target region specified; each node knows its own location, its remaining energy level, its neighbours' locations and its neighbours' remaining energy levels; this can be obtained thanks to a simple neighbour hello protocol.

The process of forwarding a packet to all the nodes in the target region consists of two steps (Figure 7.18). The first one aims at forwarding the packets towards the target region: GEAR uses a geographical and energy-aware neighbour selection process to route the packet towards the target region; the source node routes its data progressively towards the target region, trying simultaneously to balance the energy consumption among its neighbours. The second step consists in disseminating the packet within the region: under most conditions, a recursive geographic forwarding algorithm is used to disseminate the packet within the region; under some low-density conditions, restricted flooding is used.

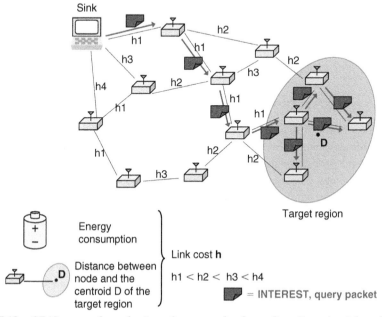

**Figure 7.18**   *GEAR protocol: mechanism of query packet forwarding. From the sink to the target region they are forwarded minimizing a defined link cost h; inside the target region flooding is used*

In order to explain the heuristic followed in the first step, assume that the node $N$ has to send a packet with target region $R$ with centroid $D$. $N$ achieve the trade-off between routing towards the target region and balancing the energy consumption of the forwarder nodes by minimizing a function cost, called learned cost, that is a combination of consumed energy to reach the region $R$ and distance to region $R$. Each node $N$ maintains a state $h(N, R)$ called learned cost to region $R$ for the node $N$; any node periodically updates its $h(N, R)$ to its neighbours. If a node does not have information about the state $h(N_i, R)$ for a neighbour $N_i$, it computes the estimated cost $c(N_i, R)$ as a default value, defined as follows: $c(N_i, R) = \alpha d(N_i, R) + (1 - \alpha)e(N_i)$, where $\alpha$ is a tunable parameter, $d(N_i, R)$ is the distance from $N_i$ to $D$, normalized by the largest such distance among all neighbours of $N$, and $e(N_i)$ is the consumed energy at node $N_i$ normalized by the largest consumed energy among neighbours of $N$. The node $N$ selects as next forwarder the neighbour $N_{min}$ that minimizes the cost function $h(N_{min}, R) + C(N, N_{min})$, where $C(N, N_{min})$ represents the cost of transmitting a packet from $N$ to $N_{min}$. Now that a node has a learned cost state or a default estimated cost function for each neighbour, we describe the forwarding actions at node $N$.

Once the packet is inside the target region, a simple flooding scheme can be used to flood the packet inside region $R$. However, flooding is expensive in terms of energy consumption, especially in high-density networks. Therefore, a recursive geographic forwarding approach is used to disseminate the packet inside target region $R$. Suppose node $N - i$ receives a packet for region $R$, and finds itself inside $R$. $N_i$ creates four new copies of the packet bound to four sub-regions of $R$. This recursive splitting and forwarding procedure is repeated until the current node is the only one inside this sub-region. The criteria to determine this is when the farthest point of the region is within a node transmission range, but none of its neighbours are inside the region. When no node is inside the sub-region, the packet is dropped altogether.

### 7.3.16   GeRaF protocol

Geographic random forwarding (GeRaF) (Zorzi & Rao, 2003a; Zorzi & Rao, 2003b) is a forwarding scheme based on geographical location of nodes and random selection of the relaying node through a contention among receivers. GeRaF integrates routing, MAC and topology management into a single layer, by allowing nodes to be put into energy-saving sleep state.

A basic assumption of the protocol is that each node knows its own position and the position of the destination node, that is, the sink. Basically, once a node has a packet to send, it broadcasts it specifying its own location and the location of the intended destination. All active listening nodes in the coverage area receive this packet and define their own priority in trying to act as a relay, based on how close they are to the destination. The selected relay continues this process, thereby geographically routing the packet towards destination, without the need of any routing table.

More specifically, when a node wants to transmit a packet, it broadcasts a message that will be heard by all active nodes in its coverage area. Each of these nodes can determine its own distance from the sink and it decides if it can act as a relay for this packet. This is done by first dividing the coverage area in two parts: the relay region, which contains all points of the area within range closer to the destination than the transmitting node, and the non-relay region, which contains all other points within range. Nodes in the non-relay region are never selected as relays. Further, the relay region is divided into $N_p$ priority regions, based on the distance from the destination: region $A_i$ contains all points in the coverage region whose distance from the final destination is $D - 1 + (i - 1)/N_p \leq d \leq D - 1 + i/N_p$, where $D$ is the distance from the transmitting node and the sink, $d$ is the remaining distance after one hop and $i = 1, \ldots, N_p$. Nodes in the region $A_1$, which are those closest to the destination, can contend for the medium firstly. If no nodes are found in $A_1$, then nodes in $A_2$ contend, and so on until there are nodes in the relay region (Figure 7.19).

GeRaF adopts sensors equipped with two radios: the message exchange occurs in the data frequency while a busy tone frequency is used for busy tone only.

When a sleeping node has a packet to send, it enters the active state and monitors both frequencies for T seconds. If either frequency is busy, the node backs off and reschedules an attempt at a later time. If both frequencies are sensed idle the node broadcasts an RTS message, which contains the location of the intended destination and its location. Then the transmitting node listens in the subsequent slots for CTS messages from potential relays. In each of the CTS slots following the RTS message, three different cases may occur (1) if only one CTS message is received, the node can start transmission of the data packet; (2) if it receives no CTS, it sends a CONTINUE message and it listens again for CTS, timing out after a defined number of empty CTS slots; (3) if it hears a collided signal it sends a COLLISION message to start a collision resolution algorithm and then it listens again for CTS. If after the packet transmission an ACK packet is correctly received the communication is completed and the node can go back to sleep.

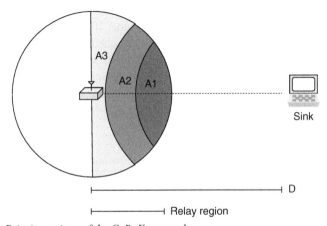

**Figure 7.19** *Priority regions of the GeRaF protocol*

Each node can periodically wake up and put itself in the listening mode, independently of other nodes. If nothing happens throughout a defined listening time, the node goes back to sleep. On the other hand, if the node detects the start of a transmission, it goes into the receiving state. Upon detecting the start of a message, node starts receiving and it activates the busy tone on the busy tone frequency. If no valid RTS is received, the node goes back to the listening state, where it stays for the originally scheduled duration. On the other hand, if a valid RTS is received, the node reads the information in it and determines its own priority as a relay, as described before. In the first CTS slot after the RTS, all nodes in $A_1$ will send a CTS message, while all others will be silent. All nodes will then listen for the message from the transmitter in the latter part of the CTS slot. If a packet start is heard (which contains the identification of the node which sent the CTS), only the designated node will continue to receive, whereas all others will go back to sleep. If in the second part of the first CTS slot a CONTINUE message is heard, it means that there are no nodes in $A_1$, and all nodes in $A_2$ will contend in the second CTS slot. If an ABORT message is received, the transmitter has reached the maximum allowed number of CTS slots and the transaction is aborted. If, on the other hand, a COLLISION message is received, this means that more than one CTS was generated in the CTS slot that was solved by a collision resolution algorithm.

### 7.3.17   Rugin protocol

GeRaF has good performance but it uses a two-radio channel. The work of Rugin & Mazzini, 2004, presents a MAC-routing integrated algorithm for WSNs able to preserve the GeRaF capacity of working in random environment but using a single-channel radio.

Generally, nodes of the network are in an idle state. Each node periodically wakes up and it puts itself in a listening mode. If nothing happens throughout the listening time, the node goes back to sleep. The ratio between sleeping and listening time, when the node is in the idle state, is considered a fixed parameter of the network, called duty cycle.

When an idle node has a packet to send, it makes transition to the transmit state and it monitors the radio channel for a fixed period of time. If an activity on the channel is sensed during this interval, the node backs off and reschedules another attempt at a later time. Otherwise, the node broadcasts a probe message, which contains its address and the address of the destination node. After the transmission of the probe, the transmitter listens to the channel in an implicit slotted time structure for a fixed number of slots. Receivers reply in these slots, so that the transmitter can then select the best relay among the nodes that have replied in these slots.

The selection of the next relay follows a proper algorithm that takes into account information related both to the remaining energy of the nodes and their positions (Figure 7.20): the transmitter chooses the node that minimizes the value of $I_C = \alpha I_E + \beta I_P$; $I_E$ is a function inversely proportional to the node residual energy and can assume value in the range [0, 1]; $I_P$ is a function decreasing as the relay position is more close to the

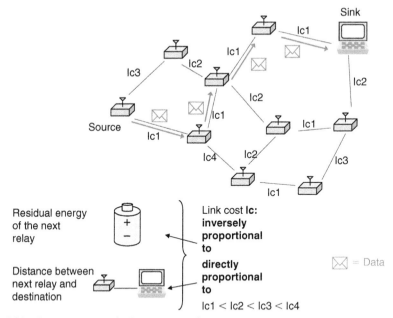

**Figure 7.20** *Data routing in the Rugin protocol*

destination and can assume value in the range [0, 1]; $\alpha$ and $\beta$ are two constants used for setting the relative weight between energy and hop length. When the next relay has been chosen the transmitter sends a broadcast selectrelay message which contains the selected relay address and the data message addressed directly to that node. At the end of the data transmission, the transmission node checks the radio channel by waiting for the receiver probe message, that testifies successful to delivery, and then it can power off its radio going into the idle state; otherwise such a data transmission is rescheduled.

When a listening node receives a probe packet, it analyzes the address of the probe's sender and it decides coherently with the routing algorithm to turn on the idle state and power off or to participate in the communication. In case a node decides to participate, it transmits an alive packet in a slot randomly chosen between the available ones and it powers off. At the end of the slotted window, the node wakes it up to listen for the selectrelay packet. If the node has been selected by the transmitter, it remains on receiving the data and then it sends a probe packet to the transmitter to confirm the correct reception of the data.

# 8

# Localization and time synchronization techniques for WSANs

## 8.1 Introduction

Time synchronization and localization represent fundamental issues for several WSN applications. They are jointly discussed in this chapter since, as will be clear later, they are strictly related to each other.

Sensed data without time and position information is often meaningless. For example, in habitat environments monitored sensed events must be ordered both in time and space to permit a correct interpretation. If nodes lack a common timescale (i.e., they are not synchronized) the final estimate will be inaccurate.

Time synchronization and positioning are also essential for basic mechanisms composing the WSN to work efficiently. In fact, properly designed MAC scheduling algorithms that can reduce packed collisions or power-saving strategies (wake-up sleeping times in low duty cycle working nodes) are feasible if nodes are synchronized, thus leading to a substantial energy conservation. In addition, as will be shown in the following text, a proper time synchronization is required to achieve high-ranging accuracies in positioning techniques. As explained in Chapter 7, the position information may be advantageously used to improve the performance of location-based routing protocols as well.

For outdoor localization the GPS is the most used system today (Spilker, 1978), not only to retrieve position information but also as a reference base time source. Unfortunately, battery drain, cost and size constraints preclude the utilization of GPS for several of the nodes in many WSN applications. Moreover, in indoor or cluttered environments its use is precluded entirely. Depending on application constraints, only a small fraction of nodes might be equipped with GPS or are placed in known positions (*anchor nodes* or *beacons*).

The other nodes with unknown position are referred to as *unknown nodes* (or *agents*) and must estimate their position by interacting with the anchor nodes. As will be shown later, when a direct interaction with a sufficient number of anchor nodes is possible, *single-hop* algorithms can be adopted. Otherwise, cooperation between nodes is required to propagate, in a *multihop* fashion, the anchor node positions information to those nodes which cannot establish a direct interaction with anchor nodes.

In certain scenarios none of the nodes is aware of its position (*anchor-free* scenario). Moreover, in many applications the knowledge of absolute coordinates is not necessary (e.g., ad hoc battlefield and rescue systems). In these cases only relative coordinates are estimated (sometimes called *virtual coordinates*) and ad hoc positioning algorithms have to be designed.

## 8.2 Time measurements

Usually, nodes are equipped with a local oscillator from which an internal clock reference is derived to measure the real time $t$. Unfortunately, all oscillators are subjected to frequency drifts due to various physical effects (see Figure 8.1). Hence, only an estimation $\hat{t} = \mathbb{C}(t)$ of the real time $t$ can be obtained.

The frequency of an oscillator changes over the time, however it can be approximated with good accuracy to be constant if the time intervals under measurement are small. In that case

$$\mathbb{C}(t) = (1 + \delta)t + \mu, \tag{8.1}$$

where $\delta$ is the *clock drift* relative to the correct rate and $\mu$ is the *clock offset*. The rate of a perfect clock, $d\mathbb{C}(t)/dt$, would equal 1 (i.e., $\delta = 0$). The clock performance is often expressed in terms of part per million (ppm), defined as the maximum number of extra (or missed) clock counts over a total of $10^6$ counts, that is, $\delta \cdot 10^6$.

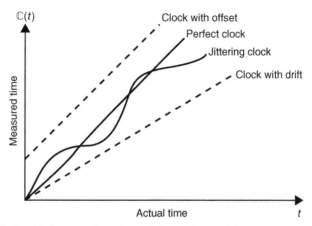

**Figure 8.1** *Relationship between the estimated time and actual time*

Suppose a node has to generate a time delay of $\tau_d$ seconds, the effective generated delay $\tau_{d_{eff}}$ in the presence of a clock drift $\delta$ is

$$\tau_{d_{eff}} = \frac{\tau_d}{1+\delta}. \tag{8.2}$$

In case a node has to measure a time interval of true duration $\tau = t_2 - t_1$ seconds, the corresponding estimated value $\hat{\tau}$ would be

$$\hat{\tau} = \mathbb{C}(t_2) - \mathbb{C}(t_1) = \tau(1+\delta). \tag{8.3}$$

In both cases there is no dependence on the clock offset $\mu$.

Consider two nodes, whose oscillators will run at slightly different frequencies, causing the clock values to gradually diverge from each other. This divergence is called *clock skew*. Network synchronization algorithms try to correct the clock skew by exchanging messages, as will be shown in Section 8.7.

## 8.3   Distance measurements (ranging)

Positioning techniques are based on measurement of certain physical quantities from which, mainly, the final position estimation accuracy and system complexity depend. According to the node's hardware capabilities, different kinds of measurements can be available based on RF, DC electromagnetic field, infrared and ultrasound (Hightower & Borriello, 2001).

### *Proximity*

The simplest way to obtain useful measurements for positioning is *proximity* where the mere connectivity information is used to estimate node position. The key advantage of this technique is that it does not require any dedicated hardware and time synchronization among nodes since the connection information is almost available in wireless devices.

Making the common assumption that nodes are randomly located in the Poisson plane in the presence of a deterministic propagation scenario (disk channel model), where $r_0$ is the transmission range corresponding to a certain maximum tolerable path loss $L_{th}$, the connection event between a pair of nodes $A$ and $B$ changes the probability distribution of the distance between a pair of connected nodes, which can be written as

$$f_R(r) = \frac{2r}{r_0^2} \tag{8.4}$$

for $0 \le r \le r_0$ and 0 otherwise.

In a more realistic scenario, where shadowing is experienced according to the model (3.11) introduced in Chapter 3, the connection event between $A$ and $B$ changes the probability distribution of the distance $r$ between a pair of connected nodes, which can be written as

$$f_R(r) = r \cdot e^{\frac{-2}{k_1}(L_{th} - k_0 - \sigma^2/k_1)} \operatorname{erfc}\left(\frac{k_0 - L_{th} + k_1 \ln r}{\sqrt{2}\sigma}\right) \tag{8.5}$$

where parameters $k_0$, $k_1$ and $\sigma$ are defined in Chapter 3.

## Received signal strength (RSS)

Based on the consideration that, in general, the further away the node the weaker the received signal, it is possible to obtain an estimate of the distance between two nodes (*ranging*) by measuring the received signal strength (RSS). Theoretical and empirical models are used to translate the difference (in dB) between the transmitted signal strength (assumed known) and the received signal strength into a range estimate. RSS ranging does not require time synchronization between nodes.

Generally, propagation effects cause small-scale slow and fast fading components (Parsons, 1992). For ranging the extraction of only large-scale fluctuations are desirable. With wideband signals the mean received power can be calculated by summing the powers of the multipath in the power delay profile. With narrowband signals, received power experiences large fluctuations over a local area and averaging should be used to estimate the mean received power.

A widely used model adopted to describe the attenuation of the radio signal in indoor environments is the WAF-based propagation model (3.8) described in Chapter 3 (Motley & Keenan 1988; Rappaport, 1996).

Unfortunately, signal issues such as refraction, reflection, shadowing, and multipath cause the attenuation to correlate poorly with distance resulting in inaccurate and imprecise distance estimates. In Pavani, Costa, Mazzotti, Dardari & Conti, 2006 an example of parameter tuning is given for a realistic indoor scenario.

Usually, the statistical model (3.11) representing the log-normal shadowing effect is adopted. In this case the Cramér-Rao lower bound (CRLB) for a distance estimate $\hat{d}$ from RSS measurements provides the following inequality related to the estimate variance (Gezici, Tian, Giannakis, Kobayashi, Molisch, Poor & Sahinoglu, 2005)

$$\mathrm{Var}(\hat{d}) \geq \frac{d^2 \cdot \sigma^2}{k_1^2}, \qquad (8.6)$$

where $d$ is the distance between the two nodes, $\sigma$ is the spread factor of the shadowing phenomena and $k_1$ is given in (3.11). It can be observed that the best achievable limit depends only on channel parameters and not on signal characteristics.

## Time-of-arrival (ToA)

Considering that the electromagnetic waves travel at the light speed $c = 3 \cdot 10^8$ m/s (this as first approximation), the distance information between a couple of nodes $A$ and $B$ can be obtained from the measurement of the propagation delay or time-of-flight (ToF) $\tau_p = d/c$, where $d$ is the actual distance between $A$ and $B$.

In a first simple scheme (*one-way ranging*), node $A$ emits at time $t_1$ a packet to a receiving node $B$. The packet contains the timestamp $t_1$ at which the transmission started. Node $B$ receives the packet at time $t_2$. If the nodes were perfectly synchronized to a common reference clock (i.e., sharing the same time reference and time base), it is clear that $\tau_p$ would be calculated at node $B$ as $\tau_p = t_2 - t_1$ and the distance estimated.

Consider now the more realistic case where nodes are not perfectly synchronized. Suppose that node $A$ and $B$ have clock drifts $\delta_A$, $\delta_B$ and offsets $\mu_A$, $\mu_B$, respectively. According to node $A$'s local time, the packet is transmitted at time $t_1^{(A)} = \mathbb{C}_A(t_1)$ (included as a timestamp in the packet) and it is received at node $B$'s local time $t_2^{(B)} = \mathbb{C}_B(t_2)$. Node $B$ calculates the estimated propagation delay as

$$\hat{\tau}_p = t_2^{(B)} - t_1^{(A)} = \tau_p \cdot (1 + \delta_A) + t_2 \cdot (\delta_B - \delta_A) + \mu_B - \mu_A. \tag{8.7}$$

As can be noticed in (8.7), $\hat{\tau}_p$ could be significantly different from the true value $\tau_p$ if stringent synchronization constraints are not satisfied as happen in many practical cases.

This problem becomes negligible when ultrasound devices are adopted. Considering that the acoustic waves propagation speed ($\approx 340\,\text{m/s}$) is much lower than the light speed, synchronization errors can be several orders of magnitude smaller than the typical propagation delay values, making this technique very attractive for some specific applications (Hightower & Borriello, 2001).

A second scheme which requires less stringent synchronization constraints is *two-way ranging*. In this scheme node $A$ (see Figure 8.2) emits a packet to node $B$ which, after a response delay $\tau_d$, gives an answer by transmitting back a second acknowledge (ACK) packet to node $A$. The round-triptime (RTT) between the node $A$ transmission and response receiving instants is

$$\text{RTT} = 2\tau_p + \tau_d. \tag{8.8}$$

Starting from the measurement of the RTT it is possible to estimate the distance $d$ between node $A$ and $B$. In this case clocks are not required to have the same time reference since the effect of different clock offsets is eliminated by the difference operation. However, relative clock drifts still affect the ranging accuracy. In fact, with reference to

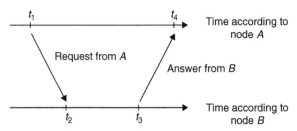

**Figure 8.2**   *Two-way ranging between two nodes*

(8.2) and (8.3), the effective response delay introduced by node $B$ is $\tau_d/(1 + \delta_B)$, whereas the estimated RTT, according to node $A$'s time scale, is

$$\widehat{RTT} = 2\tau_p(1 + \delta_A) + \frac{\tau_d(1 + \delta_A)}{(1 + \delta_B)}. \tag{8.9}$$

In the absence of other information, node $A$ derives the estimation of the propagation time, $\hat{\tau}_p$, by equating (8.9) with the supposed round-trip time $2\hat{\tau}_p + \tau_d$ leading to

$$\hat{\tau}_p = \tau_p(1 + \delta_A) + \frac{\tau_d(\delta_A - \delta_B)}{2(1 + \delta_B)}. \tag{8.10}$$

Defining $\epsilon = \delta_A - \delta_B$, the error on ranging estimate is

$$\hat{\tau}_p - \tau_p = \tau_p\delta_A + \frac{\epsilon\tau_d}{2(1 + \delta_A - \epsilon)}. \tag{8.11}$$

In Figure 8.3 the ranging error is plotted as a function of $\epsilon$ considering a maximum $\tau_p$ of 100 ns (about 30 metres) and a pessimistic value $\delta_A = 10^{-5}$, that is, 10 ppm. The curves correspond to different values of the response delay $\tau_d$. Starting from this figure it is possible to obtain a specification concerning $\epsilon$ and $\tau_d$. For example, a target ranging error of 33 ps (about 1 cm) can be satisfied for $\epsilon$ up to $10^{-5}$ if the response delay is below 10 μs. For a less pessimistic value of $\epsilon = 10^{-6}$, the constraint on the response delay relaxes to $\tau_d \approx 100$ μs.

The accuracy obtained in measuring the RTT can be reduced by adopting high-precision oscillators (not convenient in low-cost WSNs) or, better, by adopting suitable time synchronization techniques similar to those explained in Section 8.7 but implemented at MAC level to avoid further delays at upper protocol layers.

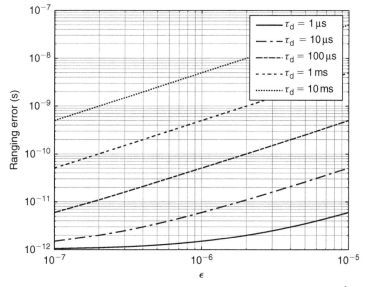

**Figure 8.3** *Ranging error in two-way ranging due to relative clock drifts.* $\delta_A = 10^{-5}$, $\tau_p = 100$ ns

In the above ranging error derivation, a perfect detection and estimation of the packet time-of-arrival (ToA) has been implicitly assumed. However, the detection of the exact arrival time of the transmitted packet in the presence of noise and multipath is a challenging problem which will be discussed in Section 8.3.1 and that could significantly degrade the ranging error.

## Time difference-of-arrival (TDoA)

In a first scheme (see Figure 8.4), multiple signals are broadcasted from synchronized nodes at distinct known locations (beacons). The receiver with unknown position measures the time difference-of-arrival (TDoA) and solves for the ToF. This technique is the same adopted by GPS. In a second scheme (see Figure 8.5), a reference signal is broadcasted from the unknown node and received at several known locations with synchronized receivers (anchor nodes). The receivers share their estimated ToA times, compute the TDoA, and solve for the ToF. Typically, receivers are synchronized through a wired network connection.

To calculate the position of the unknown node, atleast three anchors with known position and two TDoA measurements are at least required. A typical approach uses a geometric

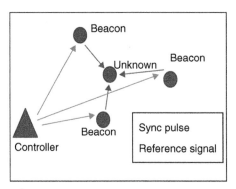

**Figure 8.4**    *TDoA scheme* 1

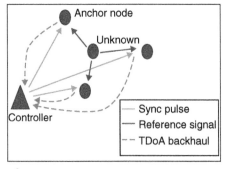

**Figure 8.5**    *TDoA scheme* 2

interpretation to calculate the intersection of two or more hyperbolas. In fact, each sensor pair gives a hyperbola which represents the set of points at a constant range-difference (time-difference) from two sensors.

## Angle-of-arrival (AoA)

Instead of providing information about distance among nodes, angle-of-arrival (AoA) measurements provide the information about node direction with respect to neighbouring nodes. The AoA on an incoming radio signal can be estimated by using multiple antennas with known separation (antenna array) and measuring the ToA of the signal at each antenna.

Given the differences in arrival times and the array geometry, it is possible to estimate the direction of propagation of a radio-frequency wave incident on the antenna array. AoA does not require the precise time synchronization needed for ToA and TDoA techniques. Two angle measurements are required to determine node position (*triangulation*). In NLOS environments, the measured AoA might not correspond to the direct path component of the received signal and large angle estimation errors can occur. Due to the presence of multiple antenna elements, AoA techniques could be too expensive in terms of cost and device dimensions for WSN applications.

## 8.3.1   ToA detection and estimation

We have seen that a key quantity in distance estimation accuracy is the ToA estimation. In the presence of noise and multipath, the detection and the accurate estimation of the exact arrival time of the transmitted packet is a fundamental issue in the ranging process.

First of all we investigate the unbeatable performance limit in ToA estimation for AWGN channel. We will see that the ranging accuracy increases as the signal bandwidth increases. That justifies the adoption of wide or UWB signalling when sub-metre accuracies are required, especially in indoor environments. Second, the ToA estimation in harsh multipath environments using the UWB technology is addressed by explaining low complexity threshold-based algorithms suitable for WSN applications.

### Theoretical performance limits in ToA estimation

We now investigate the fundamental limits in ToA estimation by considering a simple scenario where a pulse $p(t)$, with duration $T_p$, is transmitted and received undistorted, with energy $E_p$, through an AWGN channel. In general, $p(t)$ can be a part of a multiple access signalling such as direct sequence or time hopping. The received signal is

$$r(t) = \sqrt{E_p}\, p(t - \tau_{\text{toa}}) + n(t), \qquad (8.12)$$

where $n(t)$ is AWGN with zero mean and two-sided power spectral density $N_0/2$. The parameter $\tau_{\text{toa}}$ describes the ToA to be estimated based on the received signal $r(t)$ observed over the interval $[0, T)$.

In the classical ToA estimation scheme the received signal is first processed by a filter matched to the pulse $p(t)$. The ToA estimation is then accomplished simply by observing the instant corresponding to the maximum peak at the output of the MF over the observation interval (Van Trees, 1968).

This scheme yields a maximum likelihood (ML) estimate, which is asymptotically efficient, that is, for high SNR the estimated ToA, $\hat{\tau}_{\text{toa}}$, becomes unbiased with variance equal to the CRLB. The CRLB gives the bound on the estimation mean square error (MSE) of any unbiased estimation $\hat{\tau}_{\text{toa}}$ of $\tau_{\text{toa}}$, that is,

$$\text{Var}(\hat{\tau}_{\text{toa}}) = \mathbb{E}\{(\hat{\tau}_{\text{toa}} - \tau_{\text{toa}})^2\} \geq \text{CRLB}. \tag{8.13}$$

The CRLB is given by

$$\text{CRLB} = \frac{N_0/2}{(2\pi)^2 E_p \beta^2} = \frac{1}{8\pi^2 \, \text{SNR} \, \beta^2}, \tag{8.14}$$

where we have introduced the signal-to-noise ratio $\text{SNR} = E_p/N_0$ and the parameter $\beta^2$ represents the second moment of the spectrum $P(f)$ of $p(t)$ defined by

$$\beta^2 \triangleq \frac{\int_{-\infty}^{\infty} f^2 \, |P(f)|^2 \, df}{E_p}. \tag{8.15}$$

Notice that the denominator of (8.14) is proportional to the energy in the signal where the proportionality constant $\beta^2$ depends on the shape of the pulse through (8.15). Contrary to RSS-based methods, the ranging capability depends on the signal structure. This reveals that having large values of $\beta^2$, that is, a signal with wide bandwidth, is beneficial for ranging. By considering, for example, the typical exponential path loss model in (3.4), the SNR results to be proportional to $d^{-\alpha}$, then the distance estimation MSE can be lower bounded by

$$\text{Var}(\hat{d}) \geq \sigma_0^2 \, d^\alpha, \tag{8.16}$$

where $\sigma_0^2$ is the estimation error experienced at 1 metre.

As previously mentioned, the adoption of UWB signals is particularly attractive since it makes centimetric ranging potentially feasible. In UWB systems, the $n$th derivative of the basic Gaussian pulse in (6.1) is usually adopted. For this $n$th-order derivative of the Gaussian pulse it is easy to show that the second moment $\beta_n^2$, according to (8.15), reduces to

$$\beta_n^2 = \frac{(2n+1)}{2\pi \, \tau_p^2}, \tag{8.17}$$

where $\tau_p$ is related to the pulse width. From (8.17) we can observe that a better asymptotic ToA estimation performance can be achieved by increasing $n$ or decreasing $\tau_p$.

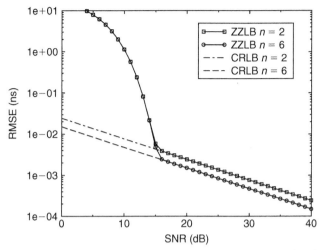

**Figure 8.6** *CRLB and ZZLB as a function of the SNR for different order derivative of the Gaussian pulse.* $\tau_p = 1$ *ns*

As the order $n$ of differentiation increases, from (8.17) it is easy to verify that the asymptotic gap between the CRLB obtained by adopting the $(n + 1)$th instead of the $n$th Gaussian derivative pulse is

$$\frac{\text{CRLB}_{n+1}}{\text{CRLB}_n} = \frac{\beta_n^2}{\beta_{n+1}^2} = \frac{2n + 1}{2n + 3}, \tag{8.18}$$

which tends to 1 for large $n$.

In Figure 8.6, the root mean square error (RMSE) related to the CRLB for different values on $n$ is shown in AWGN. For comparison, the Ziv-Zakai lower bound (ZZLB) is reported which gives an improved lower bound for low and medium SNR values (Chazan, Zakai & Ziv, 1975; Dardari, Chong & Win, 2006b). The true $\tau_{\text{toa}}$ has been considered uniformly distributed in the observation interval $[0, T)$ with $T = 100$ ns.

### ToA estimation in multipath conditions

In the presence of multipath phenomena, the previous scheme based on MF is, in general, no longer optimum. In fact, the received signal is composed of a sum of a large number of echoes components and an ambiguity problem in distinguishing the first arriving path among echoes and noise peaks arises. In case an erroneous peak is detected as the first one, a large error in ToA estimation will occur. The detection of the first path is made more challenging in NLOS condition where the first path, if not completely blocked, might not be the strongest one (Win & Scholtz, 2002b; Chong & Yong, 2005). As a consequence, different ToA estimation algorithms have to be introduced and the ranging performance can be significantly worse than that obtained in AWGN condition (Lee & Scholtz, 2002; Falsi, Dardari, Mucchi & Win, 2006). However, the complexity must be kept as low as possible to be attractive for WSNs.

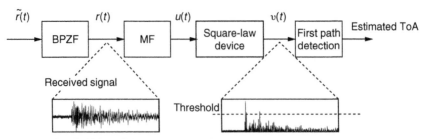

**Figure 8.7**   *Threshold-based ToA estimator with MF*

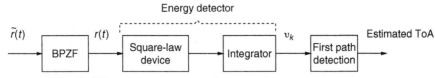

**Figure 8.8**   *Threshold-based ToA estimator with ED*

A simple technique to detect the first arriving path in such a harsh propagation environment is to compare the output of an MF or an energy detector ED with a threshold whose value has to be optimized according to the operating condition (Guvenc & Sahinoglu, 2005a; Guvenc & Sahinoglu, 2005b; Cheong, Rabbachin, Montillet, Yu & Oppermann, 2005; Rabbachin, Oppermann & Denis, 2006; Dardari, Chong & Win, 2006a; Dardari, Conti, Ferner & Win, 2008). The adoption of threshold-based estimators is attractive because complexity and computational constraints are often critical issues in WSNs and it can be implemented in the analog domain. In addition, the adoption of more complex estimators do not always correspond to a significant performance improvement, especially in dense multipath conditions. A comparison of different ToA estimation techniques based on results using real measured data can be found in Falsi et al., 2006.

With reference to the MF-based scheme in Figure 8.7, one possible procedure is a two-stage process. First, detect the portion of the observation interval in which the first path is located by comparing the output of a square-law device, inserted after the MF to remove sign ambiguity of the path amplitude, to a fixed threshold $\eta$ (Van Trees, 1968). Then make a fine delay estimation by peak searching (if particular complexity constraints are present, the refinement stage can be skipped). The output of this maxima search, subtracted by the MF delay, is taken as the estimate $\hat{\tau}_{\text{toa}}$ of the ToA. A second procedure (Figure 8.8) is to consider an ED, with integration interval equal to $\tau_s$, whose output is compared to the threshold $\eta$ (Guvenc & Sahinoglu, 2005a; Dardari, Chong & Win, 2006a). The first threshold crossing event is taken as the estimate of the ToA. In both cases the performance of the detection process, hence of the ToA estimation, depends on the choice of the threshold $\eta$ which plays important role in the optimum design of threshold-based ToA estimators. It has to remarked that the implementation of the ED scheme is less complex than the MF scheme since it can operate at sub-Nyquist sampling rates (usually $\tau_s$ is larger than 1 ns) thus requiring less demanding analog-to-digital converters. Other detection techniques are proposed and investigated in Guvenc & Sahinoglu, 2005a; Guvenc & Sahinoglu, 2005b.

The following results are based on the methodology developed in Dardari & Win, 2006; Dardari, Chong & Win, 2006a; Dardari, Chong & Win, 2008, and have been obtained by considering a dense multipath channel based on one of the IEEE 802.15.4a channels with exponential power delay profile given by (3.19) (Cassioli et al., 2002; Dardari & Win, 2006). We choose, if not otherwise specified, $\Delta = 2$ ns, $\tau_s = 2$ ns (ED integration interval), $L = 32$, $\epsilon = 6$ ns and $m_l = 2$ for our numerical examples. The observation interval is $T = 120$ ns and the 6th derivative Gaussian monocycle pulse with $\tau_p = 1$ ns is considered.

Figure 8.9 shows the RMSE of the MF and ED based estimators as a function of the normalized threshold (TNR) TNR $\triangleq \eta/N_0$ for SNR = 20, 30 dB. It can be seen that there is an optimum value of the threshold, depending on the SNR, which minimizes the RMSE. It can also be seen that the choice of the threshold $\eta$, through TNR, is more critical when operating at low SNR both for the MF and ED estimators.

In Dardari, Chong & Win, 2008 a simple sub-optimal criteria to determine the optimum TNR based on the evaluation of the probability of early detection is proposed. It is shown that the early detection effect is the dominant one, thus an alternative approximated way to evaluate the optimum value of the threshold is to calculate the value of the threshold for that the probability of early detection becomes negligible.

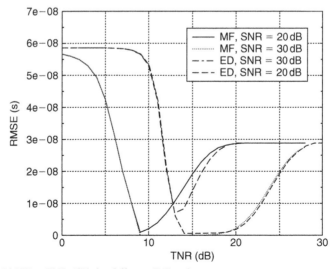

**Figure 8.9** *RMSE vs TNR (dB) for different SNR values*

The probability of early detection $P_{ed}$ is given by

$$P_{ed} = 1 + \frac{(1 - q_o)^{N_{ToA}} - 1}{N_{ToA} q_o}, \qquad (8.19)$$

where $q_o$ is defined, respectively for the MF and ED estimators, by

$$q_o^{(MF)} \triangleq Q(\sqrt{2TNR}) \qquad (8.20)$$

and

$$q_o^{(ED)} \triangleq \exp(-TNR) \sum_{i=0}^{M/2-1} \frac{TNR^i}{i!}, \qquad (8.21)$$

where $M = 2\tau_s W$, $N_{ToA} = T/\Delta$, $W$ is the signal bandwidth and $Q(\cdot)$ is the Gaussian probability integral. This expression can be used to evaluate the threshold TNR corresponding to a target $P_{ed}$ (e.g., $10^{-4}$). In Dardari, Chong & Win, 2008 it is shown that the adoption of the TNR evaluated through (8.19) does not lead to significant performance degradation with the main advantage to not require any a priori channel knowledge.

Figure 8.10 shows the RMSE as a function of SNR obtained by considering the optimal value of TNR that minimizes the MSE for each SNR. In this figure, we consider the cases where the first path is subjected to Nakagami-$m$ fading with $m_1 = 2$ and $m_1 = 10$, respectively. The former case is typical for NLOS situations, whereas the latter case is typical

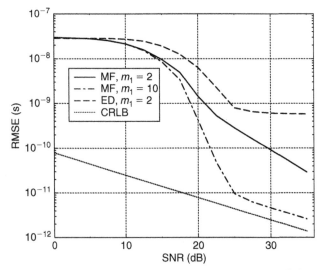

**Figure 8.10**   *RMSE vs the SNR using optimum values of $\eta$. Comparison with the CRLB*

for strong LOS situations with essentially no fading on the direct path. The MF-based estimator achieves a significant better accuracy than the ED estimator, especially for SNR > 25 dB or when in LOS situations. The results are compared to the CRLB. As can be seen in the figure, at high SNR the ED performance shows a floor equal to $\tau_s^2/12$, whereas the MF performance tends to the CRLB with a behaviour depending on the fading severity of the first path. In the presence of severe fading in the NLOS case, the ToA estimator performance tends slowly to the CRLB due to ambiguities in detecting the right peak. For low SNR the received signal is highly unreliable and the ToA estimation error is on the order of the observation interval width. For medium SNR the performance is far from that predicted by the CRLB, that is, it is dominated by large errors due to the ambiguity in peak selection. Results confirm that MF-based receivers are more suitable to be deployed for short-range LOS applications when SNR is relatively high and high ranging accuracy is desired. On the other hand, for longer range applications with low SNR, ED-based estimators are desirable in order to reduce the implementation complexity and cost.

### ToA estimation in the presence of narrowband and wideband interference

In many practical operating conditions, interference is invariably present. Experience has shown that both narrowband interference (NBI) and wideband interference (WBI), can considerably degrade the ToA estimation performance.[1]

Only a few papers deal with the effect of the interference in ToA estimation. For example, Dardari, Giorgetti & Win, 2007; Shainoglu & Guvenc, 2006 consider nonlinear filtering techniques to mitigate the interference for ED-based ToA estimators. In particular, after the de-hopping process, the observed energy samples can be arranged in a matrix form where each element $v_{n,k}$, for $n = 0, \ldots, N_t - 1; k = 0, \ldots, K - 1$, contains the energy collected in the $k$th time slot of the $n$th hopping interval. A total of $N_t$ hopping intervals and $K$ time slots are considered. Examples of energy matrices collected at the output of the ED for the cases where the receiver is affected by NBI and WBI are shown in Figures 8.11 and 8.12, respectively. Figure 8.11 shows that the energy samples corresponding to the desired signal components are time aligned and partially buried under a floor caused by NBI. The floor value corresponds to the energy collected in each interval $\tau_s$ due to NBI and noise. This floor may increase the probability of early detection, and thus degrading the performance of the estimator. Unlike the NBI case, note in Figure 8.12 that there is no energy floor. However, the detection may be compromised by contributions coming from WBI. The key characteristic to be observed is that, unlike the desired signal, the contributions from WBI are not time aligned due to the TH scheme.

The behaviour of the energy matrix in the presence of interference suggests that filtering techniques can be used to mitigate the effect of interference on the detection of the first path. Therefore, prior to the ToA estimator, the collected samples can be pre-processed

---

[1] Wideband interference is due to the presence of other users in the system, i.e., MUI.

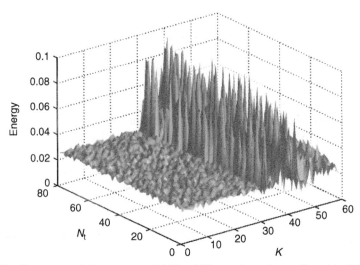

**Figure 8.11**    *Energy matrix for a system with* T = 120 *ns and* $\tau_s$ = 2 *ns, affected by NBI*

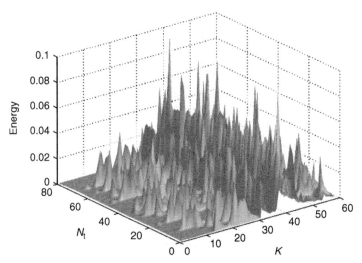

**Figure 8.12**    *Energy matrix for a system with* T = 120 *ns and* $\tau_s$ = 2 *ns, affected by WBI*

(filtered) with the purpose to reduce the interference effects and to improve the detection of the first path. This can be carried out by introducing a generic transformation $\mathbb{T}[\cdot]$ (see Figure 8.13), whose output $\{z_k\}_{k=0}^{K-1}$ is used by the ToA estimator for determining the ToA of the first path. In general,

$$z_k = \mathbb{T}[\{v_{n,k}\}] \quad k = 0, \ldots, K - 1 \tag{8.22}$$

where $\mathbb{T}[\cdot]$ can be either a 2D linear or a nonlinear transformation.

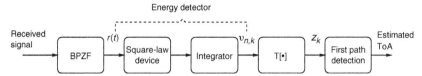

**Figure 8.13**  *Threshold-based ToA estimator with 2D filtering to mitigate the NBI and WBI*

Different filtering schemes can be adopted. The conventional way to obtain the decision vector $\{z_k\}$ is by simple column averaging (*averaging filter*), that is,

$$z_k = \sum_{n=0}^{N_t-1} v_{n,k} \quad k = 0, \ldots, K - 1. \tag{8.23}$$

Nonlinear processing can be also applied to each matrix column to reduce the impulse behaviour contributed by WBI. For example, each column of $\{v_{n,k}\}$ can be processed (*min filter*) as follows

$$z_k = \sum_{n=0}^{N_t-H} \min\{v_{n,k}, v_{n+1,k}, \ldots, v_{n+H-1,k}\}, \tag{8.24}$$

where $k = 0, \ldots, K - 1$, and $H$ is the length of the filter (Shainoglu & Guvenc, 2006).

With the purpose to improve the estimation performance in the presence of both NBI and WBI, Dardari, Giorgetti & Win, 2007 propose a scheme where the *min filter* is first applied to each matrix column as in (8.24) to produce the intermediate vector $\{\tilde{z}_k\}$, which is further processed (*differential filter*) as follows

$$z_k = \tilde{z}_k - \tilde{z}_{k+1}, \quad k = 0, \ldots, K - 1, \tag{8.25}$$

with $\tilde{z}_K = 0$.[2] The purpose of (8.25) is to reduce the presence of floors, typically caused by NBI (see Figure 8.11), and to emphasize the beginning of the multipath (see Figure 8.12).

Once the vector $\{z_k\}$ is produced, a procedure to detect the first path is required. One possible way to obtain an estimate for $\tau$ is to detect the time slot within the observation interval that contains the first path by comparing each element of $\{z_k\}$ to a fixed threshold (Dardari, Chong & Win, 2006a). The first threshold crossing event is taken as the estimate of the ToA (*simple thresholding*).

In the following some numerical results are provided to show the effectiveness in miti-gating the NBI and WBI of the nonlinear filtering techniques described. With reference to Dardari, Giorgetti & Win, 2007, we consider a square root raised cosine transmitted pulse with roll-off factor 0.6, pulse duration parameter $\tau_p = 0.8$ ns, and centre frequency $f_0 = 4.0$ GHz. According to the spectrum of the pulse, we adopt an ideal BPZF with bandwidth $W = 1.6$ GHz. The ED integration time is $\tau_s = 2$ ns. The performance of the ToA estimators are based on the preamble with length $N_{\text{sym}}$ symbols, each consists of

---

[2] Differential filtering is typically used in image processing to emphasize the presence of edges.

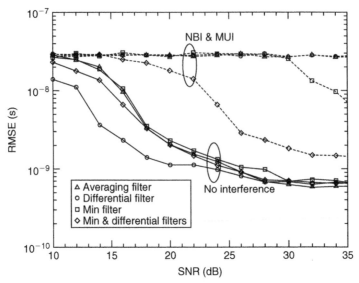

**Figure 8.14**   *Performance of the threshold-based estimator with $N_{sym} = 400$ and different 2D filtering techniques in the presence of both NBI, interference-to-noise ratio INR = 35 dB, and WBI, SIR = −15 dB*

$N_s = 4$ pulses per symbol, with a frame duration $T_f = 120$ ns and $K = 60$ slots. With the aim of evaluating the effectiveness of the algorithms in a harsh environment, we consider the presence of both NBI and WBI. In particular, NBI is modelled as a tone with frequency at $f_I = 3.5$ GHz subjected to Rayleigh fading. Regarding the WBI, we consider additional user (i.e., $U = 2$), and we define SIR $\triangleq E_s^{(1)}/E_s^{(2)}$, where $E_s^{(1)}$ and $E_s^{(2)}$ are the received energy per symbol related to user 1 and 2, respectively. The results presented are obtained for the IEEE 802.15.4a CM4 channel model characterizing NLOS indoor propagation in a large office environment (Molisch et al., 2006).

The filtering techniques considered are compared in Figure 8.14 to assess their performance, in terms of RMSE of the estimator, in a realistic scenario that accounts for multipath, NBI, and WBI. For the *min filter*, we adopted a length of $H = 5$. It can be seen in Figure 8.14 that the performance of the *averaging, min*, and *differential filters* degrades drastically in the presence of NBI and WBI. On the contrary, due to its capability to mitigate the NBI and MUI, the cascade of a min filter and a differential filter scheme provides substantial performance gain with respect to other schemes due to its capability to mitigate the combination of both types of interference.

## 8.4   Position estimation

The purpose of any positioning algorithm is, given a set of measurements (e.g., distance, angle, connectivity), to find the locations of the nodes with unknown positions (unknown nodes). Positioning occurs in two steps. First nodes measurements are obtained, then the measurements are combined using positioning techniques to deduce the location on the unknown nodes.

We can classify position estimation techniques in:

- *Anchor-based* Some nodes know their locations, either by a GPS or as pre-specified. The distance estimate to anchors can be obtained by direct interaction (*single-hop*), or indirectly by means of intermediate nodes (*multi-hop*).
- *Anchor-free* None of nodes knows its position. Only relative coordinates (virtual coordinates) can be found.
- *Range-based* Measurements provide some sort of distance/angle information among nodes.
- *Range-free*. Only connectivity information are used.

In comparing different algorithms, typical performance indexes are: the *precision*, related to the dispersion of the position estimation error (generally modelled by a Gaussian probability distribution), and the *accuracy*, the degree to which the random variation is centred on the true value. Other performance indexes are the *robustness* of the algorithm to some errors, such as range measurement errors, and the *coverage*, the percentage of nodes with estimated position.

## 8.4.1 Single-hop localization

*Multilateration*
Consider the problem of determining the position $(x, y)$ of an unknown node by using distance estimates $d_i$ between the unknown node and a set of $N$ anchor nodes (beacons) placed at known coordinates $(x_i, y_i)$, with $i = 1, 2, \ldots N$. These estimates can be obtained, for example, through ToA or RSS measurements.

The classical method for deriving the node position is *(multi) lateration*. In the presence of ideal distance estimates, the $i$th beacon defines a circle centred in $(x_i, y_i)$ with radius $d_i$ (see Figure 8.15). The intersection of the circles corresponds to the position of the target node. In a two-dimensional space, at least three beacons are required. More in detail, the position estimation can be obtained through the following system of equations

$$(x_1 - x)^2 + (y_1 - y)^2 = d_1^2$$
$$\vdots$$
$$(x_N - x)^2 + (y_N - y)^2 = d_N^2 \tag{8.26}$$

System (8.26) can be linearized by subtracting the last equation from the first $N - 1$ equations, thus arriving at a proper system of linear equations given by the following matrix form

$$\mathbf{A} \cdot \mathbf{p} = \mathbf{b}, \tag{8.27}$$

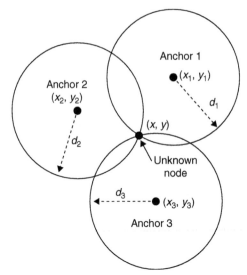

**Figure 8.15**    *Example of multilateration*

where

$$\mathbf{A} \triangleq \begin{pmatrix} 2(x_1 - x_N) & 2(y_1 - y_N) \\ \vdots & \vdots \\ 2(x_{N-1} - x_N) & 2(y_{N-1} - y_N) \end{pmatrix} \tag{8.28}$$

$$\mathbf{b} \triangleq \begin{pmatrix} x_1^2 - x_N^2 + y_1^2 - y_N^2 + d_N^2 - d_1^2 \\ \vdots \\ x_{N-1}^2 - x_N^2 + y_{N-1}^2 - y_N^2 + d_N^2 - d_{N-1}^2 \end{pmatrix} \tag{8.29}$$

and

$$\mathbf{p} \triangleq \begin{pmatrix} x \\ y \end{pmatrix}. \tag{8.30}$$

In a real scenario where estimation errors are present, (8.27) may be inconsistent, that is, circles do not intersect in one point. In addition, when $N > 3$ the system of equations is overdefined and it can be solved through a standard nonlinear least-square (LS) approach, that is,

$$\hat{\mathbf{p}}^{(LS)} = (\mathbf{A}^T \mathbf{A})^{-1} \mathbf{A}^T \mathbf{b}, \tag{8.31}$$

with the assumption that $\mathbf{A}^T\mathbf{A}$ is non-singular, where superscript $T$ denotes the transpose.

In Sayed, Tarighat & Khajehnouri, 2005 a similar approach to obtain the position estimate from AoA and TDoA measurements as well as hybrid measurements (e.g., joint AoA and ToA) is explained.

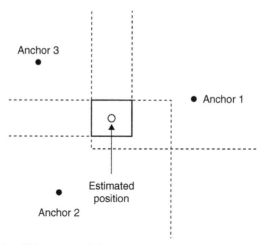

**Figure 8.16**    *Example of Min-Max multilateration*

## Low complexity multi-lateration: Min-Max

Solving (8.31) is quite expensive since complex matrix floating point operations are required and often are not available in typical WSN devices.

A much simpler method, presented as a part of the *N*-hop multi-lateration algorithm by Savvides et al. (Savvides, Park & Srivastava 2002), is *Min-Max*. The idea is to construct a bounding box starting from each known position $(x_i, y_i)$ and distance measurement $d_i$ (see Figure 8.16). In particular, the bounding box corners of node *i* are

$$(x_i - d_i, y_i - d_i) \times (x_i + d_i, y_i + d_i). \tag{8.32}$$

The estimated position is obtained as the centre of the intersection of these bounding boxes computed by taking the maximum of all coordinate minimums and the minimum of all maximums, that is,

$$[\max_i(x_i - d_i), \max_i(y_i - d_i)] \times [\min_i(x_i + d_i), \min_i(y_i + d_i)]. \tag{8.33}$$

The final position is evaluated as the average of both corner coordinates. The advantage of the Min-Max method is that it requires only low complexity sum and compare operations. Some experimental results and performance comparisons in realistic environments can be found in Pavani, Costa, Mazzotti, Dardari & Conti, 2006; Pavani, Dardari, Conti & Andrisano, 2007; Severi, Liva, Chiani & Dardari, 2007.

## Theoretical performance limits: position error bound

The range measurements $r_i$ between the unknown node and the *i*th anchor node are in practice affected by errors. We have seen, for example, that in AWGN the optimum ToA estimator is characterized by an asymptotic MSE given by the CRLB as expressed by (8.14) which is a decreasing function of the SNR. In addition, it is known from estimation theory that the ML estimation error tends asymptotically to the Gaussian distribution.

If we denote by $d_i$ the true distance, the measured range $r_i$ can be expressed as

$$r_i = d_i + \epsilon_i, \tag{8.34}$$

where $\epsilon_i$ is a random Gaussian noise with zero-mean and variance $\sigma_i^2$ accounting for the ranging estimation error. As done in (8.16), we model the dependence of the variance of $\epsilon_i$ on the distance $d_i$ as $\sigma_i^2 \equiv \sigma^2(d_i) = \sigma_0^2 d_i^\alpha$, where $\alpha$ is the path loss exponent and $\sigma_0^2$ is the variance at 1 metre.

Let $\mathbf{p} = (x, y)^T$ be the unknown node's coordinates and $(x_i, y_i)$ the coordinates of the $i$th anchor node with $i = 1, 2, \ldots N$. The PDF of the $i$th range measurement, conditioned to the true position $\mathbf{p}$ of the unknown node, is therefore given by

$$f_i(r_i|\mathbf{p}) = \frac{1}{\sqrt{2\pi}\sigma(d_i(\mathbf{p}))} \exp\left(-\frac{(r_i - d_i(\mathbf{p}))^2}{2\sigma^2(d_i(\mathbf{p}))}\right). \tag{8.35}$$

Let $\hat{\mathbf{p}} = (\hat{x}, \hat{y})^T$ be any position estimation. The CRLB gives

$$\sqrt{\mathbb{E}_r \|\mathbf{p} - \hat{\mathbf{p}}\|^2} \geq \sqrt{\mathbb{T}\{\mathbf{J}^{-1}\}}, \tag{8.36}$$

where $\mathbf{J}$ is the Fisher information matrix (FIM) (Van Trees, 1968)

$$\mathbf{J} = \mathbb{E}_r\{[\nabla_\mathbf{p} \ln(f(\mathbf{r}|\mathbf{p}))][\nabla_\mathbf{p} \ln(f(\mathbf{r}|\mathbf{p}))]^T\}, \tag{8.37}$$

and $f(\mathbf{r}|\mathbf{p})$ is the PDF of $\mathbf{r}$, the vector of range measurements, conditioned on $\mathbf{p}$.[3] Thus $\sqrt{\mathbb{T}\{\mathbf{J}^{-1}\}}$ is a lower bound on the standard deviation of any position estimator. In the remaining we refer to this expression as the position error bound (PEB)

$$\text{PEB}(x, y) \triangleq \sqrt{\mathbb{T}\{\mathbf{J}^{-1}\}}. \tag{8.38}$$

The PEB is a fundamental limit on the accuracy of any localization method.

We calculate the PEB in the case where the measurements are assumed to be independent. We have

$$f(\mathbf{r}|\mathbf{p}) = \prod_{i=1}^N f_i(r_i|\mathbf{p}), \tag{8.39}$$

where $f_i(r_i|\mathbf{p})$ is given by (8.35). After a few algebraic manipulations we obtain the FIM as (Jourdan, Dardari & Win, 2006b)

$$\mathbf{J} = \sum_{i=1}^N A(d_i)\mathbf{M}(\theta_i), \tag{8.40}$$

---

[3] The notation $\mathbb{E}_r\{\cdot\}$ denotes the expectation operator with respect to $\mathbf{r}$, $\mathbb{T}\{\cdot\}$ is the trace of a square matrix, and $\nabla_\mathbf{p}\{\cdot\}$ denotes the gradient of a scalar with respect to $\mathbf{p}$.

where $\theta_i$ is the angle between the unknown node and the $i$th anchor node measured with respect to the horizontal,

$$M(\theta) \triangleq \begin{bmatrix} \cos^2\theta & \cos\theta\sin\theta \\ \cos\theta\sin\theta & \sin^2\theta \end{bmatrix}, \qquad (8.41)$$

and

$$A(d) \triangleq \frac{1}{\sigma^2(d)} + \frac{\alpha^2}{2d^2}. \qquad (8.42)$$

The expression (8.40) provides us with some useful insights. First, note that $M(\theta_i)$ contains geometric information about the relative position of the unknown node with respect to the $i$th anchor node. The FIM is therefore a weighted sum of this geometric information, where the weights $A(d_i)$ depend on $d_i$. These weights return the quality of the range measurements and thus capture how much new information each measurement brings.

From (8.42) we see that when $d$ goes to infinity (so that the range measurement variance goes to infinity) the weights $A(d_i)$ tend to zero. This is consistent with our intuition that the larger the range estimation variance, the less valuable the corresponding range information will be in determining the unknown node's position: the corresponding $M(\theta_i)$ in (8.40) will receive a low weight and the contribution from the $i$th anchor node will be small. The weights $A(d_i)$ therefore quantify the importance of the information coming from the $i$th anchor node. This implies that the information from anchor nodes that are far away (large range measurement variance) will not contribute much to the FIM and the localization accuracy is mainly affected by local nodes.

We now use the analytical expression for the FIM to obtain the PEB. Since the FIM is a $2 \times 2$ matrix, its inverse is easily obtained and can be plugged into (8.38) to obtain

$$\text{PEB}(x, y) = \sqrt{\frac{\sum_{i=1}^{N} A_i}{\left(\sum_{i=1}^{N} A_i c_i^2\right)\left(\sum_{i=1}^{N} A_i s_i^2\right) - \left(\sum_{i=1}^{N} A_i c_i s_i\right)^2}}, \qquad (8.43)$$

where $A_i \triangleq A(d_i)$, $c_i \triangleq \cos\theta_i$, and $s_i \triangleq \sin\theta_i$. We stress that the limit on the localization accuracy given in (8.43) depends on the distance between the unknown node and the anchor nodes. If the variance was not dependent on the distance ($\alpha = 0$) then, according to (8.42), $A_i = 1/\sigma_0^2$ for all $i$ and the PEB is simply equal to the product of the measurement standard deviation $\sigma_0$ a quantity named geometric dilution of precision (GDOP) usually adopted in GPS literature (Levanon, 2000).

For a practical system we may be interested in the quality of localization not just at one point, but over an area. Let us map the value of the PEB throughout a square area for

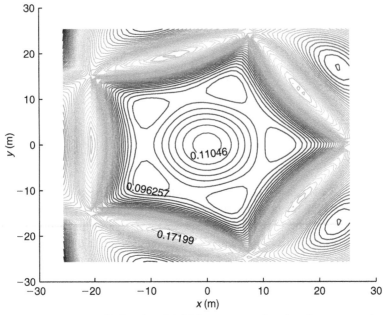

**Figure 8.17**   *Contour map of PEB when 5 LOS beacons are placed at the vertices of a polygon with* R = 25 m, α = 3, $\sigma_0^2$ = 0.001 m²

5 LOS beacons placed at the vertices of a polygon of radius $R = 25$ m (Figure 8.17). This contour plot reveals that the centre of the polygon is no longer the location with minimum PEB, contrary to the common conclusions in the literature based on a model where the range measurement variance does not depend on distance between agent and corresponding beacons (Levanon, 2000). In other words, when the beacons are so arranged, the unknown node should not expect to have optimal localization accuracy in the centre of the polygon. The situation becomes all the more complicated when NLOS beacons are included. Infact, due to multipath and extra propagation delays (caused by obstacles such as walls), range measurements could be positively biased thus affecting the performance achievable. This effect has been accounted in Jourdan et al., 2006b; Jourdan, Dardari & Win, 2006a.

*The maximum likelihood (ML) estimate*
The ML estimate of the position $\mathbf{p} = (x, y)$ of one unknown node, assuming uniform node distribution (no a priori information) and given the vector of range measurements $\mathbf{r}$, is

$$\hat{\mathbf{p}}^{(ML)} = \arg \max_{\mathbf{p}} f(\mathbf{r}|\mathbf{p}), \tag{8.44}$$

where $f(\mathbf{r}|\mathbf{p})$ is given by (8.39).

When the measurement errors are Gaussian distributed, as assumed in (8.39), the ML estimator is optimal. In cases where RSS measurements are adopted, (8.44) takes the form (Dardari & Conti, 2004)

$$\hat{\mathbf{p}}^{(ML)} = \arg \min_{\mathbf{p}} \sum_{i=1}^{N} (l_i - L_0(d_i(\mathbf{p})))^2, \tag{8.45}$$

where $d_i$ is the distance between the unknown node and the $i$th anchor node, $L_0(d)$ is the nominal path loss given by (3.4) and $l_i$ are the path loss measurements in dB which can be easily derived from the RSS measurements once the transmitted power is known.

For low ranging errors, the ML estimate is efficient, that is, tends to the PEB. However, it does not give the optimal solution if the ranging model (i.e., the path loss model) is not accurate. In addition, finding the minimum of the function (8.45) may require extensive centralized computations, especially if internode cooperation is taken into account. To face this problem (Dardari & Conti, 2004) a hierarchial ML algorithm is proposed and applied to RSS-based range measurements.

## 8.4.2 Multihop localization

Generally, a low number of anchor nodes is appreciated due to cost and feasibility constraints, hence single-hop localization could fail in cases of unknown nodes that are not able to interact with a sufficient number (at least three) of anchor nodes. As a consequence, cooperation among nodes is required to estimate node positions through *multihop cooperative* localization algorithms.

A quantitative comparison between several multihop cooperative algorithms is reported in Langendoen & Reijers, 2003a. Here a common 3-phase structure is identified:

- *Phase 1* Determine the distances between unknowns and anchor nodes.
- *Phase 2* Derive for each node a position from its anchor distances (using, for example, multilateration or Min-Max algorithms).
- *Phase 3* Refine the node positions using information about the distance to, and positions of, neighbouring nodes.

Another approach is to consider fully iterative distributed algorithms, where nodes surrounding anchor nodes cooperatively establish position estimates that are successively propagated to more distant nodes, allowing them to estimate their position without direct anchor node visibility (Savarese, Rabaey & Beutel, 2001). At each iteration step, once a node with unknown position $(x, y)$ bears $N$ nodes with known or estimated positions, it would be able to estimate its position starting from the measured distances $d_i$ and known positions $(x_i, y_i)$ if $N \geq 3$.

### N-hop multilateration
The *N*-hop multilateration algorithm has been introduced by Savvides et al., 2002. The distance to the anchors is simply determined by adding the ranges encountered at each

hop during the network flood. In particular, the anchors send a beacon message including their identity, position and path length accumulator set to 0. Each receiving node adds the measured range from the previous node to the path length field and broadcasts the new message to the other nodes. If multiple messages about the same anchor are received, the node keeps and forwards only the one containing the minimum value of path length.

One of the main disadvantages of this approach is that range errors accumulate over multiple hops. The cumulative error becomes significant in the presence of large networks with few anchors or poor ranging hardware (e.g., based on RSS measurements).

### DV-hop

The DV-hop algorithm is similar to the $N$-hop multilateration and it was proposed in Niculescu & Nath, 2001. Beacon packets are flooded by anchor nodes throughout the network. Each receiving node maintains the minimum counter value per anchor node of all beacons it receives and ignores those beacons with higher hop-count values as done in the classical distance vector routing scheme. In this way each node in the network has a rough distance information, in terms of hops, to every anchor node. To enable the conversion from number of hops and physical distance, anchor nodes evaluate the average single-hop distance, $d_{hop}$, starting from the hop count information and known position of all other anchors inside the network. In particular, anchor node $i$ estimates $d_{hop_i}$ using the following formula

$$d_{hop_i} = \frac{\sum_j \sqrt{(x_i - x_j)^2 + (y_i - y_j)^2}}{\sum_j h_{i,j}}, \qquad (8.46)$$

where $(x_j, y_j)$ and $h_{i,j}$ are, respectively, the position of the $j$th anchor node and the distance, in hops, from anchor $i$ to anchor $j$. Once calculated, anchors broadcast the estimated average hop size information. Unknown nodes can evaluate the estimated distance to anchor node $i$ by multiplying the counted hops by the average hop size $d_{hop_i}$. Finally, those unknown nodes which obtain the distance estimation to at least three anchors can estimate their location by using multilateration (e.g., the simple Min-Max algorithm).

### Performance comparison

In Langendoen & Reijers, 2003a an extensive simulation campaign is presented where several combinations as regards phases 1 and 2 are compared in terms of coverage and localization accuracy. The scenario considered consists of a network of 225 nodes placed in a square with sides of 100 units. The anchor nodes represent a fraction of 5% and are placed in a regular grid. The link budget has been set to have radio transmission range of 14 units (no shadowing effects are considered). On average, less than one anchor node is seen by each node and the connectivity degree is about 12–13 nodes. The standard deviation of the (Gaussian) range error is set to 10% of the transmission range.

As an example, Figure 8.18. gives the average position error of the six combinations, proposed in Langendoen & Reijers, 2003a, for phases 1 and 2 for varying range error (top),

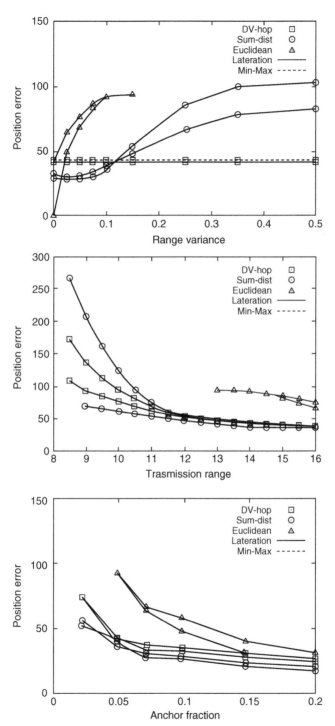

**Figure 8.18** *Performance comparison between multihop positioning algorithms (Langendoen & Reijers,* 2003)

transmission range (middle), and anchor fraction (bottom). The solid lines denote the multilateration variants whereas the dashed lines denote the Min-Max variants. The first observation is that the sensitivity to the anchor fraction is quite similar for all combinations and that all combinations are quite sensitive to the transmission range (connectivity). The Euclidean/lateration combination clearly outperforms the others in the absence of range errors. On the other hand, the Min-Max is more insensitive to distance errors than multilateration, but it requires a good anchor placement. It can be concluded that no particular algorithm gives the best performance but different behaviours against range errors, anchor fraction, placement and coverage have been found.

As a further example, we analyze the performance of two fully iterative algorithms based, respectively, on multilateration through (8.31) and Min-Max. In Figure 8.19 a simulation comparison between the two algorithms is reported in the same scenario considered in the previous example. The precision is expressed in terms of standard deviation of the estimated position error against the ranging error standard deviation.

The multilateration gives the better precision only if the range measurement is very accurate (e.g., zero measurement error or negligible with respect to the average nodes distance). As the ranging accuracy becomes more modest (as happens using RSS ranging techniques), the Min-Max technique shows a good robustness whereas multilateration fails. In the case where anchors are placed in a regular grid (controlled positioning) instead of random positions, some performance increasing is obtained, especially for multilateration. As far as the coverage (defined as the percentage of nodes able to get any location information) is concerned, no significant difference is present between

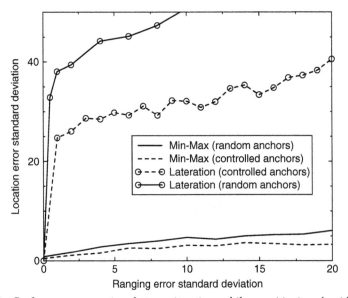

**Figure 8.19** *Performance comparison between iterative multihop positioning algorithms*

multilateration and Min-Max techniques. In this example about 90% and 50% of coverage is reached for random and controlled anchor positioning, respectively. The fully distributed iterative Min-Max technique for estimating the node position is very attractive in the WSN scenario considered because of its robustness and its low complexity since it does not require any floating point multiplication or division as needed by multilateration.

### Performance limits
The paper by Savvides, Garber, Moses & Srivastava, 2005 is a first attempt to derive analytically the fundamental behaviour, in terms of PEB, of multihop localization in the presence of measurement errors. The bounds expressions obtained are used to investigate the error trends with respect to network density, network scaling and percentage of anchor nodes independently on the specific localization algorithm. For example, results show that the localization error decreases rapidly until 6–10 neighbours are heard by each node. Another important result found is that multihop positioning is scalable since localization error is quite insensitive to increases in the number of anchor nodes.

## 8.4.3 Range-free localization

The position information can be inferred also in the presence of simple proximity information. Positioning algorithms based on proximity are also called *range-free* localization algorithms.

Consider the scenario shown in Figure 8.20 where $m$ anchor nodes are present with coordinates $\mathbf{b} = (x_1, y_1, x_2, y_2, \ldots, x_m, y_m)$ and the positions $\mathbf{x} = (x_{m+1}, y_{m+1}, \ldots, x_n)$ of the remainder $n - m$ nodes are unknown. The problem is to find $\mathbf{x}$ such that the proximity constraints are satisfied. The radio connectivity model, which considers a circle with a fixed radius $r_0$ as in (4.3) (i.e., the disk model), is the simplest way to model the proximity constraint.

As the number of constraints increases, the feasible region of solutions for $\mathbf{x}$, given by the intersection of individual constraints, becomes smaller. As an example, in Figure 8.21. the feasible set (shaded region) of solutions of a unknown node (white node) is shown for an increasing number of anchor nodes (black nodes), $m = 1, 2, 3$. From scenario (a) to (c), the intersection region decreases for each added constraint.

In Bulusu, Heidemann & Estrin, 2000 anchor nodes with overlapping region of coverage are placed in a regular grid with distance $d$ and act as beacons. A generic unknown node $i$ infers proximity to a collection $\mathbb{A}$ of anchor nodes for which the connectivity metric exceed a certain threshold (no cooperation among unknown nodes is considered here). The unknown node localizes itself to the region given by the intersection of the connectivity regions of radius $r_0$ and centred on anchor node positions. The centroid of this region is taken as position estimate $(\hat{x}_i, \hat{y}_i)$ of the generic unknown node $i$

$$(\hat{x}_i, \hat{y}_i) = \left( \frac{1}{k} \sum_{k \in C} x_k, \frac{1}{k} \sum_{k \in C} y_k \right). \tag{8.47}$$

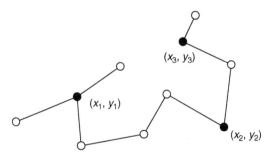

**Figure 8.20**   *Proximity-based positioning*

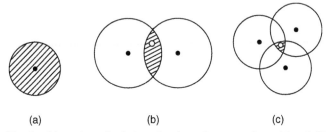

(a)                    (b)                    (c)

**Figure 8.21**   *The feasible region of solutions for the unknown node position (white node) as the number of anchor nodes (black nodes) increases*

Obviously, by increasing the ratio $r_0/d$ (i.e., the range overlap of the anchor nodes), the accuracy of the location estimate improves.

The more general case where a reduced anchor nodes range overlap and internode constraint (cooperation among nodes) exist is more challenging. The work in Doherty, Pister & Ghaoui, 2001 presents a centralized methodology to solve this problem as a linear or semi-definite program. It is shown that the position estimation error can be dramatically reduced as the network connectivity increases.

The DV-hop scheme, described in Section 8.4.2 and proposed in Niculescu & Nath, 2001, is actually a range-free positioning algorithm since distance estimation between unknown and anchor nodes is performed by hop counting.

In He, Huang, Blum, Stankovic & Abdelzaher, 2003 a range-free positioning algorithm called APIT is proposed and compared with some of the previous described algorithms in realistic environments. Results show that APIT performs better when more realistic radio channel models and random node placement are considered and low communication overhead is desired.

## 8.5   Anchor-free localization

Most of the work present in the literature concerning localization in WSNs starts from the assumption that a non-negligible fraction of nodes in the networks are a priori aware of their position (anchor nodes). In many applications the anchor nodes are not available or only relative coordinates among nodes are of interest.

The anchor-free localization problem can be formulated as follows: given a set of nodes with unknown position and range measurements among neighbours' nodes, determine the (relative) position coordinates of every node in the network.

Unfortunately, this problem is NP-hard. In addition, distributed algorithms are appreciated. Some heuristic algorithms have been proposed to face this problem. For example (see Figure 8.22), cooperative localization is analogous to finding the resting point of masses (representing the nodes) connected by springs (with length proportional to distance measurements) (Patwari, Ash, Kyperountas, Hero, Moses & Correal, 2005). Springs exert forces on the nodes which move until stabilization. The equilibrium point of masses represents a minimum-energy localization estimate. Force-directed relaxation methods can be used to converge toward a minimum-energy configuration. However, such methods are susceptible to severe false local minima. From that it is important to start from a reasonably good initial estimation.

In Priyantha, Balakrishnan, Demaine & Teller, 2003 a fully decentralized algorithm called anchor-free localization (AFL) is proposed. It is composed of two phases. First, an initial layout is computed to alleviate the false minimum problem. The strategy followed is based on the observation that many false minima are caused because nodes operating on local information converge falsely to configurations where groups of nodes are topologically folded with respect to the true configuration. AFL seeks to configure nodes into a 'fold-free' configuration. The second phase consists of the force-directed relaxation method.

The algorithm has been tested by simulation for different node connectivity and densities as well as distance estimation error conditions. It gives better precision and convergence results than incremental algorithms as the one presented in Savarese et al., 2001.

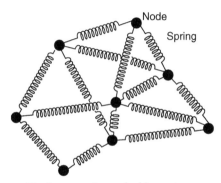

**Figure 8.22** *Anchor-free positioning as spring embedder system*

## 8.6  Position tracking

Up to now, we considered the problem of localizing a node without accounting for the knowledge (if any) of the node's previous position. Given the actual and previous estimated positions, besides forbidden location configurations (e.g., inside buildings), there are also forbidden time configurations, for example the node cannot suddenly move from one room to another one. Position tracking algorithms, in addition to range measurements, take into account the system memory as well as a node mobility model to achieve better position estimation accuracy.

The Bayesian filtering offers a powerful mathematical tool for the positioning problem through WSN (Fox, Hightower, Liao, Schulz & Borriello, 2003). Here the localization problem is modelled as a dynamic system where the vector state $\underline{s}_n$, at discrete time $n$, represents the coordinates $(x_n, y_n)$ of the unknown node. In particular, the state at time $n$ is

$$\underline{s}_n = f(\underline{s}_{n-1}, \underline{s}_{n-2}, \ldots, \underline{s}_1; \underline{r}_n), \tag{8.48}$$

which is a function of previous estimated positions and actual range measurements vector $\underline{r}_n$. Specifically, the vector $\underline{r}_n = [r_n^{(1)}, r_n^{(2)}, \ldots, r_n^{(K_n)}]$ is the vector of distance estimates between the unknown node and the $K_n$ anchor nodes at time $n$. At time $n$ we can define the following function

$$Bel(\underline{s}_n) \sim p(\underline{r}_n \mid \underline{s}_n) \sum_{\underline{s}_{n-1}} p(\underline{s}_n \mid \underline{s}_{n-1}) \cdot Bel(\underline{s}_{n-1}) \tag{8.49}$$

which is called Belief function and represents the posteriori probability distribution over the random variable $\underline{s}_n$ (Fox et al., 2003). Through the Belief function it is possible to identify the most likely state at time $n$ among all possible states.

Two elements operate in the Belief function (see Figure 8.23.). The first one is the *mobility model* $p(\underline{s}_n \mid \underline{s}_{n-1})$ that represents the dynamic model for the system. It gives

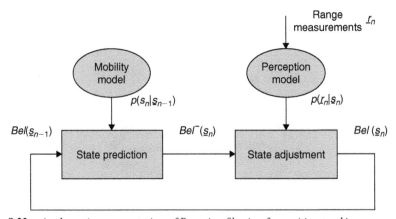

**Figure 8.23**    *A schematic representation of Bayesian filtering for position tracking*

the description of variation of the state $\underline{s}_{n-1} \rightarrow \underline{s}_n$, that is, the statistical description of unknown node movement. The second one is the term $p(\underline{r}_n|\underline{s}_n)$, that provides statistical information on the position at time $n$ starting from the measure vector collected at time $n$. This is the *perception model* and operates as an updater for the system state. In fact it updates and corrects the forecast obtained from the elaboration of the system dynamic model.

Once the mobility and perception models are properly defined, the Belief function allows to answer the following question: *What is the probability that, having the probability of previous positions and the range estimates vector at current time, the unknown node position is $\underline{s}_n = (x_n, y_n)$?* In other words, it allows for estimating the probability map from which it is possible to extract the most likely estimation of the current state of the system.

In Chapter 10 a case study example based on this approach is provided.

## 8.7   Time synchronization

Random fluctuations in the network dynamics such as propagation delay, channel access, operative system internal time scheduling, make the synchronization task in a wireless network challenging. Network time synchronization techniques rely on some sort of message exchange between nodes. When a node sends a packet carrying a timestamp to an intended node, the packet is received after a certain variable amount of time which prevents the receiver from understanding the real time difference among local clocks and then to accurately synchronizing to the sender node.

An important parameter for evaluating network time (or clock) synchronization algorithms is the *clock skew*. It provides the maximal absolute difference between the time perceived by nodes.

The main affecting factor of clock skew is the uncertainty about the accumulated delay of delivering messages between the nodes that, in general, increases with the number of hops.

The purpose of any time synchronization scheme is to equalize the nodes' logical clocks, that is, to minimize the clock skew among nodes. This is generally achieved through message exchange between nodes.

The time synchronization requirements can be classified into three basic models according to application needs (Ganeriwal, Kumar & Srivastava, 2003). The simplest form of synchronization is *event ordering* where the purpose of the synchronization algorithm is to tell if an event $E1$ has occurred before of after another event $E2$ (Romer, 2003).

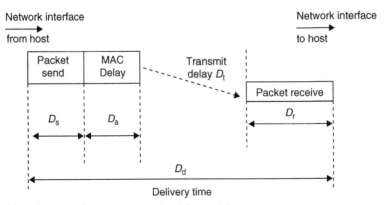

**Figure 8.24**   *Sources of error in the packet delivery delay*

A second type of synchronization requirement is focused on maintaining relative clocks. All nodes' oscillators run independently but, through message exchanges, the information about relative drift and offset between network clocks is available. In this way the local time of a node can always be converted to the local time of another node. Most of pro-posed time synchronization algorithms address this model (Elson, Girod & Estrin, 2002).

The third type of synchronization model is also the most complex (*always on* model). In fact it requires that all nodes maintain a clock synchronized to a reference clock in the network to preserve a global timescale throughout the network (Ganeriwal et al., 2003; Hong & Scaglione, 2005).

The source of errors in network time synchronization can be listed into four basic com-ponents (see Figure 8.24) (Sivrikaya & Yener, 2004). (1) The *send time, $D_s$*, for message construction at the transmitter which includes the overhead of the operative system, the time spent to transfer the message to the physical interface for transmission and buffer-ing delays. (2) The access time, $D_a$, that accounts for MAC delay. For example, in TDMA systems, $D_a$ is the time to access a slots, whereas in CSMA systems $D_a$ is the waiting time of free transmitting channels. Therefore, $D_a$ is closely related to network loads. (3) The *transmit delay, $D_t$*, spent to transmit the message from the sender to the receiver. (4) The *receive time, $D_r$*, that is, the time needed by the receiver to receive the message and trans-fer it to the microcontroller of the node.

These sources of error introduce non-determinism in the network dynamics, thus mak-ing the task of synchronization algorithms challenging. We denote with *delivery time, $D_d$*, the total time spent by the packet between the network interfaces of the sender and the receiver, that is,

$$D_d = D_s + D_a + D_t + D_r. \tag{8.50}$$

In Zhang & Deng, 2005 a statistical characterization of $D_d$ is given based on measurement on real hardware.

## 8.7.1  Network time synchronization

### Performance bounds

The work of Lundelius-Welch & Lynch, 1984 presents upper and lower bounds for clock synchronization in the case of a fully connected network composed of $n$ nodes. It is proved that, even if the clocks all run at the same rate as real time (non-drifting clocks), an uncertainty (*jitter*) of $u = D_d^{(\max)} - D_d^{(\min)}$ in the message delivery time makes it impossible to synchronize the clocks of $n$ nodes any more closely than

$$u \cdot \left(1 - \frac{1}{n}\right), \tag{8.51}$$

where the $D_d^{(\max)}$ and $D_d^{(\min)}$ are, respectively, the maximum and minimum message delivery times among two nodes.

By naming the uncertainty $u_{i,j}$ between nodes $i$ and $j$ as *distance*, this result has been extended in Biaz & Welch, 2001 to show that, for any network time synchronization algorithm, the worst case clock skew between some pair of nodes in the network is $\Omega(D)$,[4] where $D = \max_{i,j} u_{i,j}$ is called *diameter of the network*. In the case of equal distances $u$ among nodes, the diameter is simply $D = u \cdot h_{\max}$, where $h_{\max}$ is the maximum number of hops necessary to cross the entire network. Most of the algorithms (e.g., Srikanth & Toueg, 1987) achieve a worst case skew of $O(D)$. A different approach to evaluate time synchronization performance limits based on estimation theory is presented in Severi & Dardari, 2008.

### Two-way nodes synchronization

The basic procedure of most synchronization algorithms is a two-way message exchange between a pair of nodes $A$ and $B$. It is exploited, for example, by the network time protocol (NTP) which is widely used for clock synchronization on the Internet (Mills, 1991). The purpose is to estimate the clock offset between $A$ and $B$. Assuming constant propagation delay and clock drift in the small message exchange time period, consider the scheme in Figure 8.25. Node $A$ initiates the synchronization procedure by sending a synchronization packet at time $t_1$ (according to its own local time reference). This packet contains the value of $t_1$ (timestamp). Node $B$ receives the packet at $t_2 = t_1 + S + D_d^{(A,B)}$, where $S$ is the clock skew between $A$ and $B$ and $D_d^{(A,B)}$ is the delivery time of the packet from $A$ to $B$. $B$ responds at time $t_3$ with an acknowledge packet, which includes the timestamps $t_2$ and $t_3$. This packet is received by node $A$ at time $t_4 = t_3 - S + D_d^{(B,A)}$, where $D_d^{(B,A)}$ is the delivery time from $B$ to $A$. By means of the knowledge about $t_1$, $t_4$ (according to node $A$'s local time reference) and $t_2$, $t_3$ (according to node $B$'s local time reference),

---

[4]The symbol $\Omega(x)$ means a lower bound proportional to $x$.

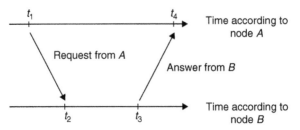

**Figure 8.25**   *Basic time synchronization procedure: two-way message exchange between a pair of nodes*

node $A$ can estimate the clock skew $S$ as follows and try to synchronize itself to $B$

$$\hat{S} = \frac{(t_2 - t_1) - (t_4 - t_3)}{2} = \frac{2S + D_{\mathrm{d}}^{(A,B)} - D_{\mathrm{d}}^{(B,A)}}{2}$$

$$= S + \frac{u}{2} \tag{8.52}$$

where $u = D_{\mathrm{d}}^{(A,B)} - D_{\mathrm{d}}^{(B,A)}$ is the uncertainty in the delivering time which determines the residual skew $u/2$ among $A$ and $B$.

In general, several trials are performed, because of the randomness of $D_{\mathrm{d}}$, by taking the least value or the average value as estimate.

### Reference broadcast synchronization (RBS)

The reference broadcast synchronization (RBS) scheme was proposed in Elson & Estrin, 2001; Elson et al., 2002 for WSNs. The key idea of this scheme is that a set of receivers are synchronized with one another to act as reference node. In particular, the reference node sends a beacon packet which does not include any timestamp (see Figure 8.26). Its ToA is used by receiving nodes as the reference point to estimate their relative clock offsets through the exchange of their receiving ToA. With reference to the example of Figure 8.26, the estimated clock skew among nodes $A$ and $B$ is

$$\hat{S} = t_A - t_B = D_{\mathrm{t}}^{(R,A)} - D_{\mathrm{t}}^{(R,B)} + D_{\mathrm{r}}^{(A)} - D_{\mathrm{r}}^{(B)}, \tag{8.53}$$

with

$$t_A = t_R + D_{\mathrm{s}}^{(R)} + D_{\mathrm{a}}^{(R)} + D_{\mathrm{t}}^{(R,A)} + D_{\mathrm{r}}^{(A)}$$
$$t_B = t_R + D_{\mathrm{s}}^{(R)} + D_{\mathrm{a}}^{(R)} + D_{\mathrm{t}}^{(R,B)} + D_{\mathrm{r}}^{(B)}. \tag{8.54}$$

As can be noted, the final precision is affected only by the receivers' non-determinism and not by the sender's non-determinism $D_{\mathrm{s}}^{(R)} + D_{\mathrm{a}}^{(R)}$ (as happens in typical two-way message-exchange-based schemes). The authors argue that RBS achieves much better precision than traditional synchronization methods. By adding more reference nodes the precision can be further increased. It has been demonstrated to have a precision within 11 μs on real hardware based on Berkeley's Motes. In addition, it has been shown, by simulation, that by increasing the number of reference nodes up to 30, the precision of 1.6 μs is achievable when synchronizing a pair of nodes.

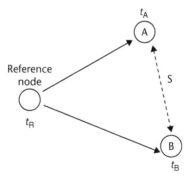

**Figure 8.26**  *The basic principle of the RBS scheme*

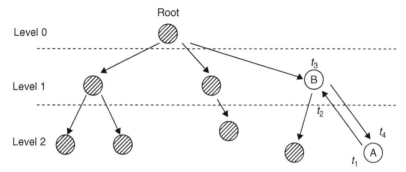

**Figure 8.27**  *An example of the hierarchial topology created by the TSPN*

## Time-synch protocol for sensor networks (TSPN)

Another interesting network time synchronization scheme for WSNs is the time-synch protocol for sensor networks (TSPN) (Ganeriwal et al., 2003). This scheme works in two phases: *level discovery* and *synchronization*. During the first phase a hierarchial topology in the network is created, where each node is assigned a level (see Figure 8.27). This usually happens once at network deployment. A node in the network acts as the root (level 0) because it is the sink or by self-election among nodes.

In the second phase, the root node issues a *time synch* packet which triggers a random timer at all level 1 nodes. After the timer is expired, the generic level 1 node asks its parent for a two-way node synchronization to estimate the RTT and then its clock offset. With reference to the example of Figure 8.27, the estimated clock skew among nodes $A$ and $B$ belonging to levels 1 and 2 respectively follows directly from (8.52).

Child nodes at level 2, hearing a *time synch* packet, also start a random timer (with longer values to permit level 1 nodes to conclude their synchronization) and initiate a two-way node synchronization phase with a level 1 node. This procedure makes at the end all nodes synchronized to the root node.

The authors tested the TSPN protocol on Berkley's Motes by inserting times stamping packets at the MAC level to reduce uncertainty at the sender level. Precisions of 16.9 μs are claimed.

## Gradient clock synchronization

In many WSN applications, nearby nodes in the network need to cooperate to perform some tasks, and nodes that are far away interact much less frequently. Hence, only nearby nodes need to have highly synchronized clocks. Far nodes can tolerate greater clock skew. Thus, for these applications, the maximum acceptable clock skew between two nodes forms a gradient in their distance.

In Fan & Lynch, 2004 the *gradient property* for clock synchronization is introduced. The gradient property requires that the skew between two nodes forms a gradient with respect to the distance between the nodes. However, typical network time synchronization algorithms do not satisfy the gradient property since they ensure that any pair of nodes have $O(D)$ clock skew, where $D$ is the diameter of the network as defined previously. Hence, they do not guarantee a gradient in the clock skew, because even nodes that are one hop distance apart can have $O(D)$ skew.

In Fan & Lynch, 2004 it is required that the skew between any two nodes' logical clocks be bounded by a non-decreasing function of the distance between the two nodes. The main result is that the worst case clock skew between two nodes at distance $u$ (defined at the beginning of this paragraph) or less from each other is

$$\Omega\left(u + \frac{\log D}{\log D \cdot \log D}\right). \tag{8.55}$$

From this result we can argue that clock synchronization is not a local property, in the sense that the clock skew between two nodes depends not only on the distance between the nodes, but also on the size of the network.

## Other algorithms

A different interesting approach for network time synchronization is presented by Hong & Scaglione, 2005, where a bio-inspired protocol for large-scale, sensor networks is proposed. The main idea is to emulate the simple strategies adopted by the biological agents. The strategy synchronizes pulsing devices that are led to emit their pulses periodically and simultaneously.

Other network time synchronization schemes are discussed in Maroti, Kusy, Simon & Ledeczi, 2004; Sichitiu & Veerarittiphan, 2003; Su & Akyildiz, 2005; Zhang & Deng, 2005 and in various surveys (Sivrikaya & Yener, 2004; Sundararaman, Buy & Kshemkalyani, 2005; Elson & Romer, 2003). Experimental tests on IEEE 802.15.4a platform can be found in Cox, Jovanov & Milenkovic, 2005.

# 9

# Signal processing and data fusion techniques for WSANs

Signal processing is one of the key aspects affecting both the performance and the design of WSANs. In fact, in the last five to ten years, this is reflected in a large literature, as well as in an increasing number of sessions at international conferences and numerous national and international projects funded on this topic. As practical examples, brief summaries of some projects, for what concerns activities related to WSANs in which the authors have been involved, are reported in the Section 9.4.3.

For a survey discussion on distributed signal processing in sensor networks, see, for example, Luo, Gastpar, Liu & Swami, 2006 and the related issue in *Signal Processing Magazine*, Special Issue, 2006. The low cost and complexity requirements for each sensor affect the design of WSANs: while each sensor alone is typically able to perform simple local computation, short-range and low data-rate communications, once sensors are properly deployed in large numbers across the space they can form an intelligent network and cooperate to perform complex tasks. WSANs of this type are attractive for context-aware applications such as environmental monitoring, immersive guide, healthcare monitoring, home applications, surveillance, precision agriculture, space exploration, intelligent transportation, etc.

To fully exploit the potential of sensor networks, it is essential to develop energy-efficient and bandwidth-efficient signal processing algorithms that can also be implemented in a fully distributed manner. Distributed signal processing in a WSN has a communication aspect not present in the traditional centralized signal processing framework, thus it differs in several important aspects.

- Sensor measurements are collected in a distributed fashion across the network. This necessitates data sharing via intersensors communication. Given a low energy budget per sensor, it is unrealistic for sensors to communicate all their full-precision data samples with one another. Thus, local data compression becomes a part of the distributed signal processing design. In contrast, in a traditional signal processing framework where data is centrally collected, there is no need for distributed data compression.

- The design of optimal distributed signal processing algorithms depends on the models used to describe: the nodes' connectivity, the nodes' distribution, the knowledge of sensor noise distributions, the qualities of intersensors communication channels, and the underlying application metrics. Distributed signal processing over a wireless sensor network requires proper coordination and planning of sensor computation as well as careful exploitation of the limited communication capability per sensor. In other words, distributed signal processing in sensor networks has communication aspects which are not present in the most of traditional signal processing frameworks.
- In a WSN, sensors may enter or leave the network dynamically, resulting in unpredictable changes in network size and topology. This can be due to failure between intersensors communication (propagation conditions, interference or non-available communication channels), duty cycling, drained batteries or nodes' damages. This dynamism requires the necessity for distributed signal processing algorithms to be robust to the changes in network topology or size. These algorithms and protocols must also be robust to poor time synchronization across the network and to inaccurate knowledge of sensor locations.

There are many theoretical challenges such as establishing models, metrics, bounds, and algorithms for distributed multimodal sensor fusion, distributed management of sensor networks including auto-configuration, energy-efficient application-specific protocol designs, formal techniques for the study of architectures and protocols, representation of information requirements, and sensor network capabilities on a common mathematical framework that would enable efficient information filtering (Luo et al., 2006). From these aspects, it is clear that the design of sensor networks under energy, bandwidth, and application-specific constraints spans all layers of the protocol stacks and it is very important to have a common framework enabling all these points be taken into account even if with different approximations degrees.

In this view an example of cross-layer methodology of WSN design for environmental monitoring will be shown in the following with particular emphasis on the impact of distributed digital signal processing (DDSP) on the spatial process estimation error on one side and network lifetime on the other side.

Depending on the process under monitoring and the goal of the WSAN, such as detection of distributed binary events and spatial process estimation, several techniques can be pursued with envisaging of centralized and distributed processing.

## 9.1 Distributed detection

In most applications, the intelligent fusion of information from geographically dispersed sensor nodes, commonly known as distributed data fusion, is an important issue. A related

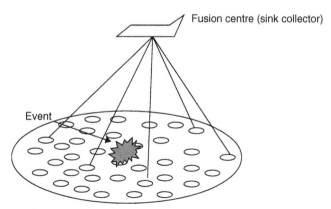

**Figure 9.1** *Example of distributed detection scenario*

problem is the binary decentralized (or distributed) detection problem, where a large number of identical sensor nodes deployed randomly over a wide region, together with a global detector or fusion centre (FC), cooperatively undertake the task of identifying the presence or absence of a phenomenon of interest (PoI) (see Figure 9.1). Specifically, each node takes a local decision about the presence or absence of the PoI and sends its decision to the FC which is responsible for the final decision based on the information gathered from local sensors.

Two problems have to be considered: the design of the decision rule at the FC and the design of the local sensor signal processing strategies. In the case of perfect knowledge of system parameters the design of the decision rule at the FC is a well-established task. The design of the local sensor decision rule, that is, the likelihood ratio test (LRT) threshold in binary detection, is more challenging due to the distributed nature of the system. In fact, the optimal choice of each sensor LRT threshold is coupled to each other node threshold, although nodes are not in general fully connected due to propagation effects or energy constraints. Recently it has been demonstrated that under the asymptotic regime (i.e., large number of nodes), the identical LRT threshold rule at the sensors provides the optimal error exponent if local sensor observations are independent and identically distributed (Sung, Tong & Swami, 2005). For a more complete overview of decentralized detection the reader is recommend to read Varshney, 1997; *Signal Processing Magazine, Special Issue*, 2006.

Unlike in classical decentralized detection problems (Blum, Kassam & Poor, 1997; Varshney, 1997), greater challenges exist in a WSN setting. There are stringent power constraints for each node, and communication channels from nodes to the FC are severely bandwidth-constrained. In addition, the communication channels are no longer lossless (e.g, fading, noise and, possibly, interference are present), and the observation at each sensor node is spatially varying (Sung et al., 2005; Niu & Varshney, 2005).

Recently, there has been great interest in cooperative communication (Winters, 1987; Sendonaris, Erkip & Aazhang, 2003b; Sendonaris, Erkip & Aazhang, 2003a; Laneman &

Wornell, 2003; Laneman, Tse & Wornell, 2004; Bletsas, Shin, Win & Lippman, 2006; Lucchi, Giorgetti & Chiani, 2005). One may also exploit diversity associated with spatially distributed users, or simply *cooperative diversity*, in WSNs. In these networks, multiple sensor nodes pool their resources in a distributed manner to enhance the reliability of the transmission link. Specifically, in the context of decentralized detection, cooperation allows sensor nodes to exchange information and to continuously update their local decisions until consensus is reached across the nodes (Swaszek & Willett, 1995; Pados, Halford, Kazakos & P. Papantoni-Kazakos, 1995; Hong, Scaglione & Varshney, 2005; Dyck, 2001; Quek, Dardari & Win, 2006c; Quek, Dardari & Win, 2006b; Quek, Dardari & Win, 2006a). For example, cooperation in decentralized detection can be accomplished via the use of Parley algorithm (Swaszek & Willett, 1995). This algorithm has been shown to converge to a global decision after sufficient number of iterations when certain conditions are met. However, without a fully-connected network and given that the sensor observations are spatially varying, Parley algorithm may result in convergence to a wrong decision at most of the nodes.

In the following a practical example of distributed detection system design is given. We consider a scenario in which sensor observations, conditioned on the alternate hypothesis, are independent but not identically distributed across the sensor nodes. We compare two different fusion architectures, namely, the parallel fusion architecture (PFA) and the cooperative fusion architecture (CFA). For such bandwidth-constrained WSNs, where each sensor node is restricted to send a 1-bit information to the fusion centre, we describe a consensus flooding protocol for CFA and analyze its average energy consumption.

For more details refer to Quek, Dardari & Win, 2007.

### 9.1.1 The sensing model

First, we model the PoI as an isotropic signal source with path-loss factor $\alpha_f$. This model is general and captures PoI, such as leakage of some contaminating chemical in industrial settings, a moving armoured vehicle in a battlefield, or a source of a radioactive material (Haschberger, Bundschuh & Tank, 1996; Hawkes & Nehorai, 2003; Nemzek, Dreicer, Torney & Warnock, 2004; Brennan, Mielke & Torney, 2005). The path-loss factor $\alpha_f$ will depend on the type of signal considered (chemical contamination, sound, radioactive radiation, etc.). Thus, the received signal strength at a distance $d$ away from the PoI is given by

$$P(d) = \frac{P_0}{d^{\alpha_f}}, \tag{9.1}$$

where $P_0$ is the signal strength of the PoI measured at 1 metre from the location of the PoI.

The location of the sensor nodes can be a direct consequence of certain random deployment strategies. For example, sensor nodes may be air-dropped (see Figure 9.1) or launched via artillery in battlefields or unknown environments. Under this scenario, the spatial distribution of the nodes over the region can be modelled by a homogeneous Poisson point process (PPP) with intensity $\rho$. According to (4.1), the probability that there are $n_t$ sensor nodes within region $A$ of size $|A|$ is given by

$$\mathbb{P}\{N_t = n_t\} = \frac{\lambda_t^{n_t} \exp(-\lambda_t)}{n_t!}, \quad n_t \geq 0, \tag{9.2}$$

where $N_t$ is a Poisson r.v. with mean $\lambda_t = \mathbb{E}\{N_t\} = \rho \cdot |A|$. We assume that the sensor observations are independent conditioned on whether the PoI is present or absent. In particular, when conditioned on the presence of the PoI, the sensor observations are not identically distributed across the nodes, that is, the observations at the nodes are spatially varying. In this case, the independent observation at each sensor node after appropriate sampling and processing is given by

$$y_n = \begin{cases} z_n, & \text{when PoI is absent} \\ \sqrt{P(d_n)} + z_n, & \text{when PoI is present,} \end{cases} \tag{9.3}$$

where $n = 1, \ldots, N_t$, $z_n$ is the independent observation noise across the nodes distributed according to a zero-mean Gaussian distribution with variance $\sigma_z^2$, that is, $z_n \sim \mathcal{N}(0, \sigma_z^2)$, and $P(d_n)$ is the received signal strength at the $n$th node with a distance $d_n$ away from the PoI given by (9.1).

Thus, we can formulate the above-mentioned decentralized detection problem as a binary hypothesis testing problem with the following hypotheses:

$$\begin{array}{l} H_0: \text{PoI absent} \\ H_1: \text{PoI present.} \end{array} \tag{9.4}$$

For simplicity, we assume that the PoI is located at the centre of region $A$ when conditioned on $H_1$ so that border effects can be neglected. The FC's task is to decide whether the PoI is present in the WSN based on the information collected from the sensor nodes.

## 9.1.2 Parallel fusion architecture

In the PFA all the nodes make their local decisions independently without cooperating with one another. Since we are considering bandwidth-constrained WSNs, that is, the communication channels between the sensor nodes and the FC are bandwidth-constrained, each sensor is restricted to sending a 1-bit information to the FC.

Consequently, local decisions are quantized as follows:

$$u_n = \begin{cases} -1, & \text{when } \hat{H}(y_n) = H_0 \\ +1, & \text{when } \hat{H}(y_n) = H_1, \end{cases} \tag{9.5}$$

and $\hat{H}(y_n)$ is the decision made at the $n$th node. The detection performance of the $n$th node can be characterized by its corresponding probability of false-alarm and probability of detection, denoted by $P_f^{(n)}$ and $P_d^{(n)}$ respectively. The probability of false-alarm is given by

$$P_f^{(n)} = \mathbb{P}\{y_n \geq \zeta_n | H_0\} = Q\left(\frac{\zeta_n}{\sigma_z}\right), \tag{9.6}$$

where $Q(\cdot)$ denotes the Gaussian $Q$-function and $\zeta_n$ is the local decision threshold of the $n$-th node. The probability of detection at the $n$th node is then given by

$$P_d^{(n)} = \mathbb{P}\{y_n \geq \zeta_n | H_1, P(d_n)\} = P_d(\zeta_n, d_n), \tag{9.7}$$

where $P_d(\zeta, d)$ is defined as follows:

$$P_d(\zeta, d) \triangleq Q\left(\frac{\zeta - \sqrt{P(d)}}{\sigma_z}\right). \tag{9.8}$$

During the data-retrieval period, the FC will trigger the nodes within its activation range by sending a beacon signal. All the nodes that are within this activated region $A$ then send their local decisions to the FC.

Assuming ideal communication channels between nodes and the FC, the optimal LRT in the case of conditionally independent decisions can be performed by taking the weighted sum of the incoming local decisions $u_n$ and comparing it with a threshold as follows (Varshney, 1997)

$$\sum_{n=1}^{N_t} u_n \log\left[\frac{P_d^{(n)}\,(1 - P_f^{(n)})}{P_f^{(n)}\,(1 - P_d^{(n)})}\right] \underset{H_0}{\overset{H_1}{\underset{\leq}{\gtrless}}} \log\left[\frac{\mathbb{P}\{H_0\}}{\mathbb{P}\{H_1\}} \prod_{n=1}^{N_t} \frac{(1 - P_f^{(n)})}{(1 - P_d^{(n)})}\right], \tag{9.9}$$

where $\mathbb{P}\{H_0\}$ and $\mathbb{P}\{H_1\}$ are, respectively, the a priori probabilities of the null and alternate hypotheses which are assumed to be known at the FC. Note that the above expression is no more optimal in the presence of not ideal channel conditions. An extension of (9.9) is presented in Chen, Tong & Varshney, 2006 in the case of full or partial channel state information is available at the FC.

In a more realistic scenario, local decisions $\{u_n\}$ are transmitted over noisy parallel channels to the FC, as shown in Figure 9.2. Without loss of generality, we consider an equivalent discrete-time communication model.

The received signal at the FC from the $n$-th sensor node is given by

$$r_n = \sqrt{aE_b}\,u_n + w_n, \tag{9.10}$$

where $E_b$ is the transmitted energy per bit, and $a$ accounts for the up-link path loss, which is assumed to be identical for all nodes.[1] The channel noise, $w_n$, is modelled as a zero-mean Gaussian r.v. with variance $N_0/2$ and it is assumed to be i.i.d. across the

---

[1] When the FC is located at an altitude significantly higher than the radius of the sensor field, this is a reasonable assumption.

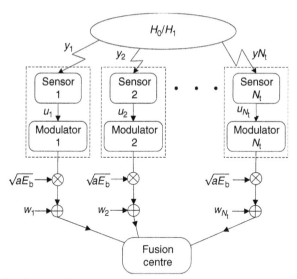

**Figure 9.2** *Parallel fusion architecture*

nodes. We can define SNR $= aE_b/N_0$ as the received SNR from each node at the fusion centre. The goal of the FC is to make a global decision about the hypotheses based on the $N_t$ received observations given by $\mathbf{r} = [r_1, \ldots, r_{N_t}]^T$.

Given $N_t = n_t$, $a$, $\{P_d^{(n)}\}$, and $\{P_f^{(n)}\}$, the optimal fusion rule is given by

$$\Lambda(\mathbf{r}) = \log \left[ \prod_{n=1}^{n_t} \frac{p_{r_n|H_1}(r_n|H_1)}{p_{r_n|H_0}(r_n|H_0)} \right] \underset{H_0}{\overset{H_1}{\underset{\leq}{\gtrless}}} \tau, \tag{9.11}$$

where $p_{r_n|H_1}(r_n|H_1)$ and $p_{r_n|H_0}(r_n|H_0)$ are the p.d.f. of $r_n$ conditioned on $H_1$ and $H_0$, respectively. As pointed out in Chen, Jiang, Kasetkasem & Varshney, 2004; Niu, Chen & Varshney, 2006, the above fusion rule is not easily computable at the FC, particularly for bandwidth-constrained WSNs, since it requires each node to send its $P_d^{(n)}$ and $P_f^{(n)}$ to the FC or the FC needs to know a priori $P_d^{(n)}$ and $P_f^{(n)}$ for all $n$, which directly depends on the locations of all the nodes, the intensity of the PoI, and the local thresholds $\{\zeta_n\}$.

As a result, the FC can only rely on the sub-optimal fusion rule; in particular, we adopt the equal gain combining (EGC) fusion rule given by

$$\Lambda(\mathbf{r}) = \frac{1}{n_t} \sum_{n=1}^{n_t} r_n \underset{H_0}{\overset{H_1}{\underset{\leq}{\gtrless}}} \tau, \tag{9.12}$$

which has been shown to be robust for a wide range of SNR (Chen et al., 2004; Niu et al., 2006). In the following, we consider the scenario in which all nodes use the common local threshold $\zeta$. In this case, $P_f^{(n)} = P_f$ is fixed as a design parameter (generally $P_f \ll 1$) so that the decision threshold $\zeta$ can be evaluated according to (9.6).

Now, using the spatial Poisson distribution of the nodes and (9.8), the number of nodes that can detect the PoI when conditioned on $H_1$, hence with local decision $u_n = +1$, is a Poisson r.v., $N_d$, with mean given by

$$\lambda_d = \mathbb{E}\{N_d\} = \rho \int_A P_d(\zeta, \|\mathbf{x} - \mathbf{x}_{\text{PoI}}\|) d\mathbf{x}, \tag{9.13}$$

where $\mathbf{x}$ and $\mathbf{x}_{\text{PoI}}$ denote the locations of sensor node and PoI, respectively. Conditioned on $N_t$, the ratio $\lambda_d/\lambda_t$ is the average percentage of nodes in region $A$ which successfully detects the PoI. Using $P_d(\zeta, d)$ given by (9.8), the integral in (9.13) can be evaluated numerically. Note that the expression in (9.13) is general and is applicable for general PoI and sensor measurement models, as long as $P_d(\zeta, d)$ is well-defined.

On the other hand, by the spatial Poisson distribution of the nodes, the false-alarmed nodes can be obtained by thinning the original sensor nodes with thinning probability $(1 - P_f)$. Hence, the number of false-alarmed nodes, $N_f$, is also Poisson distributed with mean given by

$$\lambda_f = \mathbb{E}\{N_f\} = \lambda_t P_f. \tag{9.14}$$

Note that $\lambda_f$ is generally smaller than $\lambda_d$ for small $P_f$.

We consider a Bayesian approach, whereby the a priori probabilities of the null and alternate hypotheses, $\mathbb{P}\{H_0\}$ and $\mathbb{P}\{H_1\}$, are known at the FC. Without loss of generality, we assume that the hypotheses are equally likely. The FC employs the EGC fusion rule in (9.12) with threshold $\tau = 0$. Note that the EGC fusion rule treats all the received observations equally, and $\tau = 0$ is a reasonable choice due to our antipodal signal structure of $u_n$. Utilizing the total probability law, we can write the probability of decision error at the FC as

$$P_e = \frac{1}{2} \mathbb{P}\{e|H_1\} + \frac{1}{2} \mathbb{P}\{e|H_0\} \tag{9.15}$$

The detailed analytical derivation of in $P_e$ in (9.15) can be found in Quek, Dardari & Win, 2007.

### 9.1.3   Cooperative fusion architecture

In CFA, the sensor nodes need to disseminate and agree on a common decision throughout the network via a consensus flooding protocol before sending the agreed decision to the FC as shown in Figure 9.3. Similar to (9.10), activated nodes send the agreed decision to the FC via parallel channels.

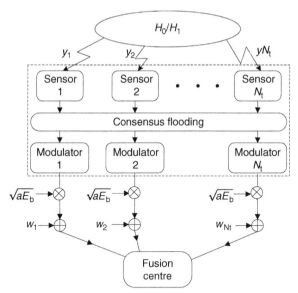

**Figures 9.3** *Cooperative fusion architecture*

## Consensus flooding protocol

When the PoI intensity is weak, Parley algorithm is likely to lead to consensus in the wrong decision since the majority of the nodes in the WSN have rejected $H_1$. The work in Dyck, 2001 attempts to relax the stringent assumptions of conventional Parley algorithm, even though it requires a self-organizing network that is capable of exchanging enormous amounts of information. This greatly limits the applicability of this algorithm in large-scale networks, especially if energy constraint is enforced at each sensor node. As such, in Quek, Dardari & Win, 2007 a consensus flooding protocol that accounts for weak PoI intensity and reduces the possibility of false-alarm flooding is proposed. We define the deliver ratio, $D_r$, as the ratio between the number of nodes that declare $u = +1$ at the end of flooding and the total number of nodes. Specifically, a voting scheme in the flooding protocol is introduced via the use of a threshold $T_h$. Through $T_h$, we can find a good trade-off between a high $D_r$ when PoI is present and a low $D_r$ when PoI is absent. The latter situation can be achieved by minimizing the possibility of false-alarm flooding. The details of the proposed consensus flooding protocol are given as follows:

**S1** The consensus flooding protocol is activated by sending a beacon signal from the FC to the sensor field.
**S2** All activated nodes make a decision based on the measured strength of the PoI intensity.
**S3** Nodes that have declared $\hat{H} = H_1$ (PoI is present) will each send a broadcast packet to neighbouring nodes only once. Each node then initiates a counter with a value of one, sets a fixed assessment delay (FAD), and proceeds to **S5**.

**S4** Nodes that have declared $\hat{H} = H_0$ (PoI is absent) will remain silent and listen to neighbouring nodes. Each node then initiates a counter with a value of zero, sets a FAD, and proceeds to **S5**.

**S5** During FAD, the counter is incremented by one for each received broadcast packet.

**S6** After FAD expires, each node compares its counter with a preset value $T_h$. If the counter is less than $T_h$, set $\hat{H} = H_0$. Otherwise, set $\hat{H} = H_1$.

**S7** If the node has changed its decision and has not broadcasted before, it will proceed to **S3**. Otherwise it will remain silent.

**S8** The consensus flooding protocol is stopped after a certain number of iterations, and all activated nodes send their decisions to the FC.

Note that the above flooding protocol differs from the conventional broadcasting protocols. Conventional broadcasting protocols are designed to maximize the delivery ratio as well as to minimize the redundant retransmissions, regardless of the correctness of the message broadcasted (Tseng, Ni, Chen & Sheu, 2002; Williams & Camp, 2002).[2] Unlike the Parley algorithm (Swaszek & Willett, 1995) or conventional broadcasting protocols (Tseng et al., 2002; Williams & Camp, 2002), this consensus flooding protocol adopts a voting scheme to enable agreement in decisions and to control false-alarm flooding. In addition, only nodes that declare $H_1$ are allowed to broadcast their decisions. When time constraint is not stringent, the FAD value and the number of protocol iterations can be chosen large enough to allow the consensus flooding protocol to terminate correctly. The choice of the threshold $T_h$ in the voting scheme essentially depends on the degree of connection, which is defined as the average number of neighbours, $\lambda_h$, each node can hear. Thus, we can parameterize this consensus flooding protocol by $(T_h, \lambda_h, D_r)$, where the parameters, $T_h$ and $\lambda_h$, are chosen to meet a given $D_r$ when conditioned on $H_1$, and to minimize the possibility of false-alarm flooding on $H_0$.

### 9.1.4 Energy efficiency analysis

The average energy consumed by each node in the PFA to convey a single information bit to the FC at a target $P_e$ is simply given by

$$E_{avg}^{(PFA)} = E_b^{(PFA)}, \tag{9.16}$$

since the PFA does not have any cooperation overhead. To execute the consensus flooding protocol in the CFA, the average number of nodes that send a broadcast packet under $H_1$ is equal to $\lambda_a + \lambda_d \cdot P_{T_h}$, where $\lambda_a = \lambda_t D_r$ is the average number of nodes which sent a broadcast packet during the flooding process and agreed on the correct decision whereas $P_{T_h}$ is the probability a node does not hear a sufficient number of neighbours' nodes (i.e., less than $T_h$) as evaluated in Quek, Dardari & Win, 2007. As a consequence, $\lambda_d \cdot P_{T_h}$ represents the average number of nodes which detected the PoI but did not participate in the

---

[2] In Tseng et al., 2002; Williams & Camp, 2002, the delivery ratio is simply defined as the ratio between the number of nodes that received the broadcasted message over the total number of nodes.

flooding process since they do not hear a sufficient number of neighbours. Combining both the transmission and the flooding energy, the average energy consumed by each node to convey a single information bit to the FC when conditioned on $H_1$ is given by

$$E_{\text{avg}|H_1}^{(CFA)} = E_b^{(CFA)} + \frac{E_{\text{flood}}}{\lambda_t}(\lambda_a + \lambda_d \cdot P_{T_h}), \tag{9.17}$$

where $E_{\text{flood}}$ is the average energy required to broadcast a packet during the cooperative phase. The consensus flooding protocol proposed can ensure that the probability of false-alarm flooding is lower than $P_f$. As a result, the average number of nodes that send a broadcast packet is not larger than $\lambda_f$, and the average energy consumed by each node to convey a single information bit to the FC when conditioned on $H_0$ is given by

$$E_{\text{avg}|H_0}^{(CFA)} = E_b^{(CFA)} + \frac{E_{\text{flood}}\lambda_f}{\lambda_t}. \tag{9.18}$$

Combining (9.17) and (9.18), the total average energy consumed by each node in the CFA to convey a single information bit to the FC at a target $P_e$ is given by

$$
\begin{aligned}
E_{\text{avg}}^{(CFA)} &= \mathbb{P}(H_0)\left[E_b^{(CFA)} + \frac{E_{\text{flood}}\lambda_f}{\lambda_t}\right] \\
&\quad + \mathbb{P}(H_1)\left[E_b^{(CFA)} + \frac{E_{\text{flood}}}{\lambda_t}(\lambda_a + \lambda_d \cdot P_{T_h})\right] \\
&= E_b^{(CFA)}\left\{1 + \frac{\delta}{2\lambda_t}[\lambda_f + (\lambda_a + \lambda_d \cdot P_{T_h})]\right\},
\end{aligned}
\tag{9.19}
$$

where $E_{\text{flood}} = \delta E_b$. In realistic WSNs, the choice of $\delta$ depends mainly on the relationship between the up-link pathloss and the internode wireless links. In general, $\frac{E_{\text{flood}}}{2\lambda_t}[\lambda_f + (\lambda_a + \lambda_d \cdot P_{T_h})]$ accounts for the cooperation overhead since it represents the increase in the average energy consumption per node in order to cooperate.[3]

By using (9.16) and (9.19), we can then compute the average energy gain (in dB) due to cooperation as

$$
\begin{aligned}
\Delta_E &= 10\log\left(\frac{E_{\text{avg}}^{(PFA)}}{E_{\text{avg}}^{(CFA)}}\right) \\
&= 10\log\left(\frac{E_b^{(PFA)}}{E_b^{(CFA)}\left\{1 + \dfrac{\delta}{2\lambda_t}[\lambda_f + (\lambda_a + \lambda_d \cdot P_{T_h})]\right\}}\right).
\end{aligned}
\tag{9.20}
$$

---

[3] Note that we have considered implicitly that the energy required for listening is negligible.

For a target $P_e$, $\text{SNR}^{(\text{PFA})}$ and $\text{SNR}^{(\text{CFA})}$ can be determined by simply inverting the probability of error expressions for PFA and CFA respectively. Substituting these values into (9.20), we obtain

$$\Delta_E = 10 \log \left( \frac{\text{SNR}^{(\text{PFA})}}{\text{SNR}^{(\text{CFA})} \left\{ 1 + \dfrac{\delta}{2\lambda_t} [\lambda_f + (\lambda_a + \lambda_d \cdot P_{T_h})] \right\}} \right). \tag{9.21}$$

From (9.21), we can observe that the gain from cooperation depends on how much energy is spent on local data exchange. This additional required energy explicitly depends on the connectivity of the network through $P_{T_h}$, the delivery ratio of the consensus flooding protocol through $\lambda_a$, the average flooding energy through $\delta$, and the average number of active sensor nodes through $\lambda_t$.

### 9.1.5   Performance comparison between PFA and CFA

The effect of PoI intensity and flooding energy on the energy efficiency of CFA at $P_e = 1 \times 10^{-4}$ and $\lambda_t = 500$ is plotted in Figure 9.4. We can observe that as the PoI intensity increases, the average energy gain due to cooperation decreases since PFA becomes more reliable. However, this average energy gain due to cooperation can go below 0 dB at a certain region of high PoI intensity, since the increase in average energy consumption needed to execute flooding outweighs the gain in energy efficiency resulting from cooperation. To further investigate this phenomenon, we consider different

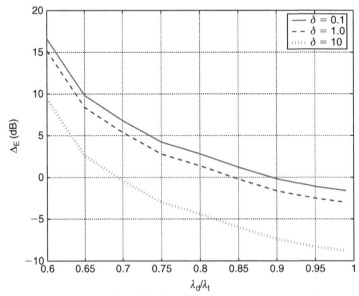

**Figure 9.4**   *Effect of PoI intensity through $\lambda_d/\lambda_t$ and flooding energy on energy efficiency of CFA when $P_e = 1 \times 10^{-4}$ and $\lambda_t = 500$. The flooding protocol parameter set used is $(T_h, \lambda_h, D_r) = (3, 11, 0.9)$*

values of $\delta$ ($\delta = 0.1$, $\delta = 1$ and $\delta = 10$). It can be seen clearly in Figure 9.4 that the average energy gain due to cooperation decreases as $\delta$ increases. Recall that $\delta$ mainly depends on the node transmitter power, bit rate and packet length. This result shows that these parameters have to be carefully designed to make $\delta$ small, in order for CFA to be more energy efficient than PFA. Depending on the PoI intensity, PFA can be more energy efficient than CFA, especially for larger values of $\delta$. For example, when $\delta = 10$, it is ore energy efficient to implement PFA in regions with PoI intensity $\lambda_d/\lambda_t \geqslant 0.7$.

In conclusion, a trade-off exists among spatial diversity gain, average energy consumption, delivery ratio of the consensus flooding protocol, network connectivity, node density, and PoI intensity in CFA. In particular, cooperation among nodes is advantageous in scenarios where the PoI intensity is weak or unknown.

## 9.2 Distributed scalar field estimation

In the following an example of cross-layer WSN design for process estimation, (e.g., as for the spatial distribution of the temperature or pressure in environmental monitoring), is proposed (see, e.g., Figure 9.5). Both a particular DDSP technique, with resampling at the FC, and centralized SP are considered following the hierarchical architecture of the self-organizing WSN, showing their impact in both network lifetime and the quality of process estimation. This approach was presented in Dardari, Conti, Buratti & Verdone, 2007.

### 9.2.1 Sampling the target process

We consider a scenario where nodes are randomly and uniformly placed with spatial density $\rho$. As will be clear later, only a subset of nodes, with density $\rho_0$, participate to the sampling process with positions assumed known. To save energy, nodes are normally in sleeping mode and periodically commute in receiving mode.

**Figure 9.5** *Example of distributed scalar field estimation scenario for environmental monitoring*

The supervisor (sink) wakes up a certain number of nodes by transmitting a sequence of packets (triggering packets) long enough to cover the activity period of each node. Only woken nodes will participate to successive communication phases. We assume energy efficiency is not an issue for the supervisor, which is equipped with more complex capabilities both in terms of signal processing and transmission power, with respect to nodes.

Let us denote as *round* the entire sequence of operations that starts with the supervisor-generated triggering event and brings to the determination of the estimate at the supervisor.

Typical environmental phenomena (e.g., temperature or pressure measurements) are slow time-varying if compared with the packet delivery time in WSNs. For this reason, we consider here a quasi-static scenario, which means that the round time is considered to be much smaller than the change rate of the observed field. In this scenario no stringent time synchronization constraints among nodes are present.

The signal to be sampled is described here through the (target) $l$-dimensional spatial random process $Z(\mathbf{s})$ ($\mathbf{s}$ being the spatial variable) with realizations $z(\mathbf{s})$. We consider the sample space as a finite region $A$ where the process is observed, centred in the supervisor. Without loss of generality, we consider $A$ a circular area with radius $R$. Hence, the actual (truncated) signal of interest is $x(\mathbf{s}) = z(\mathbf{s}) \cdot r_A(\mathbf{s})$, where

$$r_A(s) = \begin{cases} 1 & s \in A \\ 0 & \text{otherwise.} \end{cases} \tag{9.22}$$

The signal $x(\mathbf{s})$ has finite energy $E_0$ and belongs to the random process $X(\mathbf{s})$. The goal is to create an estimate of $x(\mathbf{s})$, which we denote by $\hat{X}(\mathbf{s})$. In Figure 9.6 a scheme of the whole estimation process is shown. We can define the Fourier transform, $S_X(\mathbf{v})$, the auto-correlation function, $R_X(\tau)$, and the energy spectral density, $E_X(\mathbf{v})$, of $x(\mathbf{s})$, as follows

$$S_X(\mathbf{v}) = \Im^{(l)}[x(\mathbf{s})] \tag{9.23}$$

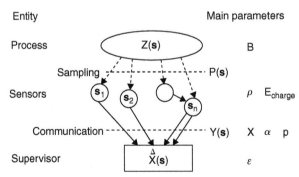

**Figure 9.6**  *Scenario considered and main quantities involved in the process estimation*

$$R_X(\tau) = \int_{\Re^l} x(\mathbf{s})x(\mathbf{s} - \tau)d\mathbf{s} \tag{9.24}$$

$$E_X(\mathbf{v}) = \Im^{(l)}[R_X(\tau)], \tag{9.25}$$

where $\tau = (\tau_1, \tau_2, \ldots \tau_l)$ and $\mathbf{v} = (v_1, v_2, \ldots v_l)$ is a spatial frequency. The operator $\Im^{(l)}[\cdot]$ represents the *l*-dimensional Fourier transform. We assume that $z(\mathbf{s})$ is bandlimited, that is, $S_Z(\mathbf{v}) = \Im^{(l)}[z(\mathbf{s})]$ does not contain significant spectral components outside $S_0$, where $S_0 = \{\mathbf{v} \text{ s.t. } (-B_0 < v_1 < B_0, -B_0 < v_2 < B_0, \ldots, -B_0 < v_l < B_0)\}$ and $B_0$ represents the bandwidth per dimension of $z(\mathbf{s})$. The Fourier transform of $x(\mathbf{s})$ is then

$$S_X(\mathbf{v}) = S_Z(\mathbf{v}) \otimes R_A(\mathbf{v}), \tag{9.26}$$

where $R_A(\mathbf{v}) = \Im^{(l)}[r_A(\mathbf{s})]$ and $\otimes$ is the convolution operator. In the practical bidimensional case, it is $l = 2$ and

$$R_A(\mathbf{v}) = \frac{R}{\|\mathbf{v}\|} J_1(2\pi R \|\mathbf{v}\|), \tag{9.27}$$

where $J_1(\cdot)$ is the Bessel function of the first kind of order one and $\|\cdot\|$ is the norm operator. Note that, due to the spatial truncation of the original signal, $x(\mathbf{s})$ is not bandlimited. However, it can be easily verified that $R_A(v) \approx 0$ when $\|v\| > \frac{1}{\pi R}$.

In general, $S_X(\mathbf{v})$, and hence $E_X(\mathbf{v})$, are $\approx 0$ outside $S$, where $S = \{\mathbf{v} \text{ s.t. } (-B < v_1 < B, -B < v_2 < B, \ldots, -B < v_l < B)\}$ and, in the two-dimensional case, $B = \frac{1}{\pi R} + B_0$.

The dimension of $S$ is $\dim(S) = \beta$, where $\beta = (2B)^l$ represents the minimum Nyquist sampling rate in the case of uniform sampling (Gardner, 1990; Masry, 1978). In practical applications we have $\frac{1}{\pi R} \le B_0$, that is, the area of observation is chosen larger than the typical process correlation distance which is proportional to $1/B_0$: the worst case corresponding to the largest bandwidth expansion is obtained when $\frac{1}{\pi R} = B_0$.

The *n*th node is located in the spatial point $\mathbf{s}_n$ and takes the sample $x(\mathbf{s}_n)$.[4] Considering the sensors randomly placed with a spatial density $\rho_0$ in the monitored environment, we can statistically describe the sequence $\{\mathbf{s}_n\}$ of spatial samples as a homogeneous PPP. The derivative of the corresponding counting process is the stationary random process $H(\mathbf{s}) = \sum_n \delta(\mathbf{s} - \mathbf{s}_n)$, having mean $\mu_H = \mathbb{E}\{H(\mathbf{s})\} = \rho_0$, and whose statistical autocorrelation function and power spectral density are given by Gardner, 1990; Marvasti, 1986

$$R_H(\tau) = \rho_0 \cdot \delta(\tau) + \rho_0^2, \tag{9.28}$$

$$S_H(\mathbf{v}) = \rho_0 + \rho_0^2 \cdot \delta(\mathbf{v}), \tag{9.29}$$

respectively, where $\delta(\cdot)$ is the Dirac pseudo function, $H(\mathbf{s})$ represents the sampling process.

---

[4] We neglect quantization errors.

## 9.2.2  Building the process estimate

Starting from the collected samples, an estimate of the target process can be determined through either a centralized or a distributed procedure: in the former case, nodes, via a self-organizing communication protocol, transmit their samples to the supervisor, which is in charge of the signal processing and the estimation (no DDSP option); in the latter, they send the samples to some elected nodes (cluster head, CH) in a clustered architecture, that perform suitable distributed signal processing and transmit the estimate to the supervisor which, by collecting different estimates, provides the final result as explained in the following (DDSP option).

In both cases, owing to communication failure, there exists a probability $p$ that a node is unable to send its information to the entity performing signal processing. In this case, the corresponding sample does not contribute to the signal estimation (sample loss). We denote by $q = 1 - p$ the probability of correct sample reception.

The set of samples received by the entity performing signal processing forms a new stationary sampling process, $P(\mathbf{s})$. Using the result derived in the Appendix of Dardari, Conti, Buratti & Verdone, 2007 (with $P_1(\mathbf{s}) = H(\mathbf{s})$), and expressions (9.28), (9.29), the statistical autocorrelation function and mean of $P(\mathbf{s})$ are $R_P(\tau) = \delta(\tau)\rho_0 q + q^2 \rho_0^2$ and $\mu_P = \mathbb{E}\{P(\mathbf{s})\} = q \cdot \rho_0$, respectively. As expected, the new process has the same characteristics of the original one with density $q \cdot \rho_0$ and power spectral density

$$S_P(\mathbf{v}) = q\rho_0 + q^2\rho_0^2 \cdot \delta(\mathbf{v}). \tag{9.30}$$

The sampled version, $Y(\mathbf{s})$, of the target signal conditioned to the realization, $x(\mathbf{s})$, can be expressed as $Y(\mathbf{s}) = x(\mathbf{s}) \cdot P(\mathbf{s})$, representing a finite energy non-stationary random process. The autocorrelation function of the generic process realization $y(\mathbf{s})$ is

$$R_y(\tau) = \int_{\mathfrak{R}^l} x(\mathbf{s}) \, x(\mathbf{s} - \tau) \, p(\mathbf{s}) \, p(\mathbf{s} - \tau) ds, \tag{9.31}$$

where signal $p(\mathbf{s})$ is a realization of the random process $P(\mathbf{s})$.

Let us define the average statistical autocorrelation function of $Y(\mathbf{s})$ as

$$\begin{aligned}
\bar{R}_Y(\tau) &= \mathbb{E}\left\{ \int_{\mathfrak{R}^l} x(\mathbf{s}) \, x(\mathbf{s} - \tau) \, P(\mathbf{s}) \, P(\mathbf{s} - \tau) ds \right\} \\
&= \int_{\mathfrak{R}^l} x(\mathbf{s}) \, x(\mathbf{s} - \tau) \, R_P(\tau) ds \\
&= R_P(\tau) \cdot R_X(\tau),
\end{aligned} \tag{9.32}$$

and the average energy spectral density

$$\bar{E}_Y(\mathbf{v}) = \mathfrak{I}^{(l)}[\bar{R}_Y(\tau)] = E_X(\mathbf{v}) \otimes S_P(\mathbf{v}). \tag{9.33}$$

From (9.30), (9.32) and (9.33) it follows

$$\bar{E}_Y(\mathbf{v}) = q^2 \rho_0^2 E_X(\mathbf{v}) + E_0 \cdot \rho_0 \cdot q. \tag{9.34}$$

We consider that the estimated signal is obtained through linear interpolation of the received set of samples $Y(\mathbf{s})$. The estimate $\hat{X}(\mathbf{s})$ can then be expressed as

$$\hat{X}(\mathbf{s}) = \phi(\mathbf{s}) \otimes Y(\mathbf{s}), \tag{9.35}$$

where $\phi(\mathbf{s})$ is the impulse response of the linear interpolator, whose transfer function is $\Phi(v) = \mathfrak{F}^{(l)}[\phi(\mathbf{s})]$.

In the following, let us consider the case of an ideal low-pass interpolator with transfer function

$$\Phi(\mathbf{v}) = \begin{cases} 1/\mu_\phi & \mathbf{v} \in \mathcal{S}^* \\ 0 & \text{otherwise,} \end{cases} \tag{9.36}$$

with $\mu\phi$ and $\mathcal{S} \subseteq \mathcal{S}^*$ described later.

### 9.2.3 Mathematical derivation of the estimation error

A good indicator of the estimate quality is the average normalized estimation error defined as the normalized MSE

$$\varepsilon = \frac{1}{E_0} \mathbb{E}\left\{ \int_{\Re^l} (\hat{X}(\mathbf{s}) - x(\mathbf{s}))^2 \, d\mathbf{s} \right\} \tag{9.37}$$

In the Appendix of Dardari, Conti, Buratti & Verdone, 2007, the expression for (9.37) is derived as a function of the spectral densities $\bar{E}_Y(\mathbf{v})$ and $E_X(\mathbf{v})$. When the ideal low-pass interpolator (9.36) is adopted, it results in

$$\varepsilon = 1 - 2\frac{\mu_P}{\mu_\phi} + \frac{1}{E_0 \mu_\phi^2} \int_{\mathcal{S}^*} \bar{E}_Y(\mathbf{v}) d\mathbf{v}. \tag{9.38}$$

By substituting (9.34) in (9.38) and considering that $\int_{\mathcal{S}^*} E_X(\mathbf{v}) d\mathbf{v} = E_0$, we can write

$$\varepsilon = 1 - 2\frac{q\rho_0}{\mu_\phi} + \frac{q\rho_0}{\mu_\phi^2}[q\rho_0 + \dim(\mathcal{S}^*)]. \tag{9.39}$$

It is worthwhile noting that, due to the characteristics of Poisson sampling, no aliasing effects arise whatever reconstruction bandwidth $\mathcal{S}^*$ is chosen, unlike in the classical uniform sampling case. The only effect is an increase of the MSE. Hence, $\mathcal{S}^*$ can be different from $\mathcal{S}$ depending on the application (see later when DDSP is adopted).

Now we find the optimum value of $\mu_\phi$, in order to minimize the estimation error $\varepsilon$; let us constrain to zero the derivative in (9.39), thus[5]

$$\frac{d\varepsilon}{d\mu_\phi} = 0 \quad \Leftrightarrow \quad \mu_\phi = q\rho_0 + \dim(\mathcal{S}*). \tag{9.40}$$

Now we can compute the value of the MSE given by

$$\varepsilon = 1 - \frac{q\rho_0}{q\rho_0 + \dim(\mathcal{S}*)} = \frac{\dim(\mathcal{S}*)}{q\rho_0 + \dim(\mathcal{S}*)} = \frac{\beta\zeta}{q\rho_0 + \beta\zeta}, \tag{9.41}$$

where $\zeta = \dim(\mathcal{S}*)/\dim(\mathcal{S})$.

Hence, proper discussion about the meaning of this solution must be given:

- When nodes are deployed, $\rho_0$ is known; however, during network life $\rho_0$ decreases owing to the fact that some nodes expire, then an estimation of $\rho_0$ is needed.
- Knowledge of $q$ requires knowledge of the sample loss probability, which changes during network life, and has to be estimated.
- Knowledge of $\dim(\mathcal{S}*)$ requires knowledge of the dimension of the sampling space and this requires information about the process to be estimated.

For the sake of comparison, we also consider two other sub-optimal cases, which are less complex, in terms of a priori knowledge about $q$ and $\dim(\mathcal{S})$ required:

$$\mu_\phi = q\rho_0 \quad \Rightarrow \quad \varepsilon = \frac{\beta\zeta}{q\rho_0} \tag{9.42}$$

$$\mu_\phi = \rho_0 \quad \Rightarrow \quad \varepsilon = (1-q)^2 + \frac{q\beta\zeta}{\rho_0} \tag{9.43}$$

It can be easily shown that results obtained through (9.42) are better than (9.43). But even if the solution given by (9.41) and (9.42) are better than (9.43), in some cases they could not be conveniently realized, due to the need for an estimation of $q$ and $\dim(\mathcal{S}*)$.

### Absence of DDSP

When DDSP is not adopted, all samples successfully received are processed by the supervisor, in order to determine the estimate $\hat{X}(\mathbf{s})$. Each sample has a probability equal to $p$ to be missing because of an unconnected node, or owing to MAC failures.

It can be shown that without DDSP the best performance is obtained by fixing $\mathcal{S}* = \mathcal{S}$, hence $\zeta = 1$. The ratio $\eta \triangleq \rho_0/\beta$ represents the oversampling factor with respect to the

---

[5] With the second derivative it is possible to check that this value for $\mu_\phi$ represents a minimum.

minimum Nyquist uniform sampling rate $\beta$. In general, we have $\eta \geqslant 1$. Expressions (9.41) to (9.43) give the estimation error as a function of the sample loss probability and $\eta$. As will can be noted, the impact of node communication failure, through the probability $p$, becomes more relevant as $\eta$ increases, the latter being strictly related to sensor density. Apart from the value of dim($\mathcal{S}$), this result does not depend on the particular realization $z(\mathbf{s})$ of the random process $Z(\mathbf{s})$. Hence, it can be extended to the whole random process. Expressions (9.41) to (9.43) can be used to measure the quality of the estimation of the random process, under observation performed at the supervisor, when no DDSP is implemented.

## Presence of DDSP

In order to reduce the overall energy consumption due to the transmission of samples, it would be useful to partially decentralize the signal processing task necessary to have an estimate of the target process. In particular, considering a clustered network architecture, samples coming from nodes are not directly collected by the supervisor but they reach the final destination through intermediate nodes (the CHs), which perform partial signal processing. At each CH, loss-less data compression techniques can be adopted thus reducing the amount of data transmitted. Typical compression techniques take advantage of the correlation among adjacent samples. In general, the compression rate depends on the process statistics, spatial correlation and the number of samples processed. A general characterization appears to be prohibitive. An interesting survey about distributed compression in sensor networks is presented in Pradham, Kusuma & Ramchandran, 2002.

Here we will focus, instead, on the possibility to compress data considering the fact that samples come from a random sampling process and making use of a uniform resampling processing at the Nyquist frequency.

Let us assume that a cluster of area $A_{ch}$ containing $n_p$ nodes ($n_p$ having mean $N_p$) is managed by a CH which is responsible to collect samples and to retransmit them to the supervisor. We assume CHs can estimate a portion $x_{ch}(\mathbf{s}) = z(\mathbf{s}) \cdot r_{A_{ch}}(\mathbf{s})$ of the target signal in the area $A_{ch}$ based on the $n_p$ samples received. The function $r_{A_{ch}}(\mathbf{s})$ is defined similarly to (9.22) considering the area $A_{ch}$, with radius $R_{ch}$, instead of $A$. The truncated signal $x_{ch}(\mathbf{s})$, managed by the CH, is characterized by an increased bandwidth with respect to the original signal $x(\mathbf{s})$. According to (9.26) and (9.27), having substituted $R$ with $R_{ch}$, the bandwidth per dimension of $x_{ch}(\mathbf{s})$ is now $B_{ddsp} = B_0 + \frac{1}{\pi R_{ch}}$. Indeed, to keep negligible the aliasing error due to truncation, the reconstruction filter bandwidth and the resampling frequency must be suitably increased, that is, dim($\mathcal{S}^*$) $= (2B_{ddsp})^l$. Considering that $A/A_{ch} \approx N_{ch}$, $l = 2$ and $R_{ch} \approx \sqrt{N_{ch}}\,R$, we have

$$\zeta = \frac{\dim(\mathcal{S}^*)}{\dim(\mathcal{S})} = (1 + \sqrt{N_{ch}})^2 \approx N_{ch}, \tag{9.44}$$

where $N_{ch}$ is the average number of CHs in the sampling area $A$ and is considered to be much larger than one. In deriving (9.44) we considered the worst case where $\frac{1}{\pi R} = B_0$.

The CH then makes a resampling of the estimated signal at the Nyquist frequency, now $\beta \cdot \zeta$, and transmits the new set of Nyquist samples to the supervisor which collects all the estimated portions of the original signal. Considering that the original set of samples comes from a node density $\rho_0$, the average number $M$ of samples composing the new set to be transmitted is

$$M = \frac{[N_p \cdot (1 - p_{\text{MAC}}) + 1]\beta\zeta}{\rho_0} = \frac{[N_p \cdot (1 - p_{\text{MAC}}) + 1]}{\delta}, \qquad (9.45)$$

where $[N_p \cdot (1 - p_{\text{MAC}}) + 1]$ is the average number of samples received by the CH plus the one generated by itself, taking the possibility that a packet is lost owing to the MAC protocol into consideration, through the loss probability $p_{\text{MAC}}$. The ratio $\delta = \eta/\zeta$ represents the signal compression factor due to resampling of the estimated signal at the Nyquist frequency.

When $\delta \gg 1$ then $M \ll N_p$, thus a drastic reduction of the transmission throughput (and energy cost) is expected. The augmented energy consumption due to signal processing at the CH in DDSP mode will be also taken into account.

We consider no transmission errors between the CHs and the supervisor, because, as stated in the following section, we assume that during a single round the wireless channel is time-invariant and thus all CHs self-elected in a specific round can hear the supervisor, otherwise they would not have been triggered. Moreover, as far as the access to the channel is concerned, a polling procedure is performed by the supervisor, so that collisions between CHs may not occur; hence, the total estimation error at the supervisor can be still evaluated through (9.41) to (9.43) by putting $\zeta = N_{\text{ch}}$ according to (9.44). As a consequence, with respect to the no DDSP case, we expect larger values of estimation error, but, as will be shown in the numerical results, a longer network lifetime, due to reduced amount of transmitted data.

### 9.2.4 The self-organizing distributed WSN

It has been previously shown that the evaluation of the MSE requires the knowledge of some network parameters such as $p$, and $N_{\text{ch}}$, that are strictly related to the way the information is delivered. In this section a typical WSN clustered architecture is analytically investigated to derive these parameters, without any aim at considering complex protocol strategies. Moreover, those parameters also depend on the propagation characteristics of the environment, suitably modelled in the following.

### 9.2.5 Physical connectivity between nodes

At the physical level, communications between nodes (either nodes or the supervisor) is allowed through transceiver devices which are assumed to have equal receiver sensitivity, $P_{r_{\min}}$. We adopt the statistical log-normal shadowing channel model (3.11) discussed in Chapter 3. Let us denote the transmitted power as $P_t$. As far as packet capture is considered, according to a simple threshold model, we assume that packets are correctly detected if received power, $P_r = P_t - L$, is larger than the receiver sensitivity $P_{r_{min}}$, otherwise they are discarded[6]. As a consequence, $L_m = P_t - P_{r_{min}}$ is the maximum allowable pathloss for connection. We consider a quasi-static scenario where the round time is much smaller than the channel coherence time. In addition, since WSNs typically adopt a low duty cycle (that means that the time between consecutive rounds is much larger than the round duration), it is reasonable to consider the channel time-invariant during a round and uncorrelated channel between rounds.

According to the channel model given by (3.11), the probability $C(d)$ that two nodes transmitting with power $P_t$ at distance $d$ are directly connected is given by (4.7) by substituting $L_{th}$ with $L_m$.

### 9.2.6  Information routing through a clustered architecture

A clustered architecture seems particulary suitable for DDSP (Van Dick, Miller & Gaithersburg, 2001). Typically, most of cluster-based algorithms consider a distributed CH self-election algorithm, such as LEACH-based algorithms investigated in De Pedri, Zanella & Verdone, 2003; Heinzelman, Chandrakasan & Balakrishnan, 2002. See Chapter 7 for more details on clustered protocols.

In a clustered architecture, the nodes triggered by the supervisor organize themselves into clusters, with one node per cluster acting as CH. Non-CH nodes transmit packets to their CHs and then CH nodes transmit the packets received plus the one generated by themselves to the supervisor via a direct link (i.e., a three-level hierarchical organization as in Figure 9.7). Therefore, being a CH is much more energy intensive than being a non-CH node. If the CHs were chosen a priori and fixed throughout the network lifetime,

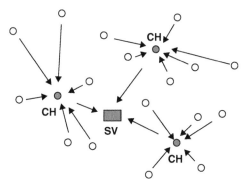

**Figure 9.7**  *Transmission flow with the clustered algorithm. Rectangle: supervisor; filled circle: CHs; circle: non-CH nodes*

---

[6]As usual, all powers and losses are expressed in logarithmic units.

these nodes would quickly use up their limited energy: once the CH runs out of energy, it is no longer operational. Thus, we assume the algorithm incorporates a randomized rotation of the CH role among the nodes in the network: at each round a node autonomously decides to elect itself CH with probability $x$. Decisions taken in different rounds are uncorrelated. In this way, the energy load of being a CH is evenly distributed among nodes during network lifetime.

## 9.2.7  Communication protocol

The simple communication protocol considered is based on the following steps, performed at each round. As far as MAC is concerned (described in the next subsection), since we assume the shadowing sample varies in different rounds, the natural choice brings to a contention-based protocol which minimizes control overhead with respect to a centralized algorithm (Buratti, Giorgetti & Verdone, 2005).

The communication protocol steps are:

*Trigger and wake up*
The supervisor transmits the triggering sequence with power $P_t = P_{su}$; a random number, $n_t$, of nodes is triggered (those receiving a power $P_r = P_{su} - L \geq P_{r_{min}}$ or, equivalently, experiencing a pathloss smaller than $L_{su} = P_{su} - P_{r_{min}}$).

As discussed in Section 4.4, by neglecting the border effects, the number $n_t$ of participating nodes is Poisson distributed with mean $N_t$ given by

$$N_t = \mathbb{E}\{n_t\} = \pi \rho e^{2(\sigma^2/k_1^2 - k_0/k_1 + L_{su}/k_1)}, \tag{9.46}$$

which follows directly from (4.3) and (4.15), where $\rho$ is the initial nodes' density. The probability that a node situated at distance $d$ from the supervisor is triggered is given by (4.7) with $L_{th} = L_{su}$. It must be noted that this probability is a decreasing function of the distance; that produces two side-effects. First, the woken node density, $\rho_w(d) = \rho C(d)$, is distance-dependent, so the process is not uniformly sampled. Second, nodes closer to the supervisor participate more frequently to the transmission than far nodes and they tend to discharge their batteries prematurely.

To overcome this situation, the protocol requires that woken nodes randomly decide whether to participate to the following phases: we assume they will, with probability $w(d) = w_0/C(d)$, otherwise they switch to sleeping mode. The constant $w_0$ defines the sampling space $A$ that will provide contributions to the target process estimation. In fact, the maximum distance $R$ will be obtained by resolving $w_0/C(R) = 1$. Moreover, it is assumed that all nodes are aware of their own position and that the supervisor, when it triggers nodes, informs them about its position and the propagation parameters characterizing the environment, that could be estimated, for example, from the RSS measurements, so that each node can compute $C(d)$.

The density of participating nodes results in

$$\rho(d) = \rho_w(d) \, w(d) = \rho w_0 = \rho_0,$$ (9.47)

which is constant for $d < R$, as desired.

*Self-election*

The participating nodes initiate the self-election phase: each of them elects itself CH with probability equal to $x$, where $x$ is a system parameter to be optimized. The number of CHs is $n_{ch}$, with mean $N_{ch} = x \cdot N_t \cdot w_0$. The correspondent CH and non-CH densities are given by $\rho_{ch} = \rho_0 \cdot x$, and $\rho_{nch} = \rho_0 \cdot (1 - x)$, respectively. To notify its election both to supervisor and surrounding nodes, each CH transmits a broadcast packet with power $P_t = P_{ch} = P_{su}$. CHs will access the channel through a contention-based mechanism, but the number of CHs is in general not very large ($x$ is usually much less than one) and collisions can be easily avoided through standard techniques. It is also assumed that the supervisor sends acknowledgments to the CHs, indicating the radio channel to be used in the following phases.

We will consider negligible the packet loss in this phase.

*Cluster selection*

Each node receiving the packet(s) sent by the CHs decides which CH to refer to (i.e., which cluster to subscribe to), based on the packet received with largest power. We assume that non-CH nodes transmit with power $P_t = P_{nch} = P_{su} \cdot \alpha$, where $\alpha \leqslant 1$ is a system parameter to be optimized. In this case the maximum tolerable pathloss becomes $L_p = L_{su} + 10 \log_{10}\alpha$. Hence, the number of CHs reachable by a generic node and the number of nodes forming a given cluster are random variables denoted as $n_{rch}$ and $n_p$, respectively; their means are $N_{rch}$ and $N_p$.

By assuming that the CHs are uniformly and randomly distributed over the infinite plane, with density $\rho_{ch}$, the number $n_{rch}$ of CHs reachable by a generic node is also Poisson distributed, with mean

$$N_{rch} = \mathbb{E}\{n_{ach}\} = \pi \rho_{ch} e^{2(\sigma^2/k_1^2 - k_0/k_1 + L_p/k_1)}.$$ (9.48)

Though the CHs can be assumed to be uniformly located owing to the distributed self-election strategy, they are not distributed over the infinite plane, as they belong to the finite set of nodes that have been triggered by the supervisor. Therefore, the above expression represents an approximation whose validity decreases for nodes farther from the supervisor. However, the approximation seems reasonable as it will be shown by simulation in the numerical results.

The number $n_p$ of nodes that subscribe to a specific CH (i.e., the number of nodes composing a cluster), can be considered as Poisson distributed with mean (Orriss & Barton, 2003)

$$N_p = \mathbb{E}\{n_p\} = \pi \rho_{nch} e^{2(\sigma^2/k_1^2 - k_0/k_1 + L_p/k_1)} \cdot \frac{1 - e^{-N_{rch}}}{N_{rch}},$$ (9.49)

where $N_{rch}$ is given by (9.48).

*Sample transmission*
Nodes generate the packet containing the samples and transmit it with power $P_t = P_{nch}$; having assumed that all sensors (both CHs and non-CHs) have the same receiver sensitivity, and that power loss is constant during one round, this packet will be correctly received by the relevant CH due to channel reciprocity, unless interference between packets takes place. The contention between nodes must be managed by the MAC protocol (see below). Intercluster collisions are avoided as the CHs use separate radio channels assigned by the supervisor during the self-election notify phase.

*Transmission to the supervisor without DDSP*
If DDSP is not implemented, then the CHs transmit the sequence of $n_p$ samples received plus the one generated by itself; under the assumptions of this paper, this packet will be correctly received by the supervisor, which will receive on average $N_s = (N_p + 1) \cdot N_{ch}$ samples, if no packet collisions occur; they can be avoided if the supervisor requires packets transmission through a polling scheme. In fact, the supervisor knows the identity of the participating CHs from the previous self-election notify phase.

*Transmission to the supervisor with DDSP*
If DDSP is implemented, then the CHs perform suitable signal processing and transmit the new $m$ samples (with mean $M$) of an estimated version of the target process; under the same assumptions as for no DDSP (see subsection above), this packet will be correctly received by the supervisor.

The probability that a node, triggered by the supervisor, is not able to connect to any CH, is denoted as $p_1$. This isolated node will not belong to any cluster, and will never send its sample. The probability of isolated node can be evaluated recalling the Poisson nature of $n_{ach}$ and (9.48)

$$p_1 = \mathrm{Prob}\{n_{rch} = 0\} = e^{-N_{rch}}. \qquad (9.50)$$

We will consider the network to be dense if $p_1$ assumes small values, that is, the network is essentially connected, which corresponds to high values of $\rho$.

## 9.2.8  Medium access control

In general, having fixed a certain MAC protocol, the information needed to evaluate the impact of such protocol on system performance, in terms of energy efficiency and sample loss probability, is the probability $p_{MAC}$ that a packet is lost due to interference problems; it is a function of the average number of transmissions per packet, $R(N)$, which depends on the specific MAC protocol adopted, of protocol parameters/contraints and of the average number of nodes $N$ competing for transmission. An exaustive evaluation of this function for different protocols is out of the scope: in the following an example will be given for a simple case.

Once $p_{MAC}$ is known, it is possible to derive the probability $p$, that a sample in the sample space is lost due to communication. In fact, the latter event occurs if the non-CH node is isolated or the packet is lost due to MAC; the probability of sample loss is thus

$$p_1 = (1 - x)(p_1 + (1 - p_1) p_{MAC}(N_p)), \qquad (9.51)$$

where the second factor accounts for MAC losses during sample transmission (we remind you that $N_p$ is the average number of nodes aggregated to each CH).

We give here an example of the evaluation of $p_{MAC}$ $(N)$ for a simple slotted random channel access protocol without retransmissions, that is, with $R(N) = 1$. The time is divided in frames and each frame is further subdivided in $Z$ slots. We have $n$ nodes, where $n$ is Poisson distributed with mean $N = \mathbb{E}\{n\}$. Each node transmits a packet randomly choosing one of the $Z$ available slots in the frame. To reduce collisions, we choose a value for $Z$ high enough with respect to $N$; moreover, since $N$ is not constant, but depends on $\rho$, $x$ and $\alpha$, it could be reasonable to set $Z$ to be dependent on $\rho$. In fact, if $Z$ is constant it might occur that for high values of $\rho$, $Z$ is too low with respect to $N$ and so there are too many collisions. To avoid this problem, we fix $Z = c \cdot N$ where $c$ is a parameter that must be suitably set.

The probability of packet loss is the probability that two or more nodes select the same slot and collide. This probability, recalling that the number of nodes, $n$, is Poisson distributed with mean $N$, is given by

$$p_{MAC}(N) = \sum_{n=2}^{\infty} \left[ 1 - \left(1 - \frac{1}{Z}\right)^{n-1} \right] \text{Prob}(n) = 1 + \frac{e^{-N} - Ze^{-N/Z}}{Z - 1} \qquad (9.52)$$

$$\leq 1 - e^{-N/Z} = 1 - e^{-\frac{1}{c}}. \qquad (9.53)$$

The upper-bound reported in the right-hand side in (9.52) is tight for $N > 10$ and shows that $p_{MAC}(N)$ mainly depends on the ratio $c = Z/N$. For example, collision probabilities below 10% require $c = 10$, which is a reasonable value that will be adopted for providing numerical results.

Note that other MAC protocols used in WSN (Ye, Heidemann & Estrin, 2002; Woo & Culler, 2001; Sohrabi, Gao, Ailawadhi & Pottie, 2000; Verdone, 2004) can be considered by properly modifying (9.52) and $R(N)$.

## 9.2.9 Energy budget

Now, let us derive the mean energy consumption of each node during one round. This is a focal point in the WSN design because it is directly connected to node lifetime. By means of the previously defined probability of CH election $x$, we obtain the mean energy consumption per round with and without DDSP. In particular, we have

$$E_{round} = w_0 \cdot [E_{non\text{-}CH}(1 - x)(1 - p_1) + xE_{CH}], \qquad (9.54)$$

where the first term represents the consumption if the node is non-CH (multiplied by the correspondent probability) and the second term the consumption for a CH. The factor $w_0$

accounts for the nodes that do not participate in the round. By considering that each non-CH consumes energy when transmitting the data packet to its CH once or more times depending on the collisions, we have

$$E_{\text{non-CH}} = E_L\, R(N_p)\, \theta_d. \tag{9.55}$$

where

- $E_L = E_H \cdot \alpha$ and $E_H$ is the energy spent to transmit a bit at power $P_{\text{su}}$
- $\theta_d$ is the size (in bits) of the data packet.

The total average energy spent per round by a CH is given by

$$E_{\text{CH}} = E_H \cdot \theta_c + (N_p(1 - p_{\text{MAC}}(N_p)) + 1)\, E_H \cdot \theta_d + f(R(N_p)), \tag{9.56}$$

for the no DDSP case and

$$E_{\text{CH}}^{(\text{ddsp})} = E_H \cdot \theta_c + M E_H \theta_d + E_{\text{ddsp}} + f(R(N_p)), \tag{9.57}$$

when DDSP is adopted.

The first term of (9.56) and (9.57) refers to the energy spent by a CH to transmit the broadcast packet to inform all other nodes and the supervisor of its role; $\theta_c$ is the size (in bits) of the broadcast packet. The second term is related to the energy consumed for the data transmission to the supervisor: in the no DDSP case the average number of packets sent by a CH is the average number of packets correctly received by its no-CHs, that is, $N_p(1 - p_{\text{MAC}}(N_p))$, plus the one generated by the CH itself; whereas in the DDSP case each CH has to transmit $M$ data packets on average (see Section 9.2.3). The generic function $f(r)$ represents the energy spent by CHs to transmit $r$ average retransmission requests. Finally, $E_{\text{ddsp}}$ quantifies the energy spent by the CH to reconstruct and resample the portion of process sensed by nodes within the cluster if DDSP is adopted. We assume this portion of energy is proportional to the number of samples processed and is given by

$$E_{\text{ddsp}} = E_H \cdot \theta_d\, \gamma\, (N_p(1 - p_{\text{MAC}}(N_p)) + 1), \tag{9.58}$$

where $\gamma$ is a parameter to be defined according to the circuital characteristics of the node; it represents the ratio between the average energy needed to process one sample and the energy required for the sample transmission. The smaller is $\gamma$, the more advantageous the DDSP strategy. According to the literature (see, e.g., Zhao, Liu, Liu, Guibas & Reich, 2003), the ratio of energy consumption for processing and communication of one bit is in the range of 0.001–0.0001. In the case of adoption of other compression techniques, (9.58) has to be modified.

Now, for a given initial battery charge, $E_{charge}$, the mean number of rounds achievable during the life of each node is given by

$$N_{round} = \frac{E_{charge}}{E_{round}}.$$ (9.59)

Note that the energy spent to receive packets is not taken into consideration, assuming that it is negligible with respect to the one used for the packet transmission; this condition occurs in some cases (Zhao et al., 2003).

## 9.2.10 Cross-layer design

Here, numerical results related to the mathematical framework proposed will be provided to highlight the interdependency of several network design issues when the performance is investigated, both in terms of process estimation quality and network lifetime. In particular, the performance is affected by a large set of parameters related to different aspects such as

- propagation ($k_0$ and $k_1$ for the path-loss model, $\sigma$ for shadowing)
- the spatial characteristics of the target process ($\beta$)
- system choices (the density of nodes $\rho$, the maximum loss $L_{su}$ for high power transmitting nodes, the initial battery energy $E_{charge}$ for each node, the energy $E_H$ consumed to transmit a bit in high power transmission mode, the ratio $\gamma$ describing energy consumption for single sample elaboration in DDSP mode, the parameter $\alpha$ characterizing energy consumption in low power transmission mode)
- the transmission protocol (the probability $x$ to become a CH).

In the following, results are given as a function of $\rho$, $x$ and $\alpha$; the other parameter settings, if not otherwise specified, are reported in Table 9.1.

In particular, the constant $w_0 = C(R)$ is chosen to be $\approx 0.5$ which corresponds to a radius $R = 1500$ m. Within this circular area, nodes participating to the following phases result in being uniformly distributed.

Mathematical results on $N_p$, $p$, $N_{round}$, and $\varepsilon$, which are based on some approximations neglecting border effects, have also been validated through Monte Carlo simulations. The difference between simulations and the mathematical model is that in the first case the reference scenario is constituted by a circular area, having radius $R = 1500$ m, whereas the second refers to an infinite plane; in both cases nodes are uniformly distributed over the area. However, in the latter case the real scenario considered is limited by the transmission range of the supervisor; therefore, simulations are used to validate such an approximate approach. Due to the clustered architecture of the WSN, the mean cluster size, $N_p$, and the sample loss probability, $p$, play an important role on the overall network

**Table 9.1**  *Values adopted for propagation, process, system and project parameters if not otherwise specified*

| | |
|---|---|
| $k_0$ (dB) | 25.1 |
| $k_1$ | 13.03 |
| $\sigma$ | 4 |
| $\beta$ (m$^{-2}$) | $4 \cdot 10^{-6}$ |
| $L_{su}$ (dB) | 120 |
| $E_{charge}$ (J) | 1 |
| $E_H$ ($\mu$J) | 3,9 |
| $\theta_c$ (bit) | 48 |
| $\theta_d$ (bit) | 1024 |
| $\gamma$ | 0.001 |
| $w_0$ | 0.462 |
| $R$ (m) | 1500 |
| $c$ | 10 |

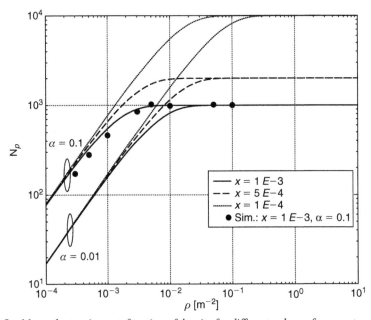

**Figure 9.8**  *Mean cluster size as a function of density for different values of parameters x and α*

performance. For this reason, both $N_p$ and $p$, evaluated through (9.49) to (9.52), are reported as a function of node density $\rho$ for different values of $x$ and $\alpha$ in Figures 9.8 and 9.9, respectively. It is interesting to observe that, as the node density increases, the cluster size tends to saturate to an asymptotic value which is a function of parameters $x$ and $\alpha$.

In Figure 9.9, it is also possible to analyze the MAC impact on the probability of sample loss $p$ for different values of the MAC parameter $c$, when $c = 10$ and $c = 100$.

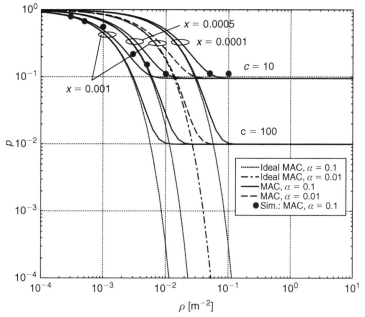

**Figure 9.9** *Probability of sample loss as a function of density for different values of parameters* x, α *and* c

According to (9.51), for low values of $\rho$ the sample loss probability is dominated by the probability $p_1$ that a node is isolated (connectivity), whereas for large values of $\rho$, it is lower bounded by the packet loss probability $p_{MAC}$ (MAC effect).

For $\rho$ approaching to infinite, we can obtain the floor value, $p_{floor}$, that is given by

$$p_{floor} = (1-x) \cdot \left[ 1 + \frac{e^{-\frac{1-x}{x}} - c \cdot \frac{1-x}{x} e^{-\frac{1}{c}}}{c \cdot \frac{1-x}{x} - 1} \right]. \tag{9.60}$$

This bound is absent if the effect of MAC is neglected (ideal MAC, $p_{MAC} = 0$, i.e. $c \to \infty$, thus $p_{floor} = 0$) and the performance depends only on the network connectivity which always increases with $\rho$. Note that for both the mean cluster size and the probability of samples loss, a very good agreement with simulations is verified.

In Figure 9.10, the MSE as a function of $\rho$, for $\alpha = 0.1$ and $x = 0.001$, is reported in the absence and in the presence of DDSP, for different choices of $\mu_\phi$. In particular, case 1 corresponds to (9.43), case 2 refers to (9.42) and case 3 is the optimum one given by (9.41). Note that the floor on the MSE is due to the floor on $p$ (see Figure 9.9). Hence,

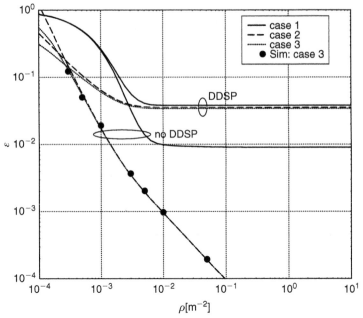

**Figure 9.10**   *MSE with and without DDSP as a function of nodes density for different values of* μ *for* α = 0.1 *and* x = 0.001 *in the three different cases for linear interpolator*

we obtain the floors on the MSE, called $\varepsilon_{\text{floor}}$. In particular, in the absence of DDSP only in case 1 exists a floor, given by

$$\varepsilon_{\text{floor}} = p_{\text{floor}}^2, \tag{9.61}$$

where $p_{\text{floor}}$ is given by (9.60).

When DDSP is adopted, all three cases provide floors, given by:

- case 1:

$$\varepsilon_{\text{floor}} = p_{\text{floor}}^2 + (1 - p_{\text{floor}}) \cdot \beta x \tilde{K} \tag{9.62}$$

     where $\tilde{K} = \pi e^{2(\sigma^2/k_1^2 - k_0/k_1 + L_{\text{su}}/k_1)}$

- case 2:

$$\varepsilon_{\text{floor}} = \frac{\beta x \tilde{K}}{1 - p_{\text{floor}}} \tag{9.63}$$

- case 3:

$$\varepsilon_{\text{floor}} = \frac{\beta x \tilde{K}}{\beta x \tilde{K} + 1 - p_{\text{floor}}} \tag{9.64}$$

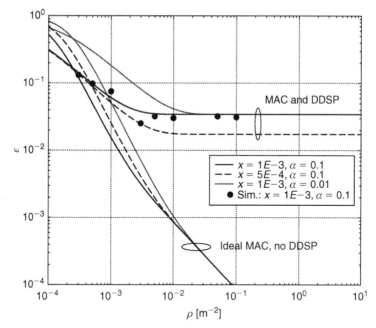

**Figure 9.11** *MSE with and without DDSP, as a function of density for different values of α and x*

With the parameters in Table 9.1, the floors' values obtained through (9.61), (9.62), (9.63) and (9.64) are, respectively, 0.009, 0.038, 0.036 and 0.034, which are perfectly verified by Figure 9.10. Again, simulation results show very good agreement with mathematical models.

In all cases the introduction of DDSP leads to a significant degradation on the MSE.

However, as will be clear in the following, there are relevant benefits of DDSP on WSN lifetime, and the trade-off between performance and lifetime will be discussed. The effect of parameters $x$ and $\alpha$ is shown in Figure 9.11 where the MSE is plotted with and without DDSP and in the presence or not of the described MAC protocol (concerning $\mu_\phi$, the case 2 is considered). Analytical and simulative results are in agreement also in this case. Note that, when ideal MAC and no DDSP are considered, the MSE, according to (9.43), can be made arbitrarily small by increasing the density $\rho$. However, this situation (ideal MAC) should be regarded as performance bound since constraints due to MAC are normally present. In the following Figures, the MAC protocol described in Section 9.2.8 is considered.

As already mentioned, the quality of process estimation is not the only focal point that drives the choice of the node density. In fact, the mean nodes' lifetime, measured in terms of the mean number of rounds achieved before they expire, is also important. It depends on a large set of parameters, such as node density and the cluster size, the energy consumed for transmitting a packet in high power mode (when the node is CH) with respect to that consumed in low power mode, the parameter $\alpha$, and the operational processing mode (DDSP or not).

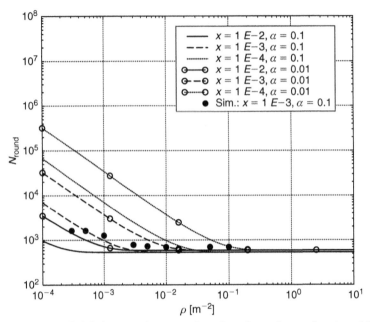

**Figure 9.12**   *Mean node life duration (in terms of number of rounds as a function of density for different values of parameters x and α without DDSP*

With DDSP it also depends on the processing consumption through the parameter $\gamma$ and the spatial correlation properties of the process under analysis, that is, $\beta$. The mean nodes' lifetime is defined in terms of the achievable average number of rounds, $N_{round}$ through (9.59).

In Figures 9.12 and 9.13, the mean number of rounds $N_{round}$ (referred to a battery charge of $E_{charge} = 1$ Joule) is reported as a function of node density for different values of $x$ and $\alpha$ in the absence (Figure 9.12) and in the presence (Figure 9.13) of DDSP. By fixing a minimum number of rounds that has to be guaranteed it is possible to obtain a constraint on the maximum tolerable node density $\rho$. Note that for dense WSNs (high values of $\rho$), the lifetime tends to an asymptotical value as a consequence of the asymptotical behaviour of $p$. It is also possible to note that DDSP strongly increases the mean lifetime of nodes.

Also for the network lifetime we evaluate the floor, $N_{round_{floor}}$, for $\rho$ approaching to infinite. The asymptotic values for $N_{round}$ in the no DDSP and in the DDSP cases follow:

- In the case of no DDSP:

$$N_{round_{floor}} = \frac{E_{charge}}{w_0\left\{(1-x)E_L\theta_d + x\left[E_H\theta_c + E_H\theta_d\left(\frac{1-p_{floor}}{x}\right)\right]\right\}}. \tag{9.65}$$

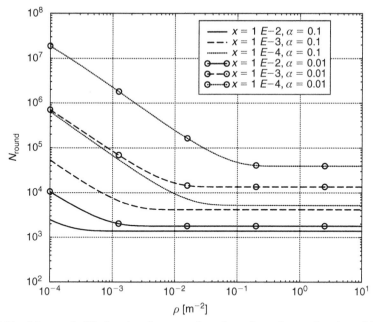

**Figure 9.13** *Mean node life duration (in terms of number of rounds as a function of density for different values of parameters x and α with DDSP.*

- In the case of DDSP:

$$N_{\text{round}_{floor}} = \frac{E_{\text{charge}}}{w_0 \left\{ (1-x)E_L\theta_d + xE_H\theta_c + xE_H\theta_d \cdot (\beta x\tilde{K} + \gamma) \cdot \left( \frac{1 - p_{\text{floor}}}{x} \right) \right\}} \cdot$$

(9.66)

As example results, in the no DDSP case with $\alpha = 0.1$, we obtain a floor $N_{\text{round}_{floor}}$ equal to 539, whereas in the DDSP case, for $x = 10^{-4}$ and $\alpha = 0.01$, we obtain $N_{\text{round}_{floor}} = 38224$, which are the same values shown in Figures 9.12 and 9.13.

The three key parameters on which the trade-off design of WSN is played are $\rho$, $x$ and $\alpha$, and the two performance metrics considered are $\varepsilon$ and $N_{\text{round}}$. A possible design criterion is the following: given a certain requirement, $\varepsilon_{\text{req}}$, on the MSE, find the values of $\rho$, $x$ and $\alpha$ such that the network lifetime, $N_{\text{round}}$, is maximized. According to this criterion, the algorithm which allows the evaluation of the three parameters can take advantage of floor expressions we derived, as in the following:

- With requirement $\varepsilon = \varepsilon_{\text{req}}$ and given $c$, $\sigma$, $k_0$, $k_1$, $L_{\text{su}}$, $\beta$, $w_0$, $\gamma$, $\theta_d$, $\theta_c$
- find $x_{\text{req}}$ s.t. $\varepsilon_{\text{floor}}(x_{\text{req}}) \cdot (1 + y\%) = \varepsilon_{\text{req}}$; (e.g., $y = 10$)

- find $\tilde{\alpha}(\rho)$ s.t. $\varepsilon(\rho, x_{req}, \tilde{\alpha}) = \varepsilon_{req}$
- find $\rho_{req} = \arg \max_{\rho} N_{round}(\rho, x_{req}, \tilde{\alpha}(\rho))$

  $\alpha_{req} = \tilde{\alpha}(\rho_{req})$

- giving $N_{round}(\rho_{req}, x_{req}, \tilde{\alpha}_{req})$

As an example design we consider the reconstruction case 1 with DDSP: by fixing $\varepsilon_{req} = 4 \cdot 10^{-2}$, from the above described project criterion, we obtain $x_{req} = 10^{-3}$, $\alpha_{req} = 2.5 \cdot 10^{-3}$, and $\rho_{req} = 3.5 \cdot 10^{-2}\ [m^{-2}]$. These parameters' values bring to a mean number of around $N_{round}$ per joule of charge of about 18000.

## 9.3   Compression techniques for WSNs

It is then apparent from previous sections that an important challenge in WSN research is the integrated design of local signal processing operations and strategies for inter-sensor communication and networking to reach a desirable trade-off among energy efficiency, simplicity, and overall system performance. For instance, to maximize battery life and reduce communication bandwidth, it is essential for each sensor to locally compress its observed data so that only low-rate intersensor communication is required (Xiao, Ribeiro, Luo & Giannakis, 2006). In fact, the most relevant power consumption is in data transmission/reception and it is convenient to process data locally before transmitting still considering the compromise due to limited processing capabilities at single node. Then, the adoption of a compression technique is considered when the sum of energy consumption for local processing plus transmission of processed data is less then the energy consumption for the transmission of unprocessed data.

The field of data compression for WSNs is gaining interest in the literarure and in this section some example techniques is discussed (see, e.g., the survey in Kimura & Latifi 2005).

### 9.3.1   Coding by ordering

The algorithm *coding by ordering* is integrated in the routing scheme data funnelling (Petrović, Shah, Ramchandran & Rabaey, 2003), which is composed of two phases (setup and data collection). The setup phase is composed of the following steps:

- The controller divides the area it wishes to monitor into small areas.
- It then initiates a directional flood towards each region.
- Each intermediate node records the cost of reaching the controller.
- When the packet reaches the region, the first node in the region that receives the packet designates itself as a border node.
- The border node adds two new fields to the packet: the cost for reaching the border node and the number of hops to the border node. If then floods the region with this modified packet.

- All nodes within the region receive packets from all the border nodes. Based on the energy required by each border node to reach the controller, they compute a schedule of border nodes where data is to be aggregated.

The phase of data collection is then composed of the following steps:

- When a sensor has a data sample it needs to send back to the controller, it uses the schedule to figure out the border node to use in the current round.
- It then waits for a time inversely proportional to the number of hops from the border node before sending out the packet.
- Along the way to the border node, the data packets are joined together until they reach the border node.
- The border node collects all the packets and then sends one packet with all the data back to the controller, using probabilistic routing.

Thus, the data funneling algorithm creates clusters with dynamic hierarchy. There is not a single CH, whose failure can be devastating to the functionality of the network. Instead, the border nodes take turns acting as the cluster head spreading out the responsibility and the load (thus the energy consumption) among them. Also, the controller can redefine the regions into which its area of interest is divided, thereby forcing the nodes to divide themselves into new clusters and elect new sets of border nodes. The controller can redefine the regions based on the data it receives from the nodes and/or the energy remaining in the nodes so as to ensure that nodes that have the greatest energy reserves act as border nodes.

The main idea behind coding by ordering is that when transmitting many unique pieces of data, and the order in which the data is sent is not important to the application (i.e., the transmitter may choose the order in which to send those pieces of data), then the choice of the order in which those pieces of data are sent can be used to convey additional information to the receiver. In fact it is possible to avoid explicitly transmitting some of those pieces of data, and use the ordering of the other information to convey the information contained in the pieces of data that were not sent. For example, in the case of the data funneling algorithm this allows the border node to choose to suppress some of the packets (i.e., choose not to include them in the superpacket), and order the other packets within the superpacket in such a way as to indicate the values contained within the suppressed packets.

Let us assume $n$ be the total number of nodes with an ID each, $m$ be the number of nodes sending a packet to a border node and to have the encoder at the border node throwing away any $l$ packets. Then, the number of possible combination of IDs for dropped nodes is $\binom{n-m+l}{l}$.

One encoding strategy is to appropriately order the remaining $m - l$ packets to indicate what values were contained in the suppressed packets. A total of $(m - l)!$ values can be indexed by ordering $m - l$ distinct objects. By assuming that each of the suppressed packets contains a payload that can take any of the $k$ possible values and an ID, which can be any value from the set of valid IDs except for the ones that belong to the packets included in the superpacket. The values contained within the suppressed packets can be regarded as symbols from a $(n - m + l) \times k$ -ary alphabet, giving $\binom{n-m+l}{l}k^l$ possible values for the suppressed packets. In order, for it to be possible for the encoder to suppress $l$ packets in this manner, the following relationship must be satisfied

$$(m - l)! \geq \binom{n - m + l}{l}k^l. \tag{9.67}$$

We are interested in the greatest $l$, that is, $l_{max}$, satisfying (9.67). The ratio $l_{max}/m$ represents the compression ratio in terms of packets. The coding by ordering technique has a good compression ratio (e.g., when $n = 2^7$, $k = 16$, and $m = 100$, then approximately 44% of packets can be dropped at the aggregation node) (Kimura & Latifi, 2005). One difficulty of utilizing this scheme is that it requires a mapping table for permutations in which size increases exponentially with the number of sensor nodes aggregated.

## 9.3.2 Pipelined in-network compression

The pipelined in-network compression (PINCO) algorithm exchanges latency for energy by first buffering and delaying nodes' measurements, and then reducing available redundancy in spatial, temporal, and spatial-temporal domain among the sensor data in a buffer by compressing them into groups of data (Arici, Gedik, Altunbasak & Liu, 2003). Each group can be decompressed independently of other groups. A very important feature of PINCO is that groups are compressed as highly flexible structures so that they can be recompressed without decompressing. Recompression exploits the likely available redundancy among groups at later stages of the WSN while propagating towards the collector. Without the need for decompression it does not impose any additional cost on the nodes with highly constrained resources. An end-to-end latency bound specified by the user when querying the network is satisfied by appropriately choosing the maximum waiting time of data in a nodes buffer. A PINCO scheme for single-valued data is an error-resilient compression scheme. Available redundancy is reduced by exploiting the commonalities among the buffered data.

The PINCO parameters have to be tweaked for specific application characteristics for good performance. Results in Arici et al., 2003 investigate the effect of PINCO parameters on two performance measures: observed end-to-end latency and the amount of dissipated energy.

## 9.3.3 Distributed compression

Many evolving low-power sensor network scenarios need to have high spatial density to enable reliable operation in the face of component node failures as well as to facilitate high spatial localization of events of interest. This induces a high level of network data redundancy, where spatially proximal sensor readings are highly correlated. In Pradham et al., 2002 a distributed technique is proposed for dense WSNs to remove this redundancy in a completely distributed manner, that is, without intersensors communications. The need for a spatially dense sensor network is driven by two requirements: (1) reliable decision-making in the face of unreliable individual components and (2) superior spatial localization of transient events of interest. This can lead to considerable system redundancy. The need to strip this redundancy is underlined by a couple of additional factors: the access to shared medium and energy saving.

From the theory of distributed source coding, if the joint distribution quantifying the correlation among nodes in the WSN is known, then by avoiding intersensors communication there is theoretically no loss in performance of the overall compression efficiency under certain conditions (Slepian & Wolf, 1973; Pradham et al., 2002).

The goal for the decoder is to reconstruct the original source using this side information as well as the reduced bitstream sent by the encoder. Let us consider, as an example, two correlated discrete-alphabet i.i.d. sources $X$ and $Y$. In the absence of distributed compression each source would transmit $H(x)$ bits (entropy of $X$) and $H(Y)$ bits, respectively, and in the presence of an aggregation node the sum of the two would be transmitted to the sink.

Typically, it is possible to efficiently encode $X$ if the encoder also knows $Y$. The important result is that it is possible to reduce the total number of bits transmitted, thus reduce the energy consumption, also when $Y$ is known only at the decoder and not at the encoder. In fact, one can still compress $X$ using only $H(X|Y)$ bits (entropy of $X$ given $Y$), the same as in the case where the encoder does know $Y$. That is, by knowing just $p(X, Y)$, the joint distribution of $X$ and $Y$, without explicitly knowing $Y$, the encoder of $X$ can perform as well as an encoder which explicitly knows $Y$ (in theory, only $H(X|Y)$ needs to be known at the encoder, not even $p(X, Y)$). Recall that

$$H(X,Y) = -\sum_{x,y} p(x, y) \log p(x, y)$$
$$H(X|Y) = H(X,Y) - H(Y) \tag{9.68}$$

with $(X, Y) \sim p(x, y)$. Since the decoder alone has access to the $Y$ process, the task is to optimally compress the $X$ process, thus form the best approximation, $\hat{X}$, to $X$ given an encoding bit budget of $R$ bits per sample. This problem can also be posed as minimizing the rate of transmission $R$ such that the reconstruction fidelity is less than a target distortion. This involves an intricate interplay of source coding, channel coding, and estimation theory.

In the particular correlation case in which the correlation is expressed by the fact that $Y$ is a noisy version of $X$ (i.e., $Y = X + N$ with $N$ zero-mean Gaussian distributed), then the least significant bits of $Y$ and $X$ are more correlated than the most significant bits and this reflects in to individuate proper unequal error protection schemes.

An example of deployment in WSN is also given in Pradham et al., 2002, here summarized: assume there are two ($X$ and $Y$) 3-bit data sets (e.g., from 8-levels quantization) and the Hamming distance between $X$ and $Y$ is no more than one bit. Since the decoder knows $Y$, and $X$ is only one Hamming distant apart from $Y$, it is not efficient to distinguish $X = 111$ from $X = 000$. For the same reason, $X = 001$ and 110, $X = 010$ and 101, $X = 011$ and 110, do not need to be distinguished from each other. These sets of two $X$ values are grouped as cosets and assigned different binary index numbers (in this case they are four):

$coset0(00) = (000, 111)$
$coset1(01) = (001, 110)$
$coset2(10) = (010, 101)$
$coset3(11) = (011, 100)$

It is important to understand that, if $Y = 110$ and $X = 010$, once $Y$ is known at the decoder, there is no reason to send three bits for $X$, but only two indicating the coset, in this case 10. Once the decoder received $Y = 110$ as a side information from $Y$ and 10 as a partial information from $X$, then it can select $X = 010$ from *coset* 2 since 110 has a Hamming distance of 2 from 110 instead of distance 1. Whether $X$ knows $Y$ or not, $X$ can still compress 3 bits of information into 2 bits.

If a node aggregates and relays the information from many nodes, this reduction of the number of transmitted bits becomes more relevant.

## 9.4    A possible classification of signal processing techniques for WSNs

The trade-offs between local or distributed data processing and transmission of raw data to an FC must be examined under the constraints of energy, bandwidth and latency. The relevant metric is application specific, such as the accuracy of target detection and tracking, the timeliness of the detection, the accuracy of classification, the quality and utility of the delivered information, etc.

### 9.4.1    Classifying signal processing techniques

To classify SP techniques for WSANs is not trivial since many techniques are available and studied in the literature. Here we present most important characteristics to which SP techniques are based on.

- The application and the information carried by each sensor, that is, the aim of the WSN (e.g., distributed detection of binary information, distributed detection of quantized information, estimation of processes distributed in space and in time etc.).
- The constraints on energy consumption and quality of service.

- The network organization (e.g., flat, hierarchical, etc.).
- The adopted routing and MAC protocols.
- The communication level of nodes, that is, the energy consumption for transmitting a bit, for receiving, for sleeping, thus breezing the node status, and, also very important, the energy consumption for elaborating a bit.

### 9.4.2   Scanning the literature

In Pottie & Kaiser, 2000 it is pointed out that as WSNs combine micro-sensor technology and low-power signal processing, computation and low-cost wireless networking take advantage of recent advances in integrated circuit technology enabling mass production. A detector is given a set of observables $\{X_j\}$ to determine which of several hypothesis $\{h_i\}$ is true. The observables may be the sampled output of the sensor and include not only the desired target but also background noise and interference from other sources. The decision concerning the presence of the target, absence, and type is usually based on estimates of a small number of parameters of these observations (as, for example, selected Fourier and wavelet transform coefficients) that provide a reduced representation of observations, the feature set $\{f_k\}$. The reliability of these parameters depends on both the number of independent observations and the SNR. As an example, according with the CRLB, which establishes the fundamental limits of estimation accuracy, the variance of a parameter estimate for a signal perturbed by white noise declines linearly with both the number of observations and the SNR. Consequently, to compute a good estimate we need either a long set of independent observations or high SNR. The choice among hypotheses is made by dividing a decision space according to proper rules on $\{h_i\}$ and deciding for the $h_l$ for which $\mathbb{P}\{h_l|\{f_k\}\} > \mathbb{P}\{h_i|\{f_k\}\}, \forall i \neq l$. The dimension of the feature set, with respect to the number of hypotheses to sort through, represents a trade-off between the complexity and the uncertainty of the decision. To build the minimum size space, we must determine the marginal improvement in the decision error rate resulting from the addition of another feature. Unfortunately, we seldom know the prior probabilities of the various hypotheses, training is often inadequate to determine the conditional probabilities, and the marginal improvement in reliability declines rapidly as more features are extracted from any given set of observables. On these facts hang many practical algorithms.

The communication costs influence the processing strategy, including our willingness to communicate and whether the processing is centralized or distributed. Optimization is mandated by the physical constraints of the WSN. Therefore, the physical layer intrudes up through the network and signal-processing layers to applications. To make concrete the effect of these constraints, assume the following (Pottie & Kaiser, 2000): a 1 GHz carrier frequency; an antenna elevation of 1/2 wavelength; a robust digital modulation, such as binary-phase-shift-keying (BPSK) transmission, $10^{-6}$ error probability, fourth power distance loss, Rayleigh fading, and an ideal (i.e., noiseless) receiver. The energy cost of transmitting 1 Kb a distance of 100 metres is approximately 3 joules. By contrast, a general-purpose processor with 100 MIPS/W power could efficiently execute 3 million instructions for the same amount of energy. If the application and infrastructure permit, it pays to process the data locally to reduce traffic volume and make use of multihop routing and advanced communications techniques, such as coding, to reduce energy costs. Indeed, exploitation of the application makes low-power design possible.

A timely snapshot of ongoing distributed signal processing research in the rapidly evolving field of WSN is given in *Signal Processing Magazine, Special Issue*, 2006.

In Chen et al., 2006, the distributed detection problem in a wireless sensor network is addressed with also a review of classical approaches. The main focus is the integration of wireless channel conditions in the design of signal processing algorithms for distributed detection. Several design examples are used to illustrate the trade-off between performance and design complexity in situations with either complete, partial, or absent channel state information. The central theme that transcends various aspects of signal processing design is that an integrated channel-aware approach needs to be taken for optimal detection performance given the available resources. It is shown that under constraints of practical applications (e.g., constraints on power, bandwidth and delay) the wireless system cannot be designed irrespective of the sensing/processing at both the local sensor and the FC level; the transceiver design and the fusion algorithm need integration. The penalty of assuming ideal transmission will become more severe when the number of sensors and the quantization levels increase.

In Xiao et al., 2006, the distributed compression-estimation for WSN is considered under bandwidth constraints. To maximize battery lifetime and reduce communication bandwidth, it is essential for each sensor to locally compress its observed data so that only low-rate intersensor communication is required. This motivates joint design of the compression-estimation module per sensor. Depending on the amount of available a priori knowledge about the sensor noise distributions, various distributed compression-estimation schemes are presented with their performance bounds for both parameter estimation and target tracking applications, all under bandwidth constraints.

Designing distributed compression-estimation algorithms in the context of a WSN differs from the traditional centralized framework in several important aspects.

- Constraints on sensor cost, bandwidth, and energy budget dictate that low-quality sensor observations may have to be aggressively quantized, down to a few bits per sample per node. Thus, estimators must be developed based on severely quantized versions of very noisy observations.
- Obtaining the complete signal models for a large number of sensors may be impractical, particularly in dynamic sensing environments. This pre-empts the application of optimum estimation algorithms and motivates distributed estimators based on partially known or unknown data/noise models.
- Sensors may enter or leave the network dynamically, resulting in unpredictable changes in network size and topology. Thus, to ensure robust operation, compression/estimation

algorithms for WSNs have to work with limited (or absent) knowledge of the network topology and/or size.

- Local compression at a sensor node depends not only on the quality of sensor observation, but also on the quality of the wireless communication channel(s) from the node. In addition, the design of distributed algorithms should be coupled with the underlying WSN topology (e.g., presence or absence of a FC).
- When an FC is present, there is no intersensor communication; communication is only between sensors and the FC. The FC collects locally processed data and produces a final estimate.
- In the absence of FC, the network itself is responsible for processing the collected information, and to this end, sensors communicate with each other through the shared wireless medium.

Hybrids are also possible in which the WSN is partitioned into clusters possibly with a hierarchical structure. Each cluster has a local FC generating intermediate estimates, which in turn are combined to obtain a final estimate.

The aim is to estimate a $p \times 1$ vector parameter $\theta$ from $K$ observations $x_k = \phi_k(\theta) + w_k$ where $\phi_k$ is a real nonlinear function and $w_k$ the noise term. Distributed estimation using a WSN entails a local compression stage in which sensors perform local quantization of their observations to obtain finite-rate messages $m_k(x_k)$. The messages are sent to the FC where a final estimate function of $(m_1, \ldots, m_K)$ is generated. Several distributed estimators can be considered and the optimal choice depends on the parametric model and the statistics of the noise.

The interplay between signal processing and communication in the context of military sensor networks is addressed in Zhao, Swami & Tong, 2006, where it is demonstrated through two case studies that the exploitation of dependencies between signal processing and networking can lead to designs with improved network performance. Two complementary categories are presented, signal processing for networking and networking for signal processing, and some typical design issues in which the networking and signal processing intertwine are investigated. While recognizing the benefit of exploiting channel state information (CSI) and residual energy information (REI) for the design of opportunistic MAC protocols, taking advantage of CSI to prioritize nodes according to CSI, thus exploiting wireless channels diversity, one cannot ignore the cost associated with obtaining this information. A protocol requiring global information on CSI and REI may encounter an unacceptable level of overhead in large-scale networks. It is thus desirable to design distributed protocols using local CSI and REI to better address the trade-off between benefit and cost. Energy-efficient signal processing techniques to acquiring CSI and REI affect this trade-off, being an example of signal processing for networking and enabling adaptive networking. The lifetime-maximizing protocols require knowledge of the network residual energy profile. This information, as well as the population/ density of functioning sensors, is also important for network maintenance. For example, the knowledge of the number of operating sensors as a function of geographical location facilitates the decision on whether and where to deploy new sensors. It is thus crucial to track the network energy profile and the sensor population. Ignoring the stringent energy constraint, the large network size, and the harsh wireless multiaccess medium leads to a trivial solution where every sensor is scheduled to report its energy level periodically. A desired solution is to piggyback the residual energy information on data packets to avoid extra transmissions solely for the purpose of network monitoring. This approach, however, provides only an energy profile sampled in space and time. The sampling pattern is determined by the network application (e.g., the spatial and temporal distributions of the random events being detected by the network) and the MAC and routing protocols used in data collections. How to infer the energy profile from the collected data samples is thus a complex signal processing problem and is coupled with the upper layer protocols. Detailed discussions on this problem and several scalable estimation algorithms for energy profile monitoring can be found in Zhao et al., 2006 and herein references. On the other hand, since WSNs are application specific, the network should be designed not for individual nodes but to optimize the application-specific metric (e.g., detection and parameter estimation, signal field reconstruction) and improve the network lifetime. The problem of cross-layer design can then be formulated as the joint optimal design of the transmitted signal format at the sensor and the statistical inference algorithm at the FC to minimize inference error.

A theoretical framework for distributed inference using graphical models and message-passing algorithms is described in Cetin, Chen, Fisher III, Ihler, Moses, Wainwright & Willsky, 2006: two applications, self-localization in sensor networks and distributed data association in multi-object tracking, are used to show how sensor network fusion problems can be cast as problems of inference over graphical models and examine how conservation of power through judicious use of communications resources provides new insights not found in the graphical model literature. The distributed inference problem is also considered in Predd, Kulkarni & Poor, 2006 but under a distributed learning framework. It is investigated how the classical robust non-parametric methods can be adapted for wireless sensor network applications, with an emphasis on their distributed implementation and convergence under bandwidth and energy constraints.

The fact that integrated approaches are needed for the design of WSNs is also emphasized in Gastpar, Vetterli & Dragotti, 2006, who investigated the interplay between sampling, source representation/coding, and communication in sensor networks. Examples and results show that, from an information theoretic perspective, deploying optimal distributed source coding and optimal channel coding separately can be overall sub-optimal in sensor networks. Thus, an integrated approach is needed for sensor network design that can take into account both source and channel conditions simultaneously.

When sensors are cameras, there is the problem of efficiently sending image samples to the FC. As example, Puri, Majumdar, Ishwar & Ramchandran, 2006 describe how video signals can be coded and compressed in a distributed and energy-efficient manner by exploiting the spatial redundancy in a broadband WSN. A specific architecture (called

PRISM) is used to illustrate the issues and challenges associated with distributed video coding.

Issues such as fusion, data aggregation, detection, estimation, and tracking are context dependent. For applications such as searching and discovering, how distributed signal processing algorithms should be designed for is described in Ermis, Alanyali & Saligrama, 2006.

Computing reliability and message delay for cooperative and distributed WSNs subject to random failure are investigated in AboElFotoh, Iyengar & Chakrabarty, 2005 for multihop networks. The self-organizing capabilities and the cooperative operation of WSNs allow for forming reliable clusters of sensors deployed near, or at, the sites of target phenomena. Reliable monitoring of a phenomenon (or event detection) depends on the collective data provided by the target cluster of sensors, and not on any individual node. The failure of one or more nodes may not cause the operational data sources to be disconnected from the data sinks (command nodes or end-user stations). However, it may increase the number of hops a data message has to go through before reaching its destination (and subsequently increase the message delay). The two related problems, computing a measure for the reliability of WSN and computing a measure for the expected and the maximum message delay, are strictly related. Given an estimation of the failure probabilities of sensors and relay nodes, in a cluster-based architecture, the WSN is modelled with a probabilistic graph. The WSN reliability is defined as the probability that an operating communication path between the sink node and at least one operational sensor in a target cluster exists. Both problems are NP-hard for arbitrary WSNs, although in this work the propagation is not carefully modelled since a deterministic law (disk model), with a maximum TR, is assumed, whereas a more proper model would also consider randomness of wireless channel.

In Chamberland & Veeravalli, 2003, a binary decentralized detection problem is investigated where a WSN provides relevant information about the state of nature to a FC. Each sensor transmits its data over a multiple-access channel. Upon reception of the information, the FC attempts to accurately reconstruct the state of nature. In the considered scenario, the sensor network is constrained by the capacity of the wireless channel over which the sensors are transmitting, and the structure of an optimal sensor configuration is studied. For the problem of detecting deterministic signals in additive Gaussian noise, it is shown that having a set of identical binary sensors is asymptotically optimal, as the number of observations per sensor goes to infinity. Thus, the gain offered by having more sensors exceeds the benefits of getting detailed information from each sensor. A thorough analysis of the Gaussian case is presented along with some extensions to other observation distributions. The estimation of a random variable based on noisy observations is a standard problem in statistics. In Chamberland & Veeravalli, 2003, the related scenario where information about a random variable is made available to a FC by a set of geographically separated sensors is investigated. Each sensor receives a sequence of observations about the state of nature and transmits a summary of its information over a wireless multiple access channel. Based on the received data, the FC produces an estimate of the state of nature $H$. The attention is focused on the special case of binary hypothesis testing, where $H$ takes on one of two possible values, where the observations across sensors are i.i.d. conditioned on $H$, and where the observation process at each sensor conditioned on $H$ is a sequence of i.i.d. random variables. If the structure of the information supplied by each sensor is predetermined, the FC faces a classical hypothesis testing problem. The probability of estimation error is then minimized by the maximum a posteriori detector. Alternatively, one can consider the problem of deciding what type of information each sensor should send to the FC as investigated in Chamberland & Veeravalli, 2003. Several different variants of this problem have been studied in the past (see, e.g., Tsitsiklis, 1993, and references herein). Notably, the class of decentralized detection problems where each sensor must select one of $D$ possible messages has received much attention. In essence, having each sensor select one of possible messages upper bounds the amount of information available at the FC. Indeed, the quantity of information relayed to the FC by a network of $L$ sensors, each sending one of $D$ possible messages, does not exceed $L$ [$\log_2 D$] bits per unit time. In the standard decentralized problem formulation, the number of sensors $L$ and the number of distinct messages $D$ are fixed beforehand. A more natural approach in context of WSNs is to constrain the capacity of the multiple access channel available to the sensors. For instance, a multiple access channel may only be able to carry $R$ bits of information per unit time. Thus, the new design problem becomes selecting $L$ and $D_l$, where $D_l$ is the number of messages admissible to sensor $S_l$, to minimize the probability of error at the FC, subject to the capacity constraint given by

$$\sum_{l=1}^{L} [\log_2 D_l] \le R. \tag{9.69}$$

### 9.4.3 International projects on WSNs and SP for WSNs: some examples

As also reported in *Signal Processing Magazine, Special Issue*, 2006, many applications are now looking for scalable, flexibly deployable, and low-maintenance, energy-efficient sensor network solutions. One such application is scientific exploration. Networks of tiny sensor nodes have been used to monitor the natural environment, such as the growth of redwood trees in California or animal habitats such as zebra herds in African game reserves. Research along this theme constantly pushes for scalability, with the vision that the sensor network will enable large-scale scientific studies of underlying phenomenologies. In a different community, government has been interested in situation awareness on the battlefield and in public spaces. A number of DARPA programme have been along this line, including the SensIT program, which focuses on information processing in sensor networks, the NEST program, which focuses on software aspects of large-scale systems, and the 'Combat Zone That Sees' program, which focuses on city-scale distributed camera networks. A carefully designed sensor network can be used for surveillance and monitoring and possibly area denial (perhaps replacing minefields), and it could be used to localize events (the E-911 problem). Sensors also could be mobile, such as robots and unmanned ground and aerial vehicles.

The authors of this book have been recently involved in international projects on WSNs, such as *NEWCOM*[7], *VICom*[8] and *EYES*[9], briefly reported in Chapter 10.

---

[7] NEWCOM, FP6-IST 2004 507325, website at https://newcom.ismb.it
[8] VICom, Virtual Immersive COMmunications, (MIUR 2002–2006), website at http://www.vicom-project.it
[9] EYES (IST-2001-34734), website at http://eyes.eu.org

# Part 3

# From theory to practice: case studies

In this final part of the book, Chapter 10 is dedicated to some case studies: real-world applications of WSANs, which are the result of experimental activities performed by the authors in the context of industrial contracts or large cooperative projects. The aim of this part is to clarify how the design guidelines provided in previous chapters can be useful when designing real-world networks. As usual when dealing with field trials and in general with experimental activities, the results are sometimes difficult to be interpreted, and discussed. However, this also provides suggestions on how to build a field trial and how to anticipate the behaviour of a network prototype. Mention of the industrial contexts of the experimental activities is given, as a necessary tribute to the availability of results.

# 10

# From theory to practice: case studies

This chapter describes the methodology used to design live WSANs, describing the scenario, the measurements done, the design methodology used, and providing samples of the measurements performed over the field. It reflects the activity performed by the authors of this book who have been recently involved in international projects briefly reported here.

This chapter concludes the book. As in all engineering activities, every idea, concept or theoretical framework must be exposed to the real world for validation. Some of the protocols and techniques presented in this book have been tested on existing platforms and in real environments, within some projects participated in by the authors. The chapter reports these experiences. First of all, a brief mention of the projects and the main participants who were involved in, is given. Then, a few case studies are presented. They do not cover all techniques and algorithms presented in the book. However, they can be considered as samples of the problems that can be encountered when moving from theory to practice.

## 10.1  The EYES project

The EYES project (IST-2001-34734) was a three years' European research project on self-organizing and collaborative energy-efficient sensor networks (from years 2002 to 2005). It addressed the convergence of distributed information processing, wireless communications, and mobile computing. The goal of the project was to develop the architecture and the technology which enable the creation of a new generation of sensors that can effectively network together so as to provide a flexible platform for the support of a large variety of mobile sensor network applications. The project had also shown the feasibility of the concepts and technologies developed with a conclusive demonstration.

The technical work focused on architectural, protocol and software issues. These areas are in fact believed to be the true bottleneck in current sensor networks. The challenges to face in developing new technologies for sensor networks are the need for the nodes to be smart, self-configurable, capable of networking together, and the inherent poverty of resources of the nodes themselves. The main thrust of the work has been therefore directed towards the development of new architectural schemes, communication protocols and algorithms at multiple layers, taking into account those specific features. In particular, schemes, which are able to work efficiently in the presence of limited energy, processing power and memory, have been developed. During the project evolution some new energy-efficient MAC and routing layer was proposed, analyzing both clustered and non-clustered network solutions based on multipath routing. Security issues have also been addressed in order to consider network resistance from external attack or sensible information sniffing. As important complement to the above described functionality, EYES has also introduced some localization schemes and analyzed the usage feasibility of different ranging methods. The EYES project had also involved external manufacturers, such as Nedap and Infineon, in order to create a new low power hardware platform useful to validate the project results.

### 10.1.1 EYES hardware description

On this section the hardware platforms developed during the EYES project are briefly discussed and analyzed.

Two EYES wireless sensor and actuator nodes, also referred to as motes, are realized on the project. The Nedap mote and the eyesIFX mote share a common structure identified by the following building blocks: an antenna, a radio, a microcontroller, non-volatile memory, sensors, actors and an energy source. The resulting block diagram is shown in Figure 10.1.

In Figures 10.2 and 10.3 the two different hardware platforms are visible.

The **antenna** is a stripline-antenna printed on the circuit board that is only offering limited gain. An alternative stub or whip antenna is present on eyesIFX devices through a SMB gold-plated connector.

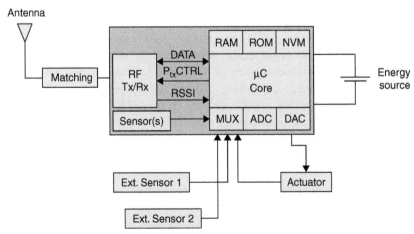

**Figure 10.1**    *Wireless sensor and actor node architecture*

The **power source** can be a primary or secondary battery. The former denotes here a cell that can not be recharged. Different chemistries are usable, such as zinc-air, lithium and alkaline, all of them differing in their energy density. Their voltage range, between 3 and 1,8 V, dictated the design of the electronics capable of coping with the wide input voltage range. Usable rechargeable alternatives can be NiMh cells or super-caps, but not LiIon cells due to cell voltage higher than 3.7 V and the power input stage that doesn't include

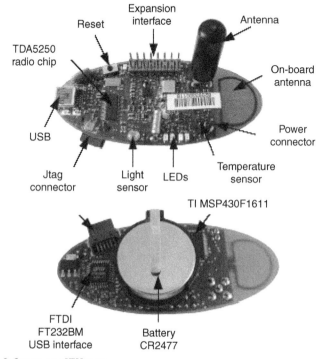

**Figure 10.2**  *Infineon eyesIFX mote*

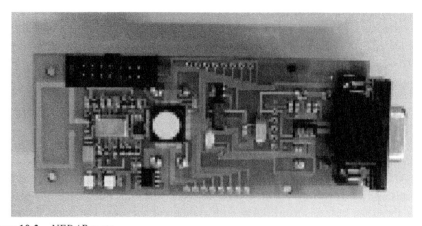

**Figure 10.3**  *NEDAP mote*

a voltage regulator in order to minimize energy loss during mote activity. Lithium-coin cell CR2106 is a default solution for eyesIFX mote, while alkaline cells are used on NEDAP devices.

The **radio** is a low-power half-duplex transceiver on 868 MHz ISM band. NEDAP mote uses a RFM TR1001 radio transceiver that utilizes amplitude modulation (AM) and eye-sIFX node mount and Infineon TDA5250 able to use both AM and frequency modulation (FM). The transceivers are directly connected to a serial port of the main microcontroller, so the bits sent on the radio channel simply match the serial frame generated. The maximum data rate is limited to 64 Kbps on both motes but due to the transceiver simplicity the speed is reduced by software implementation to decrease microcontroller overhead. Going more into details, the RFM TR1001 hybrid radio transceiver has low power consumption and has small size. The TR1001 supports transmission rates up to 115.2 Kbps. The power consumption during reception is approximately 14.4 mW, during transmission 16.0 mW, and in sleep mode is 15.0 μW. The transmitter output power is maximal 0.75 mW.

Like many other transceivers, the RFM transceiver performs best when DC-balanced data is transmitted. In practice this means that the transmission rate is reduced to 57.6 Kbps, but when choosing code words wisely, some error correction properties can be added at the same time.

Infineon TDA5250 has lower power consumption and can work on a wide range of voltage supply (5 V–2 V). In transmit mode at 3 V it needs only 12 mA; while receiving 9 mA are sufficient. For this reason, the transceiver is particularly suitable for battery-operated applications requiring very long operating times. The TDA525x transceiver family covers the 868, 315 and 434 MHz ISM bands and allows data rates of up to 64 kbit/s for Manchester-coded data transfer. The transceivers use ASK or FSK modulation, are highly integrated and thus need only very few external components. The TDA5250 deployed on the eyesIFX mote uses the 868 MHz band and has a fixed FSK modulation. The transceiver is configured via an SPI interface. This allows the microcontroller to set the configuration register (e.g., for control, filter, XTAL and wake-up configuration) and has access to a received signal strength (RSSI) value which can be read from the integrated AD-converter. The send and receive data are exchanged via the DATA pin. Another separate pin is used to do Tx/Rx mode selection.

The RFM TR1001 and TDA5250 allow their transmission power to be adjusted by their input current. A digitally controllable resistor is added to the design to modulate this input current.

The **digital section** of the node is made up of more units necessary to supply the computational power, data storage space and the programming/debug interface. The two hardware platforms that were developed in the EYES project use the same microcontroller platform, that is, the Texas Instruments MSP430 on two different versions. The Texas Instruments MSP430 microcontroller family has set the benchmark for low-power microcontrollers for a number of years. Alongside a large number of battery-operated applications, the MSP430 is particularly advantageous when it comes to preventing power loss. In addition to low power consumption in everyday use, the low-power modes allow the deactivation of various functional blocks to the point of stop mode with a power consumption of well below 1 μA. The MSP430 is unbeatable when it comes to the response time of the internal oscillator. This is stable again in less than 6 ns (2 ns with the new MSP430F2xx family) after waking up from low-power mode, enabling the controller to sleep predominantly in many applications and only be active periodically for very short intervals.

The processor used in the NEDAP sensor node is a MSP430F149. It is a 16-bit processor and it has 60 Kbytes of program memory and 2 Kbytes of data memory. When running at full speed (5 MHz), the processor consumes approximately 1.5 mW, but it also has several power-saving modes. Besides the on-chip memory and various low-power modes, the microcontroller also facilitates eight A/D 12-bits and I/O lines, that can easily be controlled by software. These lines are used as interfaces to temperature and light sensors installed on the board. However, an expansion connector is present and can be used to connect external sensors. The processor IC includes two UARTs as well. One of the UARTs will be used to drive the transceiver, the other will be shared between a serial memory and an RS232 interface for debugging. A standard JTAG connector on the board is finally used to program the device.

The eyesIFX mote instead (from the second version) uses the MSP430F161x family. With 10 kByte RAM, these derivatives are particularly suited to applications such as data logging and complex regulators. An additional D/A converter and a 3-channel DMA controller allows efficient processing of analogue measured values. All the programming and debugging phases are handled through an USB interface.

An external serial EEPROM for secondary storage is provided by an ST M25P40 or an ATMEL AT45DB041B on newer eyesIFX motes. The data flash is implemented in NOR technology, offers 100,000 write-erase cycles and data is retained for at least 20 years.

The platforms are supported by two different operating system: NEDAP node use the preemptive EYES real time operating system (PEEROS). The operating system is been written in the C language except for some hardware specific operations like flash memory control. Instead, eyesIFX mote has full support on TinyOS. TinyOS is an event-based operating system. Without initiating events (e.g., interrupts, incoming data, keystrokes or timer ticks), no code is executed and the microcontroller can go into power-saving mode. Only the functional blocks of the processor that are responsible for recognizing the possible events need to be activated. Thus a microcontroller, which flexibly activates and deactivates individual modules and which can change operational modes quickly, is beneficial for optimizing power consumption. TinyOS has replaced the programming language C with an improved version called nesC (networked embedded systems C). A normal C compiler still works in the background. The nesC precompiler generates optimized source codes for the particular application from the nesC source. These can then be interpreted by the compiler. The special structure of the nesC source texts allows the precompiler to ignore unused code, recognize dependences and conflicts, and to apply all low-power mechanisms to each function. Operating system related overheads are therefore extremely low. The TinyOS tool is available as open source.

### 10.1.2   Demo and measure

At the end of EYES project two separate demonstrations have been realized using the eyesIFX mote. A complete multihop communication framework has been developed using tinyOS 1.x operating system and some of the most interesting results are reported following this section.

The first demonstration involved a relative high number of motes and tested multihop routing capability of a network based on an highly optimized geographics random forwarding routing algorithm as described in Section 7.3.17. The implementation used a power aware routing solution with low complexity. Second and third ISO-OSI layers are integrated into the same layer to allow the information forwarding to reach the destination by selecting the relay nodes closer to it. The addressing scheme is directly linked to the nodes position and the node loops between working and idle state to preserve battery energy.

The implementation shows all the capability and also the weaknesses of this kind of algorithm. The possibility to route packet over the sensors network without the need of nodes organization was demonstrated using two emblematic cases, the first simulating a fault of same nodes while the second showing the capability of the MAC-routing to use immediately new sensors added to the network. In Figure 10.4 a measure on the paths used on the network during a 100 packets test is reported. It is interesting to observe that all the nodes are involved in routing process due to a periodic power-saving sleep strategy that randomizes the path used by a packet to reach the network sink device.

During the test on the middle of the measure area some devices were added and removed but no reconfiguration was made on the network. The results highlight the capability of GeRaF algorithm to route packet efficiently also on a randomly changing environment without significant network reconfiguration latency or extra energy consumption. The data have been collected using a Java application software connected through a USB cable to the sink device. There is a screenshot of the software in Figure 10.5. The main windows permit to monitor sensor data received from network devices and to send command to the network in order to control network motes.

The software used and node firmware is freely available for download at http://www.tlc.unife.it.

The main drawback of the implementation and, in general of GeRaF algorithm, was the need of a preprogrammed position information for each sensor used on the network; moreover an additional table with the position (or address) of the sink node is necessary to forward packets to data collector.

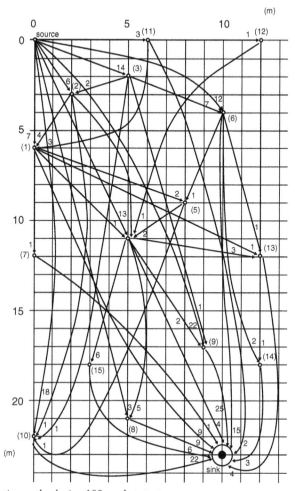

**Figure 10.4** *Routing paths during 100 packets test*

To partially solve this problem a localization algorithm was developed and implemented on sensor nodes. The methods used estimate distances through the received signal strength (RSSI), that does not require additional hardware or additional energy costs, since the amount of energy spent is just that related to a communication. The measurement of the received power, together with the knowledge of the transmitted power, lets the calculation of the signal attenuation that through a propagation model (path loss) gives an estimation of the distance between the transmitter and the receiver.

With *beacon* coordinates and estimated distances between a node and *beacons* in its coverage area, the node applies an algorithm that allows to obtain its own coordinates. In general, at least three *beacons* in a bidimensional scenario and at least four in a three-dimensional space are needed to determine the actual node position. Figure 10.6 reports the comparison on a 4-beacon-based network scenario of real position and estimated position of self-localized network devices.

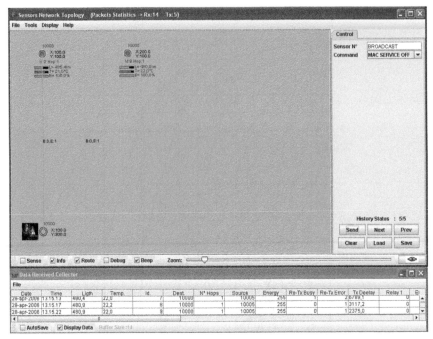

**Figure 10.5**    *Data collector software*

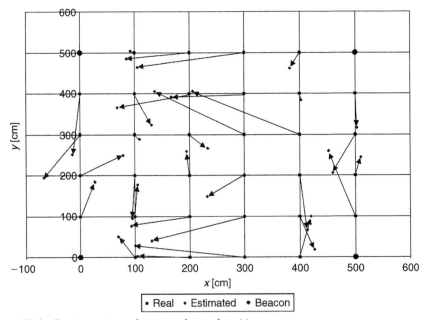

**Figure 10.6**    *Position estimated compared to real position*

The localization algorithm used is the maximum likelihood algorithm: this technique finds the node position that minimizes the differences between measured and Euclidean distances between *beacon* positions and estimated node position. For each *beacon i* inside the coverage area of the sensor under examination (node $n_0$), we can calculate the following function, that represents the difference between the node measured and the Euclidean distance from *beacon i*:

$$f_i(x_0, y_0) = d_{i0} - \sqrt{(x_i - x_0)^2 + (y_i - y_0)^2}$$

where $(x_0, y_0)$ is the estimated node position; $(x_i, y_i)$ are *beacon i* coordinates; $d_{i0}$ is the measured distance between *beacon i* and the node $n_0$; the square root term is the Euclidean distance between *beacon i* and $n_0$. The function $f_i(x_0, y_0)$ for node $n_0$ is calculated for a sufficient number of *beacon* terminals, $i = 1 \ldots B$. Node $n_0$ position is calculated as the coordinates couple $(x_0, y_0)$ that minimize (MMSE, minimum mean square estimation) the following linear combination:

$$F(x_0, y_0) = \sum_{i=1}^{B} w_i^2 f_i^2$$

where $w_i$ are weight coefficients useful to appropriate weight the functions $f_i$; on the implementation we considered $w_i = 1, i = 1 \ldots B$.

We report as an example the solution obtained with four *beacons* references, $B = 4$:

$$x_0 = \frac{\gamma_1 - \beta_1 \left( \dfrac{\gamma_2 - \gamma_1 \alpha}{\beta_2 - \beta_1 \alpha} \right)}{\alpha_1} \qquad y_0 = \frac{\gamma_2 - \gamma_1 \alpha}{\beta_2 - \beta_1 \alpha}$$

where the parameters introduced are defined as follows:

$$\alpha_1 = x_3 - x_1 \qquad \alpha_2 = x_4 - x_2 \qquad \alpha = \frac{\alpha_2}{\alpha_1}$$

$$\beta_1 = y_3 - y_1 \qquad \beta_2 = y_4 - y_2$$

$$\gamma_1 = \frac{x_3^2 + y_3^2 - x_1^2 - y_1^2 + d_1^2 - d_3^2}{2}$$

$$\gamma_2 = \frac{x_4^2 + y_4^2 - x_2^2 - y_2^2 + d_2^2 - d_4^2}{2}$$

Due to hardware differences between sensors and the very variable behaviour of the radio transmission in different environments, on second demos a dynamic ranging system was developed for tuning the ranging part of localization algorithm to each specific area and improve maximum likelihood algorithm results.

In order to limit the possible effects of hardware differences we use a calibration procedure of beacon nodes. This is done by placing a calibration node at the point where the distance to all neighbouring beacons is equal. Next, the potentiometer that controls the transmission power of each beacon is adjusted until the RSSI of all beacons is equal. It is also possible to do a calibration step before network deployment so that it has no requirements on the topology of the network.

When the calibration step is completed a table is constructed with RSSI to distance mappings. For this it uses ranges between beacons and collected RSSI information. This table can be shared with all nodes using a broadcast packet. With this table and collected beacon packets a node can estimate the ranges to the beacons. Finally, a localization algorithm estimates the position based on the ranges.

The results show an increased precision but only if distance remains limited to few metres. At higher distances the ranging methods fail on non-calibrated cases due to RSSI non-linearity. When distances-reach 4 or 5 metres with eyesIFX hardware the RSSI level collapses to a lower threshold value and low noise on the range measured is mapped on an higher distance error. Figure 10.7 reports a ranging estimated function vs distance where the performance of RSSI ranging becomes poor as the distance between beacons and unknown position device increases.

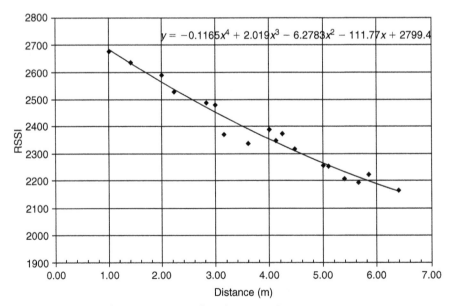

**Figure 10.7**   *Ranging function estimation based on network measures*

The results show that developed algorithms may be useful for actual applications. Areas where the sensor can be a fixed positioned are the main employ field. As an example smoke sensors for a fire alarm can take advantage of the high fault tolerance of this MAC-routing protocol. The localization method studied can also be efficiently used as a complement to GeRaF for reducing the number of sensors that need preprogrammed position information, but with the limits above reported.

## 10.2   The ambient networks project

The ambient networks (AN) project (IST-2002-2.3.1.4-507134 in the 6th framework programme) aims at an innovative, industrially exploitable new network vision based on the dynamic composition of networks to avoid adding to the growing patchwork of extensions to existing architectures. This will provide access to any network, including mobile personal networks, through instant establishment of inter-network agreements.

The project adopts the design paradigm of horizontally structured mobile systems that offer common control functions to a wide range of different applications and air interface technologies. Such a radical change requires the definition of new interfaces and a multitude of standards in key areas of future media- and context-aware, multidomain mobile networks.

The ambient networks project addresses a core component of this vision, the realization of a networking environment, which provides easy and universal access for users and enables an open marketplace for network, service and application businesses, unconstrained by focus on a particular commercial model or radio technology.

The project fully addresses the networks part of the strategic objective mobile and wireless systems beyond 3G and embraces the horizontal model as outlined in the first call in the IST part of the 6th framework programme. In this model different terrestrial access levels and technologies are combined to complement each other in an optimum way for different service requirements and radio environments. This combination or composition of network domains is a key theme of the ambient networks project providing not only technology choices but also new horizontal business opportunities. The project therefore includes specific work items to enable such composition, developing protocols and mechanisms that go beyond simple Internet-like connectivity, including open interfaces for management support and a security model, which can accommodate a dynamically changing multioperator environment.

In the following section, we address a particular case study that was included within Task 1-3. Task 1-3 was divided into three studies that deal with pure ANs and some of the common issues of the project. Low-velocity ANs were studied by creating demonstrations in sensor networks and personal area networks. Finally, there is a separate case study that combines AN concepts with vehicular area networks.

The reported demo was included on the AN project with the goal to demonstrate that although not originally thought to be AN, they could be integrated within them, exploiting different gateway functionalities that allow us to extend the scope of the ambient control space (ACS), by interacting with very specific networks. It is worth highlighting that in one of the demonstrators a wrapper of the ACS was used, in order to better assess the desired functionality.

## 10.2.1   Ambient network demo

The demo is aimed at demonstrating the advantages offered by the exploitation of WSN locationing services within an AN compatible environment. For instance, stationary WSNs could be used to estimate users' positions which could be subsequently exploited to offer location-dependent services.

In the following, we briefly illustrate the implemented components by highlighting how they interact with each other and how they make it possible to integrate an embedded and dedicated system composed by wireless sensors with an ambient network. First of all, an expandable location system is implemented through a fixed WSN. This system is used to estimate the position of mobile nodes via RSSI (received signal strength indicator) measurements.

Within the fixed WSN we have a data gathering point (in the sequel the sink), which is connected via cable to a server (WSN server). The static sensors in the network are periodically asked by the sink to send beacon messages to the mobile users which, in turn, collect these beacons and get RSSI measurements for each of them. With this procedure, every mobile node can build up a list of the closest static sensor devices. The list is subsequently used to estimate the geographical position of the mobile nodes. In fact, static sensor nodes could be placed in specific positions so as to identify a particular physical location, which could be a specific door, the entrance of a shop and so on. In addition, the aforementioned list is sent to the sink by each mobile node using an own-developed multihop wireless routing protocol.

Some mobile devices like PDAs or laptops are equipped with a WSN compatible sensor and programmed with a firmware able to collect the beacons sent by static nodes. As said above, these nodes can receive beacons and therefore report their RSSI measurements to the sink. After that, the position of these users is estimated by the WSN server, which received data from the sink through a wired connection. The WSN server verifies the correctness of the received location information, encapsulates the information into XML packets and sends it to a further server (the main server) via a TCP/IP connection.

The WSN server is exploited to gather, process, and redistribute sensor field measurements. This is a dedicated server whose aim is to interconnect the embedded WSN to external systems. The main server, instead, is AN compatible and is used here to interface WSN servers with the ambient network environment. Figure 10.8 shows all parts of the system.

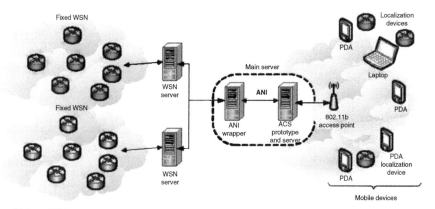

**Figure 10.8** *AN demo structures*

We have a static sensor network whose nodes are placed in strategic and specific positions (e.g., close to a shop entrance) and are used as anchors to estimate the geographical position of a given number of mobile devices (laptops or PDAs). Laptops and PDAs are equipped with a sensor node which can communicate and receive beacons from the anchors (static sensors). This communication is exploited to estimate the location of the mobile users within the network. Note that mobile devices can compose with the AN. A WSN server is used to gather information related to the location of mobile users, process it and subsequently communicate it (through an appropriate translation) to the AN. Finally, the AN exploits the location estimates to offer location-dependent services to the mobile users.

The main server integrates an ANI wrapper and reuses the BUTE PAP platform (a communication framework developed during the project) in order to handle the dynamic composition of the AN system with the mobile devices (PDAs and laptops). In addition, the main server hosts applications aimed at controlling the access rights for each user, monitoring the status of the network and managing the repository which contains the estimated geographical position for the mobile devices and the information related to routing group structures (RGs).

PDAs and laptops run an application equipped with a graphical interface to control the two main functionalities that we want to demonstrate: simulate a door opening and closing and creating group structures between mobile nodes in close proximity. These routing groups (RGs) are exploited to optimally deliver a given information to all group members. This, for instance, may be realized by communicating to all nodes in the RG using a single multicast address.

In the sequel we briefly outline the demonstration flow. Figure 10.9 depicts the elements involved in the demonstration.

We consider a hotel scenario, where guests may benefit from location-dependent services. This is achieved through localization (involving both static and mobile sensors).

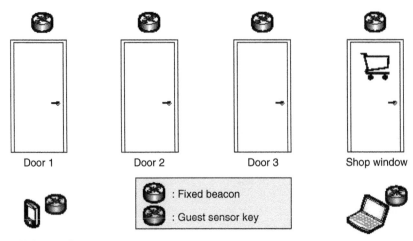

**Figure 10.9**   *AN demo setup*

For instance, a mobile user could receive advertisements depending on his/her position within the hotel shopping mall and/or automatically unlock the door of his/her hotel room by simply standing in front of it.

As a guest arrives at the hotel reception his/her PDA (or laptop) is registered, the terminal composes with the hotel AN and a special sensor device (sensor key) is given to the guest to enable his/her localization within the hotel area.

- Four sensors are placed within the demonstration area. They are positioned over different doors. Doors 1 to 3 are assumed to be hotel room doors, whereas the last door simulates the entrance of a shop.
- As the guest stands in front of his/her room, the PDA application triggers the opening of the door. Of course, access rights are checked by the hotel server in order to authorize such an action.
- The same test is subsequently repeated for a different door that, however, will not open as the user does not have the correct access rights.
- Finally, the guest stops in front of the shop entrance and, shortly, the main server collects the user position and creates a new routing group including the new customer. As new advertisements become available about some of the products in the store, these are sent to the guest using a multicast address which will be shared by all users in the RG. This increases transmission efficiency.

This demonstration has shown the possibility to integrate sensor networks capability on more complex systems and exploit the additional information into new kinds of services to the final users. The preliminary level of the AN framework definition during project phase I, has limited the integration depth of the realized system. But, also, if some functionalities are poorly implemented the realized scheme has exploited all its potentiality.

## 10.3  Wireless lamp control system

In this section we investigate a new application of wireless sensor devices. During recent years the technology has moved a number of applications to a wireless implementation. This trend is mainly driven by the high costs to realize wired connections of actuators or lighting systems but also open new possibilities on the realization of domotic houses. Here we focus the attention on emergency lighting systems.

### 10.3.1  Architecture definition

On emergency lighting systems, law requires periodic checks in order to guarantee a near to zero probability to having faulty devices. Those requirements, especially if the lighting system is particularly extended, need periodic service that can be really long and expensive. In past years in order to overcome this problem some solutions have been proposed. The simpler solution was the use of a 2-wire data bus that connects clusters of lamps to a central control unit. On alternative implementations the wiring is optimized and as in the data bus the lamp uses the same mains supply lines. The control units are connected to a central server that monitors on real time the status of the network. The server can also control actively the lamps, or in a more interesting case suggest a lamp check before a fault happens. On the sequel, we analyze the logical evolution of this network structures that moves the information exchange to a wireless connection.

The system is based on ZigBee modules fully integrated on the lamp package. On the network three groups of devices can be identified:

- bridge
- repeater
- lamp.

The bridge is the main device and has the functionality to connect the wireless network to the wired data bus and so to a control unit. The lamps are connected using multihop paths to the reference bridge and communicate only with the bridge. The network creation is initiated by the bridge and stopped manually when all lamps are added to the network. Leveraging on ZigBee standard implementation, the radio channel is automatically selected minimizing possible interference from nearest 2.4 GHz devices. A network identifier is also present and coded on wireless hardware in order to permit coexistence of different networks on the same area. Instead the repeater is a device necessary to enlarge radio coverage on radio signal shadow zone where a group of lamps is not reachable.

One of the goals of this project was the realization of a sufficiently precise prevision software able to check radio coverage and suggest or automatically estimate the best positions of bridge and eventually repeater devices. The problem is really complex because we have not only to consider propagation issues but also intrinsic limitations on ZigBee stack and control unit limits.

For example, ZigBee has a limit on the maximum number of hops and the number of children that a single device can connect due to a particular internal memory organization. These limitations are important because if radio coverage is guaranteed by neighbouring devices there is the possibility that there is no data path to the bridge due to protocol constraints.

## 10.3.2  Coverage prevision mechanisms

The prediction software uses an ACAD compatible draw as input map. Some different layers are used to store and load topology information on the considered environment. A single draw layer is used to record walls' positions with a different colour line where wall type changes. The lamps are included on different layers of homogeneous type. The software uses a multiwall propagation model as a good compromise between precision and computation complexity. This model considers only the main propagation ray between two devices and adds to the path loss an additional attenuation estimated using the number and the type of walls or obstacles crossed.

Initially, the software is required to position one or more bridge devices, so that a triangular matrix is filled with the propagation attenuation between all the network devices. A routing tree is also designed using protocol constraints on the maximum number of children and hops, in correspondence to the worst case propagation conditions. As a final phase if some lamps are not covered, the software is able to highlight the affected devices and identify in detail the problem. The network structure can be modified adding or changing position to repeaters or bridges until all the devices are covered by the radio signal.

The software prevision during first tests seem to match with good precision the real network behaviour, however a more detailed path-loss characterization is essential to guarantee good prevision on wide networks.

In order to tune the propagations model, the LQI (link quality index) parameter is considered. LQI is an index related to the packet error probability and estimated using preamble field of 802.15 PHY frame. Analyzing the measures realized on different conditions we found a high dependences from surrounding environment and LQI changes. Driven by these results we included on the prevision software the possibility to associate an additional attenuation to each network device.

## 10.3.3  Sample results

In Figure 10.10 an example of a radio coverage prevision result is reported. The second circle from the left is the bridge device and the other 3 circles represent 3 repeaters used on this case only with test purpose, all the other devices are lamps.

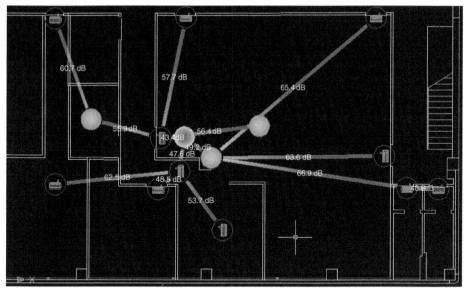

**Figure 10.10**  *Lamp radio coverage prevision*

On the high, left corner it is also possible to note the multiwall model applications; in fact the link with attenuation of 60,7 dB crossing two walls has a higher attenuation of the near link at 57,7 dB and also if distances are lower.

## 10.4  Experimental multiuser indoor localization platform based on WSN

This chapter describes the planning, algorithms and on-the-field experimental results for a multiuser indoor localization platform based on a WSN (Pavani, Costa, Mazzotti, Conti & Dardari, 2006). This activity has been performed within the project VICom referred to in Section 9.4.3.[1]

This case study is also related to concepts discussed in Chapter 8. In fact, some of the techniques and algorithms experimented are those discussed in Sections 8.3 and 8.4.

Here the feasibility of an indoor localization platform based on a WSN is investigated and verified under requirements given by real context-aware applications. To develop a localization system several aspects have to be taken into account at different levels: technology, algorithms, and specifications imposed by location-based applications. In particular,

---

[1] The authors would like to take this occasion to thanks all colleagues of others institutions involved in the project VICom, and in particular Professor F. Vatalaro, prime investigator of the project.

the following issues are addressed here: WSN deployment planning, characterization of indoor propagation, presence of low-reliability measurements, on-field characterization of the deployed platform. Two different low-complexity localization techniques are investigated and experimented. Starting from the typical context-aware application requirements we give an example of planning and tuning of the whole platform in addition to experimental results in terms of localization precision and accuracy. A case study of a complete context-aware platform (i.e., from the WSN to user applications) is presented, where the estimated position enables location-based applications and services.

Context-aware services have raised great interest in the last few years and one of the most important for context information is the user position. In an indoor environment this is a challenging task. Even if there are several techniques and technologies enabling indoor localization (Muthukrishnan, Lijding & Havinga, 2005), there is no unique answer to which is the best localization technique, since it depends on the application performance requirements and the operating environment conditions.

A common way to obtain position estimation in indoor scenarios is through RSS measurements, as done, for example, in WLANs, such as IEEE 802.11b/g (Bahl & Padmanabhan, 2000; Wang & Harder, 2006; Ekahau, n.d.). In these localization systems results tend to be not reliable for two main reasons. First, most of them require a training session to get many RF maps and, through them, try to predict users' location by comparing RSS with radio maps. The main disadvantage of these methods is their lack of scalability due to the need for extensive calibration and the lack of accuracy of RF maps due to the high dynamical characteristics of the indoor environments (for what concern, localization limits of WLAN see, e.g., Elnahrawy, Xiaoyan & Martin, 2004). The second problem of the WLAN localization approaches is the small number of access points for a given indoor environment. In the ideal case of perfect knowledge of the indoor radio channel, that is perfect estimation of the range of distances between the user and the reference points, three RF devices, deployed in three distinct positions, would be sufficient to localize a mobile user by triangulation. In realistic scenarios these ideal conditions are not met and only three reference points may not be sufficient for obtaining accurate localization in every condition. In addition, in WLANs the access points' deployment is constrained by power supply/data wired connections and focuses more on coverage maximization, rather than localization accuracy.

A good compromise between cost and localization performance is the adoption of a localization system based on WSN with RSS measurements capability. From these considerations we decided to test an indoor localization platform based on a WSN. That is, the small number of access points of the WLAN is replaced by a greater number of nodes of the WSN. Each node is smaller and cheaper than an access point and permits a deployment independent from power supply or data connections, allowing a greater flexibility and adaptability of the whole system. The use of RF environment mapping is discarded in place of a channel characterization approach that is more flexible in adapting to new scenarios and more resistant to the indoor environment dynamism.

To test our localization platform, we integrated it in a context-aware multiuser application which offers to each user an outdoor/indoor virtual visual guide as well as other location-based applications and services. Here, the platform main design issues are addressed with particular emphasis on the WSN deployment, on propagation channel modelling and on the test of localization algorithms of various complexity. On-field considerations and experimental results are then reported.

The design specification for our testbed platform is a precision less than or equal to 2 metres, in terms of MSE averaged in space and time, with an accuracy greater than or equal to 80% (fraction of time for which the target precision is reached) in several environments (e.g., indoor/outdoor, office, home, open-space, etc.).

## 10.4.1 Infrastructure and communication protocol

In our system, nodes are deployed in known locations (and, for this reason, are called *anchor nodes*) and generally listen in a low-duty-cycle sleep mode in order to save battery life. The sink nodes are located on the users' terminals in unknown positions to be estimated. Sink nodes interact with anchor nodes to obtain RSS information from nodes in electromagnetic visibility. Starting from these measurements, localization techniques can provide an estimation of the user position.

The communication protocol implemented is the following:

- The sink node, controlled by the location algorithm, begins a *broadcast phase*: it transmits a packet directed to all anchor nodes; we refer to this simply as *broadcast packet*.
- Anchor nodes reached by the broadcast packet awake and answer with a simple radio packet containing the anchor node identification number (ID) (reply packet).
- The sink node on the user terminal collects the reply packets from anchor nodes and, at the same time, estimates the RSS corresponding to each received packet. Then, all RSS from nodes in electromagnetic visibility are sent to the user terminal where the localization algorithm is performed.
- A new broadcast packet is sent every 500 milliseconds. Between two broadcast packets, that is within a localization round, all RSS are collected and sent to the localization algorithm.
- The broadcast sequence ends when a fixed number of broadcast packets are transmitted and the related sets of RSS measurements are provided to the localization algorithm. At the end of the sequence the localization algorithm estimates the user position. Then a new localization round begins.

When the anchor node is not reached by a broadcast packet for a certain period of time, it returns to the sleep state to save energy in the battery.

### 10.4.2  Localization algorithms

Two different localization algorithms have been considered: one with low complexity based on simplified multilateration, and one more complex based on Bayesian filtering (see Chapter 8 for more details). It is important to note that both algorithms do not adopt RF maps, thus strongly reducing the required time for calibration and, at the same time, granting more flexibility to environment modifications and dynamism.

*Low-complexity deterministic algorithm*
One of the first localization algorithms we realized and tested in our WSN platform is a deterministic localization algorithm characterized by a low computational complexity. It is called Min-Max (see Chapter 8) since it estimates the position of the subject within an area delimited by some maximum and minimum distances from the reference points (Langendoen & Reijers, 2003b). The Min-Max algorithm is directly derived from a standard lateration algorithm. The RSS values are properly mapped on estimated relative distances between the users and the anchor nodes, and these relative distances are then used to construct a square surrounding each anchor node.

Each square is constructed with coordinates within

$$[x^{(k)} - d^{(k)}, y^{(k)} - d^{(k)}] \times [x^{(k)} + d^{(k)}, y^{(k)} + d^{(k)}], \tag{10.1}$$

where $(x^{(k)}, y^{(k)})$ are the coordinates of the generic anchor node $k$ that answered to the broadcast packet, and $d^{(k)}$ is the relative distance between the anchor node $k$ and the object to be located. The bounding-box is obtained as given by

$$[\max(x^{(k)} - d^{(k)}), \max(y^{(k)} - d^{(k)})] \times [\min(x^{(k)} + d^{(k)}), \min(y^{(k)} + d^{(k)})]. \tag{10.2}$$

The centre of the bounding box is chosen as the object position. In Figure 10.11(a) the location method applied with three known points is shown. The idea behind the Min-Max algorithm is that standard lateration is computationally expensive and not reliable in case of distance estimation errors. This algorithm provides an accuracy near to standard triangulation, but reduces computational complexity.

*Algorithm based on Bayesian filtering*
As described in Chapter 8, the Bayesian filtering offers a powerful mathematical tool for the positioning problem through WSN (Fox et al., 2003). Here the localization problem is modelled as a dynamic system where the vector state $\underline{s}_n$, at discrete time $n$, represents the coordinates $(x_n, y_n)$ of the user. The state at time $n$ is described by (8.48), where now the vector $\underline{r}_n = [r_n^{(1)}, r_n^{(2)}, \ldots, r_n^{(K_n)}]$ is the vector of RSS in dBm received from the $K_n$ anchor nodes which answer to the broadcast packet sent by the user's node at time $n$.

It is possible to characterize two elements that operate in the model. The first one is the *mobility model* $p(\underline{s}_n | \underline{s}_{n-1})$ that represents the dynamic model for the system. It gives the

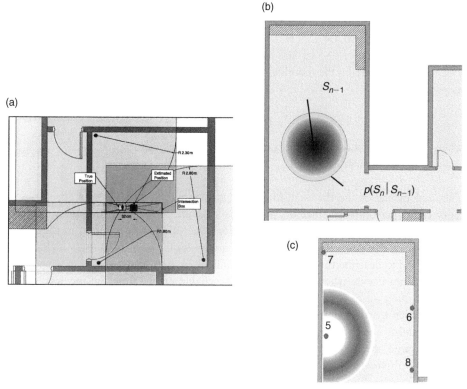

**Figure 10.11** *(a) Example of localization through Min-Max algorithm; (b) and (c) visual representation of the mobility and perception probability maps for Bayesian algorithm*

description of variation of the state $\underline{s}_{n-1} \rightarrow \underline{s}_n$, that is, the statistical description of user movement (Figure 10.11(b)). The second one is the term $p(\underline{r}_n|\underline{s}_n)$, that provides statistical information on the position at time $n$ starting from the measure vector collected at time $n$ (Figure 10.11(c)). This is the *perception model* and operates as an updater for the system state. In fact it updates and corrects the forecast obtained from the elaboration of the system dynamic model. Once the mobility and perception models have been properly constructed, from the belief function in (8.48) it is possible to extract the most likely estimation of the current state of the system.

By assuming statistical independent RSS measurements, the equation we adopted to describe the perception model is

$$p(\underline{r}_n|\underline{s}_n) = \prod_{k=1}^{K_n} p(r_n^{(k)}|\underline{s}_n), \qquad (10.3)$$

where $p(r_n^{(k)}|\underline{s}_n) = \mathcal{N}(r(d^{(k)}), \sigma_p^2)$, $\mathcal{N}(m, \sigma^2)$ represents the Gaussian distribution with mean value $m$ and variance $\sigma^2$, $r(d^{(k)})$ is the expected RSS in a point at distance $d^{(k)}$

from the $k$th anchor node (being the expected RSS dependent on the propagation model chosen), and $\sigma_p^2$ is the variance of the RSS measurements. The model for $r(d^{(k)})$ will be discussed in Section 10.4.4.

The equation describing the mobility model has been chosen to be parametric Gaussian as given by

$$p(\underline{s}_n|\underline{s}_{n-1}) = \mathcal{N}(d^2_{n-1,n}, \sigma_m^2), \tag{10.4}$$

where $\sigma_m^2$ is the motion variance, which depends on the user's speed, and $d_{n-1,n}$ is the distance between locations $\underline{s}_n$ and $\underline{s}_{n-1}$.

The implementation the Bayesian filtering algorithm within a real system enables us to enhance the performance since the probabilistic nature of this algorithm allows to take into account position and movement limitations due to obstacles present in the indoor environments. In fact, in addition to the visual map for the user interface, a logical map is integrated to describe the obstacles of the indoor environment considered. Hence, the algorithm can decide whether a position is reachable or not from the previously esti-mated position. In this way movements that imply walking through walls or obstacles are forbidden by the mobility model. Therefore, it is possible to take into account dif-ferent mobility conditions imposed by different kinds of indoor scenarios like tight cor-ridors or large open spaces. In particular, the mobility model $p(\underline{s}_n|\underline{s}_{n-1})$ is reshaped, by allowing a different variance per dimension, according to the geometrical disposition of obstacles starting from the Gaussian function in (10.4) (i.e., in a corridor the Gaussian shape is stretched along the direction of the corridor).

### 10.4.3 *The context-aware platform*

We now describe the steps carried out to develop the context-aware platform based on the WSN localization infrastructure.

The context-aware system we realized grants to the users an augmented immersive envir-onment and a model of ubiquitous computing through multiuser indoor/outdoor virtual guide services (here only indoor is considered). This example of multiuser context-aware service permits each user to localize himself in both outdoor and indoor environ-ments. This goal is achieved by the creation of a context-space with information for all users: while they share their positions and any possible public data, other special virtual entities share information about available services and about the physical environment which users are currently located in.

One of the main concepts of this testbed is the progressive discovery of the surrounding environment and the real-time adaptability to it: the management of the services that can be available to the user strictly relates to that philosophy. The presence of a service is not previously known by users' terminals, but 'announced' in the context space, thus a user is not limited to use a preset of applications.

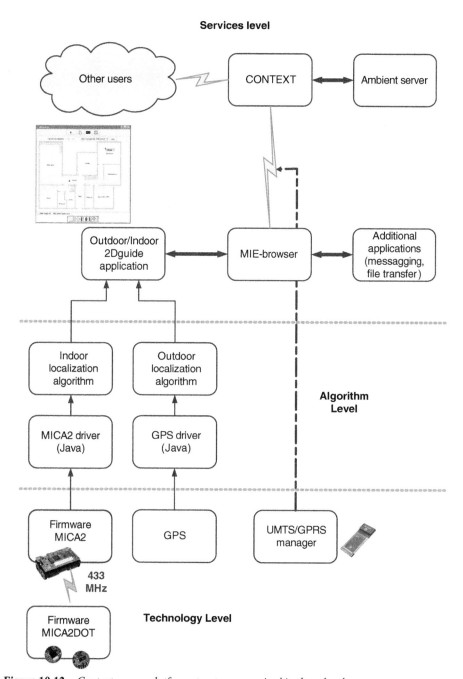

**Figure 10.12** *Context-aware platform structure organized in three levels*

The platform is three layered as depicted in Figure 10.12. The first layer (technology layer) manages all the devices used for localization and communication. The mid-layer (algorithm layer) collects data from the technology layer and processes contextual information for the upper layer, the service layer, responsible for providing interfaces and services to the user.

Due to requirements of users' mobility and immersiveness, in outdoor environments the localization task is accomplished by the GPS integrated with UMTS. In indoor environments, where the presence of obstacles hinders the use of GPS, this task is carried out by the WSN.

The system automatically recognizes, through technology interrogation, the environment surrounding the user as outdoor or indoor and downloads the correct map to visualize from the context space.

## 10.4.4  WSN configuration and deployment

The WSN is composed by crossbow mica wireless nodes (A. 2003): the MPR410 (also called MICA2) and MPR510 (also called MICA2DOT). Both these types of nodes use the ChipCon CC1000 as RF transceiver, which operates at 433 MHz, with a nominal bit rate of 19.2 Kbps and adopts the CSMA/CA protocol.

The MICA2DOT nodes are used as anchor nodes due their small size, whereas MICA2 nodes are used as sink nodes at the users' terminals. All anchor nodes are set to the minimal RF output power in transmission, that is $-20$ dBm, to reduce the energy consumption.

The deployment of anchor nodes is a key aspect when operating in realistic environments and it has been investigated in order to reach a good radio coverage and, at the same time, to limit the number of nodes deployed. The impact of nodes' distribution on the performance of WSN for environmental monitoring is described in Conti & Dardari, 2004.

We implemented and tested our localization platform in several environments presenting different geometrical and radio characteristics: in two WiLab laboratories in Bologna, Italy, and in two different floors of the Italian Ministry of Communications, Rome, Italy, to which we refer, as Ambient I (Figure 10.13(a)), Ambient II (Figure 10.13(b)), Ambient III (Figure 10.13(c)), and Ambient IV (Figure 10.13(d)), respectively. In order to minimize the setup time, some preliminary propagation characterizations of these environments have been realized through ray-tracing techniques (see Figure 10.14(a) for an example of ray-tracing characterization of Ambient I), which has been further validated by measurements campaigns.

Software simulations and on-field tests made in different environments allow us to find empirical rules on the required nodes' density to reach the target performance. As an example, we found that on average about an anchor node per 5 square metres is necessary. This density in slightly higher (a anchor node per 3–4 square metres) in tight

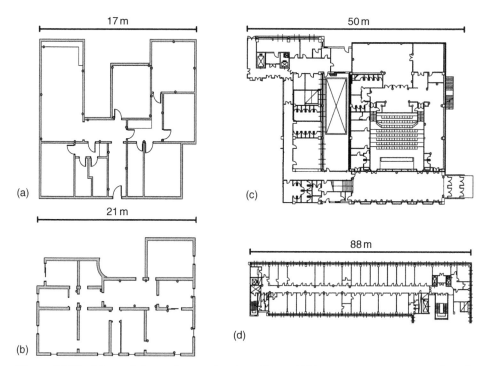

**Figure 10.13** *Maps of the environments where the on-field measurements were conducted: (a) Ambient I; (b) Ambient II; (c) Ambient III; (d) Ambient IV. The round dots represent the WSN anchor nodes deployed in the indoor environment*

corridors or offices, and a bit lower (an anchor node per 6–7 square metres) in large rooms. While performing anchor nodes installation the following empirical rules have been identified: 3–4 anchor nodes for a small to medium-sized room (up to about $4 \times 4$ square metres); 6 or more anchor nodes for bigger rooms; on corridors anchor nodes perform well if disposed alternately on the two walls at a distance of about 3 metres.

### Indoor channel characterization

The first, and very important, step of localization algorithms based on RSS is to map the RSS inputs into relative distances between the anchor nodes and the sink node. In particular, the Min-Max algorithm converts RSS into distance when constructs the bounding boxes in (10.1), whereas the Bayesian filtering algorithm uses the channel modelling within the perception model in (10.3) through the function $r(d^{(k)})$.

This conversion operation from RSS values into distances is a critical step in indoor environments. In fact the presence of furniture, objects, walls, people, and the induced reflections create a complex attenuation context for radio propagation. Thus we decided to keep the model of radio channel as simple as possible, and design robust algorithms that take possible error in the RSS-to-distance conversion into account. This choice allows us to perform a quick tuning of the channel model when the system is utilized in a particular indoor environment. At the same time the robustness of algorithms permits to setup the localization platform in new indoor environments without excessive time spent for calibration, while providing sufficiently accurate localization results. The goodness of this procedure has been verified with on-field characterization of the localization platform.

Thus, the following relationship between RSS and the distance has been considered which comes directly from (3.4) (Tam & Tran, 1995):

$$r(d) = -k_0 - k_1 \log_{10}(d), \tag{10.5}$$

where $r(d)$ is expressed in dBm and $d$ in metres. This channel model is dependent on the two parameters $k_0$ and $k_1$.

In order to mitigate fading and noise effects over the RSS samples, location algorithms do not use a single RSS sample, but use a mean value, $\bar{r}^{(k)}$, obtained over a fixed number of samples collected during the broadcast sequence:

$$\bar{r}^{(k)} = \sum_{i=1}^{n_k} \frac{r^{(k,i)}}{n_k}, \tag{10.6}$$

where $r^{(k,i)}$ is the $i$-th RSS value sampled from anchor node $k$, and $n_k$ is the number of answers from anchor node $k$ collected by the sink node in a single broadcast sequence. Note that, for a fixed number of broadcast packets in a localization sequence $N_b$, then $0 \leq n_k \leq N_b$.

The expression of the relative distance becomes

$$d^{(k)} = 10^{-\frac{\bar{r}^{(k)}+k_0}{k_1}}, \tag{10.7}$$

where $d^{(k)}$ is the relative distance between the anchor node $k$ and the sink node.

Before starting with the test on localization performance, we have carried out several experimental measurement campaigns and tests in order to estimate the optimal values for $k_0$ and $k_1$ in the test environment at the used radio frequency. In Figure 10.14(b) some results of these measurements for the indoor propagation channel characterization at 433 MHz are presented. It is possible to note how the experimental measurements follow a tendency curve, but, at the same time, present a large dispersion due to indoor environment effects on the electromagnetic propagation. By fitting experimental results

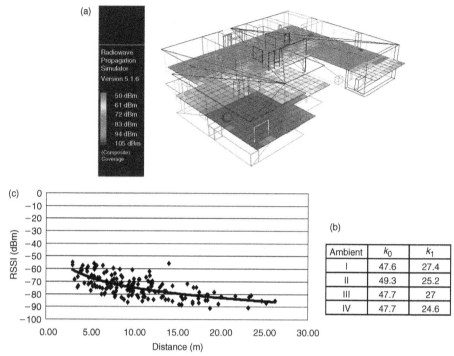

**Figure 10.14** *(a) Example of* 3D *electromagnetic field distribution obtained through ray tracing; (b) channel characterization at* 433 MHz; *(c) channel modelling coefficients for indoor environments*

with (10.5) it is possible to obtain the parameters $k_0$ and $k_1$ of the channel model (in Figure 10.14(c) the parameters in the different indoor environments are shown).

### RSS database

In order to estimate the parameters of the channel model and compare different localization techniques, an extensive measurement campaign has been carried out resulting in the construction of a vast measurements database. While testing the location platform performance in the different environments, a large amount of RSS values, more than 50,000, have been collected. This information forms a database which is usable for further investigations on the development of algorithms as well as for tuning the localization platform, thus representing one of the main results of this experimental activity.

### Experimental performance evaluation

On-field tests of localization technology and algorithms allow us to evaluate localization accuracy and precision. For the Min-Max algorithm an example of performance is presented in Figure 10.15 where the impact of configuration tuning in Ambient III is shown in terms of accuracy versus precision; it indicates that a proper choice of $k_0$ strongly affects the performance. On-field tests over the Min-Max algorithm allowed to enhance

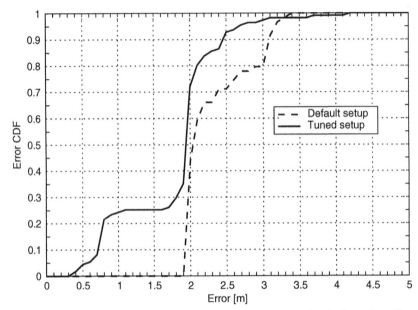

**Figure 10.15**  *Accuracy vs precision for Min-Max algorithm with different channel coefficients in Ambient III*

its robustness. The disturbed RF propagation can generate wrong distance estimation, and then non-overlapping-boxes cases that lead to unreliable positioning. To overcome this problem we modified the algorithm with an additional step in which new distance estimations obtained after the broadcast sequence are compared to the previous user's position. Since human movements are not arbitrary, some distance estimations could not be realistic, then they were discarded and wrong boxes were avoided. If the remaining boxes overlap, the position is normally obtained, otherwise the algorithm switch to a simple centre of mass technique: the user position is estimated at the centre of the three strongest anchor nodes' coordinates. In this case the positioning error slightly increases for this particular localization round.

As for the on-field test over the Bayesian filtering algorithm, an example of comparison between Min-Max and Bayesian algorithms performance is shown in Figure 10.16 where on-field experimental results (Ambient I) have been collected in the same conditions.

In Figure 10.17 another example of comparison between Min-Max and Bayesian algorithms performance is reported in terms of error cumulative distribution function. It is possible to see how the selected location test point in Ambient IV results is critical for the Min-Max algorithm thus leading to errors in the order of 5 metres. The Bayesian filter algorithm performance is improved as the localization error drops to about 2 metres fixed at an accuracy of 80%. It is evident that the superiority of the Bayesian algorithm is

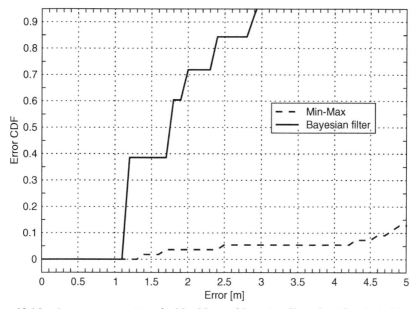

**Figure 10.16** *Accuracy vs precision for Min-Max and Bayesian filter algorithms in Ambient IV*

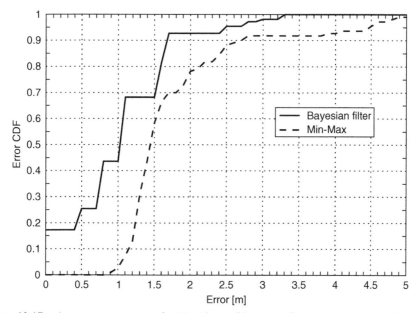

**Figure 10.17** *Accuracy vs precision for Min-Max and Bayesian filter algorithms in Ambient I*

at the expense of an increased computational complexity. With respect to the Min-Max algorithm, the Bayes localization is much more robust toward erroneous RSS measurements (i.e., wrong distance estimations) since the probabilistic approach also counteracts these events.

Finally, the performance of the Bayesian filtering algorithms has been evaluated in two successive phases in Ambient II. In the first phase the localization platform has been set up with a default configuration, without parameters tuning. In the second phase the localization platform configuration has been tuned after measurements and off-line testing on database, in order to match the propagation channel model to the particular indoor environment. In particular, Ambient II presents a high ceiling, so that values, obtained from measurements, for parameters $k_0$ and $k_1$ are quite distant from the default ones. Thus the default values ($k_0 = 47.5$ dB; $k_1 = 27.0$) in the second tests are replaced by environment specific values ($k_0 = 49.3$ dB; $k_1 = 25.2$). In Figure 10.18 the impact of configuration tuning in Ambient II is shown in terms of accuracy versus precision by reporting the error cumulative distribution function. As can be noted, after proper tuning, a location precision of a couple of metres can be achieved in most (more than 80%) of locations. This satisfies the requirements of the context-aware service that we developed.

Measures carried out using the context-aware platform allowed us to characterize the localization system in the presence of real propagation issues for several kinds of indoor environments. The performance of the localization algorithms has been optimized and tuned, with respect to the unpredictable propagation characteristics of the indoor environment, through experimental tests. At the same time, the usage of real devices enabled us to obtain a set of RSS measurements collected from a WSN deployed in the indoor environment. This collection of data permits to test and optimize the localization algorithms

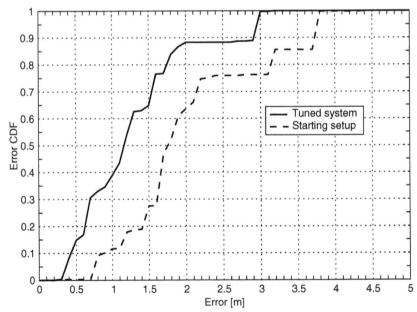

**Figure 10.18**  *Accuracy vs precision for Bayesian filter algorithm with different channel coefficients in Ambient II*

without the need of actual WSN, thus further investigation of algorithms and tuning techniques are possible from our results.

## 10.5 A positioning test-bed using UWB devices

### 10.5.1 The scenario considered

A measurement campaign has been performed to characterize the UWB channel behaviour in a typical office indoor environment at WiLAB, University of Bologna, in cooperation with the staff of Professor Moe Win from Massachusetts Institute of Technology, Cambridge. The building is made of concrete walls with thicknesses of 15 and 30 cm (see Figure 10.19). The considered environment is equipped with typical office furniture.

A positioning system composed of $N = 5$ fixed UWB beacons (tx1-5) (height 88 cm) used to localize one or more targets is deployed. Each UWB ranging device consists of one Time-domain PulseOn 210 (Time-domain) equipment operating in the 3.2–7.4 GHz 10 dB RF bandwidth.

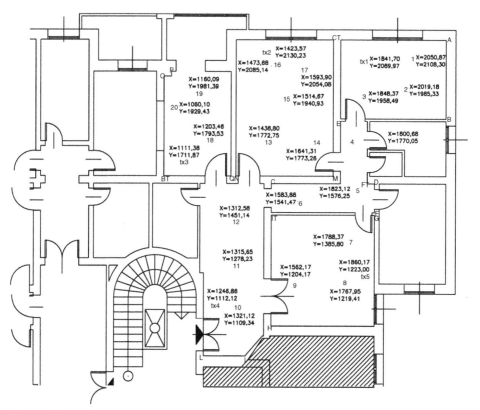

**Figure 10.19** *The scenario considered for measurements at the WiLAB laboratory of the University of Bologna. Coordinates are expressed in millimetres*

A grid of 20 possible target positions (rx1-20) at height of 76 cm has been defined from which range (distance) measurements have been taken by moving the target UWB device. For each ranging devices configuration, 1500 range measurements have been collected. Clearly, a couple of devices can be in LOS or NLOS condition depending on their relative locations.

We refer to a range measurement between two ranging devices as a direct path (DP) measurement if the range is obtained from the signal travelling along a straight line between the two points. A measurement can be non-DP if the DP signal is completely obstructed. In this case, the first signal to arrive at the receiver may come from reflected paths only. A LOS measurement is one obtained when the signal travels along an unobstructed DP, while a NLOS measurement can come from complete DP blockage or DP excess delay (in the latter case the DP is partially obstructed so that the signal has to traverse different materials, which results in additional delays). An important observation is that the effect of DP blockage and DP excess delay is the same: they both add a *positive bias* to the true range between ranging devices, so that the measured range is larger than the true value (from now on we therefore refer to such measurements as NLOS). This positive error can be as a limiting factor in UWB ranging performance, so it must be accounted for.

To this purpose, a set of ad hoc measurements have been performed to characterize the DP excess delay due to the presence of walls. In Figure 10.20, the measurement layouts are reported. In the first two layouts (Figure 10.20(a)), a simple concrete wall of thickness $d_w = 15.5$ cm and $d_W = 30$ cm, respectively, is present between two ranging devices. In the third layout (Figure 10.20(b)), two walls of thickness, respectively, of 15 cm and 30 cm are present. Ranging devices have been kept quite close to the walls to minimize the influence of multipath and better capture the DP excess delay effect. In particular, data have been collected for ranging devices located at 20, 40, 60, 80 and 100 cm from the surface of the walls. A total of 1500 range measurements have been collected for each configuration. In Table 10.1, the bias (mean) and standard deviation of the range estimation error are reported for the three layouts. As can be noted, the bias $\Delta$ due to the excess delay is approximatively equal to the thickness of the wall. The low value of the standard deviation means that the excess delay is the dominant effect in the estimation error with respect to the presence of multipath.

### A deterministic model for the bias: wall extra delay model
When the environment topology is known, a priori knowledge of the bias can be obtained starting from a first rough estimation of the target position, as will be considered later in the next section. In that case the bias value could be simply subtracted from the range measurement. The unbiased range measurements, that is, the distance estimates relative to the $i$th beacon, are therefore given by

$$d_i = r_i - b_i \quad \text{for } i = 1, 2, \ldots, N, \tag{10.8}$$

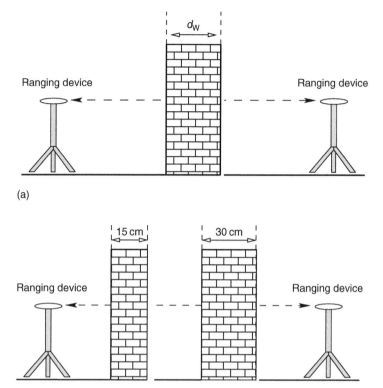

**Figure 10.20**  *The layouts considered for DP extra delay characterization. (a) one wall with thickness* $d_W = 15\,cm$ *(layout 1) and* $d_W = 30\,cm$ *(layout 2); (b) two walls with thickness* $15 + 30\,cm$ *(layout 3)*

**Table 10.1**  *Bias and standard deviation for different wall thickness*

| Layout, $d_W$ (cm) | Bias (cm) | Std dev (cm) |
|---|---|---|
| 1 wall, 15.5 | 16.4 | 3.7 |
| 1 wall, 30 | 29.5 | 3.2 |
| 2 walls, 15.5 + 30 | 45.2 | 3 |

where $r_i$ is the $i$th range measurement. The bias $b_i$ can be modelled as

$$b_i = \sum_{k=1}^{N_e^{(i)}} W_k^{(i)} \cdot \Delta_k \cdot c, \tag{10.9}$$

where $c$ is the speed of light, $W_k^{(i)}$ is the number of walls introducing the same extra delay $\Delta_k$ and $N_e^{(i)}$ is the number of different extra delay values. The total number of walls separating the ranging devices is $W^{(i)} = \sum_{k=1}^{N_e^{(i)}} W_k^{(i)}$. This model is known as the wall extra delay (WED) model (Dardari, Conti, & Lien, 2008). When all walls in the scenario have the same thickness, that is, $\Delta_k = \Delta$ for each $k$, (10.9) simplifies in

$$b_i = W^{(i)} \cdot \Delta \cdot c, \qquad\qquad (10.10)$$

### 10.5.2  Position estimation without priori information

The purpose on any positioning algorithm is, given a set of measurements (in our case the distances), to find the locations of the target(s). Positioning occurs in two steps. First ranging measurements are obtained, then the measurements are combined using positioning techniques to deduce the location of the target(s). According to the availability or not of certain priori knowledge about the environment topology and/or electromagnetic characteristics, different positioning strategies can be adopted.

The least squares (LS) solution (8.31) of the multilateration algorithm described in Chapter 8 is considered by setting $d_i = r_i$, for $i = 1, 2, \ldots, N$, where $r_i$ is the range measurement to the $i$th beacon. In Figure 10.21 the position RMSE for each location in the grid is reported in absence of any priori information about the environment topology for $N = 3$ and $N = 5$ beacons. It can be noted that the increase in the number of beacons does not necessarily correspond to a better positioning accuracy for all locations due to the fact that, in many cases, the added range measurements are subjected to a large error or the new geometric configuration is not forunate.

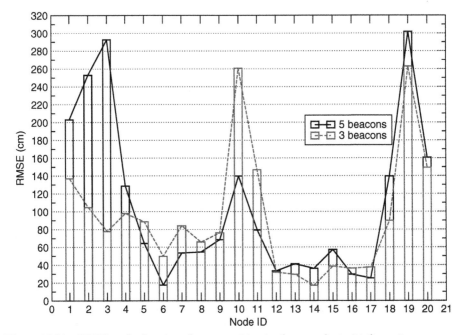

**Figure 10.21**   *RMSE as the function of target position in absence of priori information*

### 10.5.3 Positioning with priori information

Measurement results show that the NLOS configurations produces a bias which is, in many cases, the major source of error. By analyzing these data we have also seen that the bias is strictly related to the number of walls encountered by the signal. Assuming that the priori knowledge of the environment topology is available, is it possible to refine the target's position estimation once a first rough estimation has been obtained (e.g., the knowledge of which room the target is located in would be sufficient to this purpose). These considerations suggest to us the following two-steps positioning algorithm proposed in Dardari, Conti, & Lien, 2008:

- (1) A first rough position estimation $\hat{\mathbf{p}}^{(1)}$ is obtained using the LS method (8.31) by setting $d_i = r_i$.
- (2) Biases due to propagation through walls are subtracted to range measurements according to the WED model (10.9) using the first position estimate to determine the number of walls separating the target and each beacon.
- (3) A second LS position estimation $\hat{\mathbf{p}}^{(2)}$ is performed (refinement) with the corrected (unbiased) distance values.

In Figure 10.22 the position RMSE for each location in the grid is reported by using the two-steps algorithm described. By comparing Figures 10.21 and 10.22 it can be concluded that the correction of the range measurements using the WED model, in conjunction with

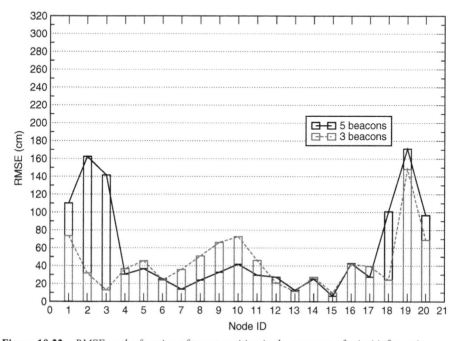

**Figure 10.22** *RMSE as the function of target position in the presence of priori information*

the knowledge of the environment topology, leads to a significant performance improvement. In general, accuracies less than 1 metre are achieved in most locations.

With the purpose to further improve position estimation accuracy, an iterative multilateration algorithm is proposed in Dardari, Conti, & Lien, 2008 in case more targets are present in the same environment and cooperate by exchanging range information not only from the beacons but also from each other.

## 10.6   Development of a multi-hop IEEE 802.15.4 network

Within the project carried out since June 2005 for two years at WiLAB in cooperation with the Italian SME (small medium enterprise) SADEL, an application scenario denoted as TCS (Tracking and Communication System) was investigated. IEEE 802.15.4 at 2.4 GHz was considered a candidate technology, owing to the requirement of low-cost devices and the need to transmit few bytes per second from each source, a condition that can be met with IEEE 802.15.4 devices.

### 10.6.1   TCS application scenario: description and requirements

The application scenario includes:

- a large number (in the order of tens) of IEEE 802.15.4 nodes deployed in a bounded environment (such as an airport, or a railway station) and connected to a backbone network using separate techniques, possibly wired (e.g., Ethernet), and
- an even larger number of mobile devices (in the order of hundreds) equipped with several types of sensors, transmitting through multiple hops their data, taken with a frequency of 1 Hz, to any infrastructure node.

The TCS application scenario includes nodes that move in an environment whose geometry is well known, at low speed, and whose location needs to be continuously tracked. To this purpose, the mobile nodes are equipped with self-localization techniques (such as GPS if outdoor) and they need to report their current position to the infrastructure every second. The nodes might also measure some physical properties of the environment they are traveling in, such as temperature, etc., also reporting every second the data measured. Examples of this scenarios are railway stations, where the nodes to be tracked are trolleys used by passengers to carry their own luggage, or the external part of airports, where the mobile nodes are the vehicles carrying food, gasoline, people, etc. to the aircraft from the buildings and vice versa; in the airport case, the mobiles might reach speeds of around 75 Kmh.

The infrastructure is composed of wireless nodes, located in fixed and properly planned positions in the environment; they all forward the data gathered to a single server, through a separate backbone network. As a result, the mobile nodes can report their data to any sinks in the area. A mobile node might reach the infrastructure through direct links, or multiple hops. However, apart from the data received from the mobile nodes, the infrastructure nodes do not exchange information about their configurations with the other nodes through the backbone network. Therefore, all radio resource management issues are completely distributed, as there is no global or even local knowledge of the network configuration at each node.

In summary, the requirements posed by the application were set as follows:

- frequency of sample generation by each source: 1 Hz
- number of bytes per sample: 20 bytes
- maximum delay between sample collection and delivery to the infrastructure: 5 s
- coverage of the area to be monitored: 90%
- maximum speed of mobile nodes: 75 Kmh
- environment to be covered: a rectangular area of no less than 0.5–0.2 square kilometres
- ability of the network to work even in the presence of interference caused by Wi-Fi hot spots.

The hardware platform selected by the SME was provided by Freescale, and was composed of boards equipped with a battery, some sensors, a RAM with 64 Kbytes, a MC9s08GT60 microcontroller and a MC1319x (x = 1 or 2 or 3) radio transceiver with maximum transmit power of 3.6 dBm.

## 10.6.2 TCS application scenario: tests

The above description of the scenario provides an indication on the main issues to be considered in order to verify whether IEEE 802.15.4 is a suitable candidate to realize the network. These issues are listed below.

*Mobility* IEEE 802.15.4 PHY layer was designed for stationary nodes. Therefore, the ability of the air interface to work in mobile conditions, even if at low speed (up to 75 Kmh), must be checked. Moreover, the association procedure in IEEE 802.15.4 devices requires some time to be completed and this is a very relevant step to be taken for each link before the true exchange of data from the mobile node to the sink can take place. Owing to the fact that the topology of the network changes very frequently because of the movements of mobile nodes, the time needed to complete the association procedure must be checked.

*Interference* Since in such environments other networks working at 2.4 GHz might co-exist (e.g., Wi-Fi hot spots), it is also very important to check the ability of IEEE 802.15.4 to work in realistic interfered conditions.

*Throughput* According to IEEE 802.15.4 nomenclature, the infrastructure nodes will play the role of the PAN coordinators, while the mobile nodes are full function devices able to forward data transmitted by other nodes. Each PAN coordinator uses one of the 16 carrier frequencies available at 2.4 GHz, according to a proper planning, or to a self-organizing distributed channel selection procedure. Many devices might be simultaneously connected to a given infrastructure node, in the order of tens. So, even if few bytes per second are transmitted by each node, the sinks (the coordinators) might gather a significant amount of data every second. IEEE 802.15.4 has a channel bit rate of 250 Kbit/s when used at 2.4 GHz, and this seems to promise that no throughput problems will be encountered. However, owing to the protocol overhead, at application layer the throughput can be significantly lower. So, this is also one of the aspects that needs to be checked.

If all the above checks are passed by the technology, then it is proven that IEEE 802.15.4 is a suitable candidate and the software to prepare the application can be completed.

It was then decided to test the technology on the field, performing field trials. The following subsections report on these tests.

### 10.6.3   TCS application scenario: mobility test

The IEEE 802.15.4 PHY layer was designed for stationary nodes. The effects of mobility can play a significant role on the performance of a digital receiver. The node movement can determine a Doppler shift of the carrier frequency with respect to the transmitted signal, proportional to the speed. Multipath components can also produce signal distortion in the presence of a Doppler shift. The overall effects of such phenomena on the performance of a digital link can not be predicted easily with theoretical tools. A field trial can show whether they represent a significant limit for the performance.

It was then decided to test the IEEE 802.15.4 nodes under mobility conditions, with one node acting as PAN coordinator (denoted as C in the following) located in a fixed position, and a device (denoted as D) running over a vehicle at different speeds, trying to associate to the PAN, and sending data. The field trial was built as follows. A large and empty street was considered as an environment for the measurements. The node C was located on one side of the street, about 70 cm above ground, and the node D was placed inside a car (at about 100 cm above ground) running at constant speed along the empty street. Both nodes were equipped with omnidirectional antennas. Pictures of both nodes are shown in Figure 10.23.

The maximum transmission range of the coordinator was measured, setting the transmit power of nodes at the nominal level of 0 dBm: received power was found to be above the receiver sensitivity for distances below about 100 m, generating a coverage length of about 200 m owing to the use of an omnidirectional antenna mounted on node C. The car was run along the street at speed $v$ which was kept constant for about 300 metres, 150 before and 150 after the point where the node C was located. Figure 10.24 sketches the field trial geometry.

**Figure 10.23**  *PAN coordinator C and device D*

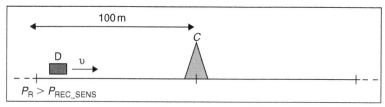

**Figure 10.24**  *Field trial geometry*

The following steps where then taken:

1. Before launching the car, the coordinator performed a frequency scan procedure to select the least interfered channel among the 16 available in the ISM band at 2.4 GHz.
2. The best carrier was selected and the coordinator was set in a status waiting for association requests, based on the non-beacon-enabled mode of IEEE 802.15.4 MAC, with acknowledge.
3. The device in the car was set in a status of cyclic search for a coordinator. Basically, the node D was trying each frequency by sending a packet and waiting for an acknowledgement by C; if no packet was sent back before a time out set at 0.5 s, the next channel was tried. Once an acknowledgement on a channel was received, the node D started immediately the transmission of data with packets having a payload of 20 bytes, counting the acknowledgements received from the coordinator.
4. The car was launched at constant speed $v$ and all packets transmitted and received by both nodes were recorded.

The experiment was repeated several times (only five for practical reasons) in order to generate average values of the measurements.

The scope of the field trial was to check whether the association procedure was successful and the data transfer efficient at various speeds.

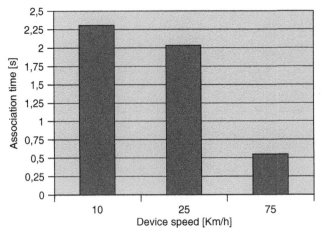

**Figure 10.25**   *Association time measured at different speeds*

Figure 10.25 shows the association time measured at different speeds. The association time is defined as the time interval between the reception of the first packet from C and the transmission of the first data packet sent by D to C. The Figure shows that the association time is shorter for larger speeds. The reason for this is in the fact that at very low speed the car keeps close to the border of the transmission range of the coordinator for some seconds; in such conditions, even if a first packet transmission was successful, a large packet error rate is experienced and some packets can be lost. Since the association procedure requires the transmission in both directions of the link of several packets, the loss of some of them determines an increased time needed to complete the procedure. When the speed was larger (above 25 Kmh in our experiments), the car moved quickly towards the coordinator, finding better channel conditions and low packet error rates after the first packet was received, thus bringing the association time to a value which is the minimum possible according to the IEEE 802.15.4 MAC procedure: about 0.5 s. From the viewpoint of the distance travelled by node D during the association procedure, the case at 25 Kmh was the worst encountered (with measures performed at 10, 25, 50, 75 Kmh): about 14 metres were run before the procedure was concluded. In other words, 14 out of 200 metres of coverage were lost (i.e., 7 %) to complete the association to the coordinator, before data transmission could take place.

Figure 10.26 reports the throughput measured during the data transmission phase at various speeds. The throughput was computed as the ratio between the amount of data transmitted successfully (packets sent by D and acknowledged by C) and the time the node D was associated to the coordinator. Both numerator and denominator of such ratio can depend on $v$: in fact, the amount of data successfully transmitted depends on the packet error rate which might increase for larger speeds, and the time the node D was associated is shorter when increasing the speed of the car. The Figure shows that the throughput does not change significantly with $v$ ranging from 10 to 75 Kmh, with the maximum value obtained for larger speeds. The reason for this might stand in the fact

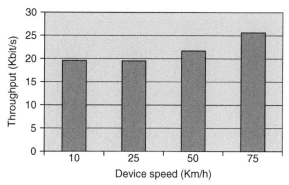

**Figure 10.26** *Throughput measured at different speeds*

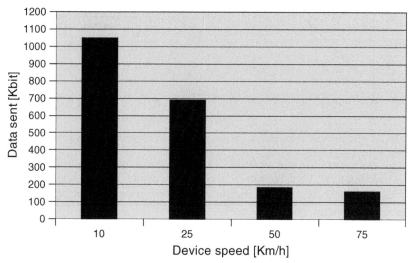

**Figure 10.27** *Amount of data successfully transmitted at different speeds*

that at 75 Kmh the time the device was associated to the coordinator was shorter. Indeed, a better figure to be considered is the amount of data successfully transmitted by node D during the time it was associated. Figure 10.27 shows such figure as a function of speed. As expected, at 75 Kmh a smaller amount of data was delivered, owing to a shorter time available for transmission.

The data throughput ranges from about 18 to 25 Kbit/s, values that are significantly below the throughput of 38 Kbit/s that can be measured with both transmitting and receiving nodes being still, with link distance equal to 1 m; Figure 10.28 shows the values of throughput measured, both in acknowledged and unacknowledged mode for the non-beacon-enabled mode of IEEE 802.15.4, in such ideal conditions when the payload size is changed. The reason for the lower throughput (18–25 instead of 38 Kbit/s) clearly can be found in the non-ideal channel conditions of mobile environments.

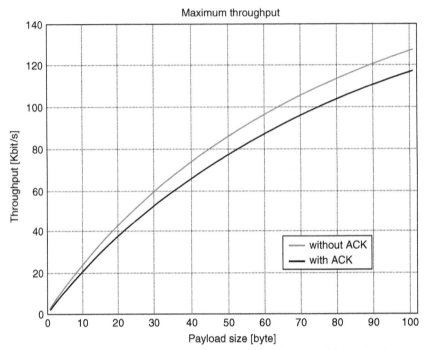

**Figure 10.28** *Throughput measured in ideal conditions, as a function of the payload size*

In all cases, however, transmission of significant amount of data was possible. This proves that IEEE 802.15.4 can be efficiently used even in a mobile environment, at least at speed no greater than 75 Kmh.

### 10.6.4 TCS application scenario: tests with interference

IEEE 802.15.4 uses ISM frequency bands that can also be occupied by other wireless systems, like for example, Bluetooth devices or IEEE 802.11a/b/g/ hot spots. Since the network to be deployed under the TCS specifications needs to work in environments like railway stations, where Wi-Fi hot spots might exist, it was important to check the ability of the IEEE 802.15.4 network to work in the presence of such sources of interference.

An IEEE 802.11 hot spot transmits a signal whose spectrum approximately spans over a bandwidth of 20 MHz. Since the IEEE 802.15.4 signal occupies a frequency channel of 5 MHz and the entire ISM band at 2.4 GHz is 80 MHz large, it is expected that even in the presence of a few hot spots, the IEEE 802.15.4 network can work efficiently, provided that a suitable channel is selected by the PAN coordinator.

Given such expectation, a test was performed in an indoor environment where two Wi-Fi hot spots were active and transmitting continuously, at different frequencies, and the packet error rate was measured on the IEEE 802.15.4 links.

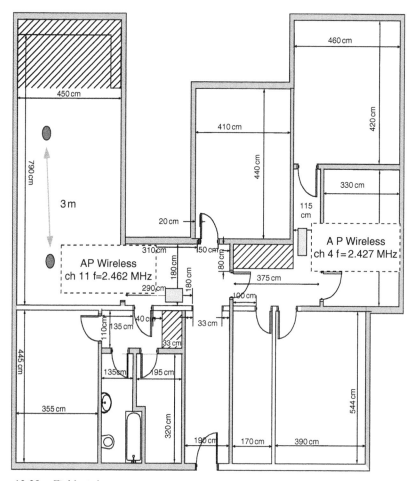

**Figure 10.29** *Field trial geometry*

Figure 10.29 shows the geometry of the field trial, and also the carrier frequencies used by the two hot spots is reported. Figure 10.30 shows the frequency channels defined for IEEE 802.15.4 and IEEE 802.11 systems. Based on these numbers, one can expect a large interference contribution on 4 + 4 802.15.4 channels, while the other eight channels should not be impacted.

Figure 10.31 shows the packet error rate measured over the IEEE 802.15.4 link, with 1000 packets transmitted. Transmit power was set at minimum (−16 dBm), nominal (0 dBm) and maximum (3.6 dBm) levels. The Figure shows that as expected a packet error rate significantly above 0 was found at the channels close to the carrier frequencies of the two hot spots, at least at minimum transmit power, while the effect of interference is almost absent when the transmit power is set at 0 dBm or above. This proves that IEEE 802.15.4 can work efficiently in an environment with active Wi-Fi hot spots, provided a suitable channel selection procedure is included that minimizes the level of interference.

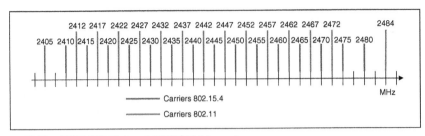

**Figure 10.30**   *IEEE 802.15.4 and IEEE 802.11 carriers*

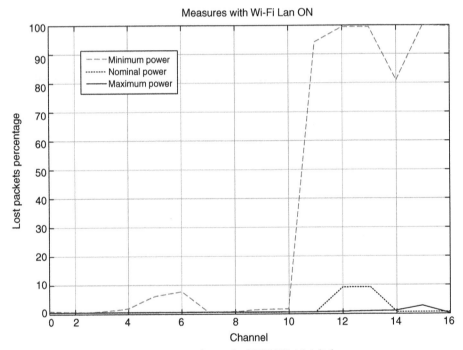

**Figure 10.31**   *Packet error rate measured over the IEEE 802.15.4 link*

An unexpected high level of packet error rate was measured at channels 15 and 16. It was found to be generated by a source of interference nominally external to the ISM band, which in fact was also generating e.m. activity within the highest portion of the ISM band.

### 10.6.5   *TCS application scenario: throughput test with multiple hops*

Even if the channel bit rate of IEEE 802.15.4 links at 2.4 GHz is 250 Kbit/s, the application layer throughput is significantly smaller because of the protocol overhead, mainly due to the MAC (sub)layer. Figure 10.28 reported above has shown that even if the payload size is maximum (about 100 bytes), the application layer throughput can not go above

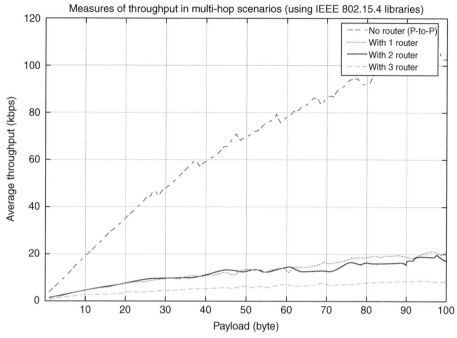

**Figure 10.32** *Throughput measured with one, two or three routers*

around 130 Kbit/s for point-to-point links. However, in the presence of multihop links, the throughput can be significantly lowered owing to the potential interference among the separate hops disturbing each other.

Figure 10.32 reports measures performed with one, two or three routers located in between a transmitter and a receiver. With one router, two hops form the link between source and destination. With two routers, three hops, etc. The Figure refers to the beacon-enabled mode of IEEE 802.15.4. It clearly shows that the throughput can be significantly lowered when multiple hops are included. With a payload size of 20 bytes, the throughput lowers from about 35 Kbit/s to about 7 Kbit/s with two routers, and even worse with three routers.

## 10.6.6 TCS with IEEE 802.15.4: selection of the transmission modes

According to the description of the application scenario, multiple sinks are deployed in the rectangular area. Each of them will play the role of PAN coordinator, serving all devices travelling the sub-area (in the following denoted as cell) they cover. Concerning the frequency channel used by coordinators, the following options can be considered.

- They all use the same channel, manually selected during their configuration
- They use different channels, suitably reused, manually selected during their configuration

- They all start a frequency scan procedure and autonomously select the least locally interfered channel.

With the third option, the mobile nodes have no knowledge of the frequency reuse pattern, as there is no a priori information on the channel frequencies used by the coordinators, and they do not exchange such information through the backbone network. As a result, once the devices de-associate from a PAN coordinator because they are leaving the cell, they need to start a new frequency scan procedure in order to find the next channel, used by the coordinator they are reaching. With the second option, the same is true unless only two frequencies are reused in the area. In fact, if only two channels are used in an alternate way (e.g., channels 4 and 8, like 4-8-4-8-4-) when moving from one coordinator towards the next one, the mobile nodes can be made aware of the fact that when they de-associate from a coordinator using channel 4, then the next channel used is 8. No frequency scan is needed, and this reduces the time needed for associating to the next coordinator in the line. With the first option, no frequency scan or reselection is needed and this simplifies the development of the software to be implemented over the IEEE 802.15.4 nodes.

However, if all nodes use the same channel, then a strong co-channel interference is present. Even if two channels are reused alternatively, the interference can be severe. Moreover, when using multiple channels, adjacent interference can also play a significant role.

To assess the ability of IEEE 802.15.4 to work in such conditions, a simulation tool was developed. It considers a rectangle having sides $a$ and $b$ with coordinators uniformly distributed along the perimeter, with a distance $D$ separating each other. Then a mobile node is placed along the perimeter, in all possible positions; the signal-to-interference ratio (SIR), $\gamma$, for the mobile-to-infrastructure link is computed for all positions. The power loss law is given by $L = k_0 + k_1 \log(d) + s$ where $d$ is link distance and $s$ is a Gaussian r.v. with zero mean and standard deviation $\sigma$; the power loss is expressed in logarithmic scale. A point over the perimeter is assumed to be covered if $\gamma$ is above a given threshold set at 6 dB.

Adjacent and co-channel interference were taken into account. Concerning the co-channel interference, we simply evaluated the power that the device receives from the coordinators transmitting on the carrier's same channel. For what concerns the adjacent one, we started from the analysis of the power spectrum of the signal generated by ZigBee devices, that is equivalent to an MSK spectrum; then, the percentage of spectrum that overlaps the adjacent channel (and that generates interference) was evaluated.

The coverage probability of the scenario is then assessed as the ratio between the number of covered points and the total number of positions considered. We denote as outage probability the complementary of the coverage probability; the target values for outage probability are below 10%. The evaluation is done for various reuse patterns, with one or more channels used.

The following set of figures shows some simulation results, obtained by fixing $a = 1000\,$m, $b = 500\,$m, $D = 100\,$m, $k_0 = 15$, $k_1 = 20$, $\sigma = 5$, transmit power set at $0\,$dBm, receiver sensitivity of $-85\,$dBm.

Figure 10.33 shows results over 200 metres of perimeter when using a single channel. The outage probability value is about 0.33. In Figure 10.34 the case with two frequencies (channels 0 and 8) is shown; in these conditions, the outage probability is about 5.5%, a value which fulfils the requirements.

By using four frequencies (channels 0, 4, 8, 12) the outage probability falls below 0.01. Figure 10.35 shows the values of $\gamma$ in this case.

These results show that with two channels alternatively reused, interference effects can be kept under control and only about 5% of the scenario suffers from excessive interference. Therefore, option two, above, is a viable solution. On the other hand, the signal-to-interference ratio is too low in one-third of the scenario if the same channel is used by all coordinators; therefore, the first option can not be implemented.

According to these discussions, a situation with two channels used alternatively seems to be the best option, and will be considered as the final choice in the remainder of the case study.

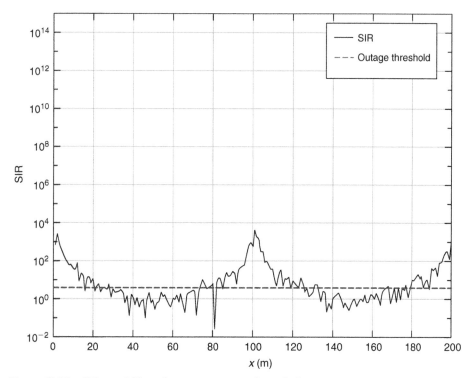

**Figure 10.33**  *SIR over 200 m of perimeter, case with single frequency*

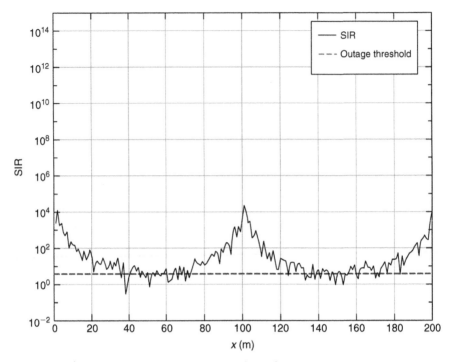

**Figure 10.34**   *SIR over 200 m of perimeter, case with two frequencies*

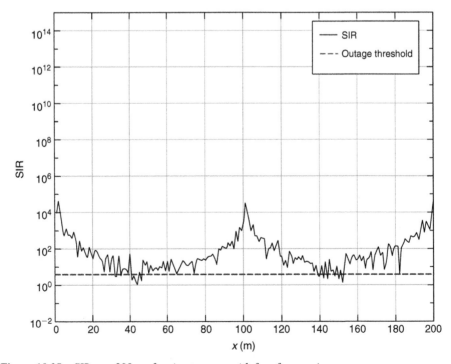

**Figure 10.35**   *SIR over 200 m of perimeter, case with four frequencies*

## 10.6.7 TCS with IEEE 802.15.4: system design

*Possible implementations: beacon enabled or non beacon-enabled; static or dynamic*

A step-by-step methodology has been used to develop the TCS project; initially the features of a communication on a mobile network based on IEEE 802.15.4 have been studied, then the possible problems related to the cellular architecture have been analyzed. In particular, some network characteristics have been examined in detail, to study and theoretically compare several possible implementations of the project:

- beacon enabled (BE) network (see Section 6.1)
- non-beacon enabled (NBE) network (see Section 6.1)
- dynamic scenario: the position of the PAN coordinators can change, they can move through the scenario (for instance, they could be located on moving vehicles)
- static scenario: the PAN coordinators are fixed.

Concerning the coordinators' dynamism, a dynamic scenario implies that the coordinators involved in the cellular architecture need to communicate among themselves to collect information on their neighbours. Unfortunately the coordinators will be probably distributed in the area in such a way that they cannot communicate with each other using 802.15.4, but they will need a different technology (e.g. Wi-Fi, Ethernet, . . . ) to exchange information. Therefore, we focus the attention on the static scenario.

For what concerns the choice between the two different IEEE 802.15.4 MAC protocol modes (BE and NBE), we need to take some of their characteristics into consideration:

- NBE: It is not possible to use the beacon packet as a synchronization signal, therefore the devices are not synchronized. If they are not associated to any coordinator, they need to scan the frequencies till they find a coordinator that hears the device's packets and starts the association procedure.
- BE: A synchronization signal (given by the beacon packet) is available, and, thanks to the periodic beacon sent by the PAN coordinators, a non-associated device will easily hear a PAN coordinator during its scan phase. Moreover, some existing MAC procedure that points out an error after three lost beacon packets is available, and could be used to realize the de-association.

Concerning the association times, through some easy calculation it can be shown that they (inclusive of both scan times in the worse case and decision times) are very similar for the BE and NBE cases. So, it can be concluded that the BE solution seems better than the NBE, since it allows the use of existing MAC procedures and simplifies association and de-association. Moreover, it has another advantage from the traffic point of view; the NBE case involves one more packet in the initial phase, and this implies more overhead.

Before implementing the whole application, some preliminary tests have been carried out to verify the reliability of the above theoretical conclusions. The experiments confirmed the drawn conclusions, but the tests with the BE mode showed one crucial software problem, not present with the NBE mode. When a device is associate to a PAN coordinator and the link quality lowers, if the communication falls down the device starts sending association requests and the coordinator stops working. Owing to this problem, the NBE MAC mode was chosen for the implementation of the TCS in a static scenario.

### Leacky Bucket

The TCS scenario foresees several coordinators deployed in the area where mobile devices can run. As soon as the quality of the communication between the devices and the PAN coordinator starts decreasing, probably the device is moving away from the coordinator, approaching the boundaries of its coverage area. In this situation, the device needs to get some actions to discover if there is any coordinator capable of providing it with a higher received power in its neighbourhood, in order to associate to it.

To realize it, a Leaky Bucket algorithm has been introduced to determine the moment in which the device is forced to a scan procedure, in order to find a possible new coordinator. If the device, through the scan, finds an available coordinator, it de-associates from its previous coordinator and tries to associate with the new one.

A counter (whose value cannot go below zero) has been introduced to implement the Leaky Bucket; it increases by one at each lost packet, whereas it decreases by one for every successfully transmitted packet (i.e., when the device receives the coordinator's acknowledge). As soon as the counter exceeds a certain threshold (N lost packets), we can assume that the losses are not related to temporary signal attenuations (due, for instance, to fast fading), but they are caused by the fact that the device is close to the end of the coverage area, and a new scan phase is forced.

The application requirements set the frequency of packet transmissions to 1 Hz; however, a moving vehicle takes less than a second to go across a zone strongly affected by fading (calculated at about 20 ms at 10 Kmh speed). Therefore, the Leaky Bucket threshold for our application will not be fixed as a consequence of the mean width of an area of maximum fading attenuation, but it will be related to the mean vehicle's speed, to the mean coverage area size, and to the mean distance among coordinators.

To give an example, let us consider several PAN coordinators distributed along a street (separated by a distance of 100 m) and characterized by a coverage length of about 100 m, and a vehicle, equipped with an IEEE 802.15.4 device, that moves along the street at a speed of 50 Kmh. The device sends to its coordinator one packet every second, and let us suppose that in a certain instant it is communicating with the coordinator C1. If we fix the Leaky Bucket threshold N to 4, and the device is approaching the end of the coverage area, the device will take 5 seconds (that is 5 consecutive lost packets) before realising it is outside the C1 range, and starting the de-association procedure. In 5 seconds the device will cover about 70 m, thus it might exit as it risks going out of the next coordinator coverage area without having the time to associate with it.

To avoid this situation, there are three possibilities: reduce the Leaky Bucket threshold, increase the frequency of the samples generation and transmission, or increase the distance between the coordinators. By appropriately choosing the values of these three parameters according to the particular scenario specifications, the optimum configuration the desired application can be obtained.

### Experimental test: scenarios

The first mobility test, described in Section 10.6.3, shows that IEEE 802.15.4 can be efficiently used to build up a mobile sensor network. Starting from those results, it was decided to make another measurement campaign to examine the actual system performance in the TCS cellular scenario and to test the association and de-association procedures in three different realistic situations. The experiments have been realized in a large street characterized by some natural obstacles (trees) and bordered by warehouses.

In the three tested scenarios, an IEEE 802.15.4 device (D), generating one sample every 0.5 s, is positioned inside a vehicle (about 100 cm above ground, see Figure 10.36), and the vehicle runs the street alternately in both directions. The device speed has been fixed to 50 Kmh in each test session, because the goal of this second test phase was to verify the impact of some parameters (e.g., the Leaky Bucket threshold, or the transmitted power) on the system performance, having previously tested the feasibility of the communication in mobility conditions up to 75 Kmh.

In the first of the three tested scenarios, the PAN coordinator (denoted by C) was positioned in a fixed location, in the centre of the street about 70 cm above ground, as we can see in Figure 10.37. It's interesting to note that the group of trees decreases the signal intensity, making the coverage length different in the two opposite directions.

The transmission power has been changed during the tests from 0.4 to 3.6 dBm; the results do not point out a deep impact of the transmission power on the system performance. This is fair, because the receiver sensitivity is −92 dBm (see Section 6.1), and a 3 dBm increase on the transmitted power does not significantly impact the received power.

**Figure 10.36** *PAN coordinator C and device D, second test phase*

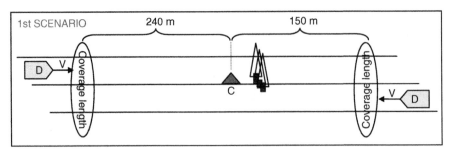

**Figure 10.37**   *First test scenario*

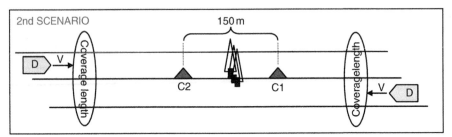

**Figure 10.38**   *Second test scenario*

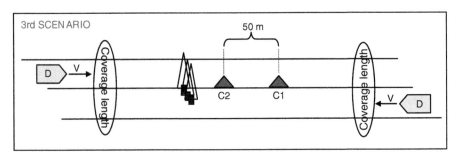

**Figure 10.39**   *Third test scenario*

The second tested scenario is characterized by two PAN coordinators separated by a distance of 150 m, denoted by C1 and C2 (see Figure 10.38). The coordinators are independent: they autonomously start the initialization procedure, and scan the channels, choosing the one less interfered. The objective of this experiment was to test the capability of the mobile device D to properly associate and de-associate in a cellular scenario (in this test we have two coordinators, C1 and C2, and consequently two cells), and to correctly deliver the most data as possible to the coordinators. In particular, we examined how the Leaky Bucket procedure threshold affects both directly the association and de-association times and indirectly PER and throughput.

The third scenario could seem similar to the second, but it gives very different results. As in the previous case, two coordinators C1 and C2 are used, but the distance between them has been decreased to 50 m (as shown in Figure 10.39), to enlarge the overlapping of the coverage areas related to the two coordinators. This enlargement on one side increases the probability that D can communicate with at least one coordinator when it is inside the overlapping zone, but on the other reduces the whole covered area. The main goal of this test was to measure the same quantities considered in the second scenario in order to compare the results, and determine the impact of the cells' overlapping on the performance of the system.

*Results*

One of the main objectives of this second phase of tests was to verify the system behaviour in a cellular scenario, and in particular to verify if the device is capable to de-associate from its coordinator in the case of low quality of received signal, and associate to the next coordinator that provides it with sufficient signal strength. The obtained results show that the device correctly executes the association and de-association procedures, successfully delivering its data to the coordinators.

To evaluate the time taken by the device to change its PAN coordinator, we refer to the second and third scenario described above. We are going to examine the time for the device de-association from the first coordinator and that for the association to the second coordinator. Through the experiments we obtained a good confirmation of the de-association times expected according to the implemented Leaky Bucket algorithm, as we can observe in Figure 10.40. The Leaky Bucket is based on the counting of lost packets; with threshold equal to four, after five consecutive lost packets the device executes the de-association procedure; this implies that, if the device sends one packet every 0.5 seconds, the times obtained with the tests are exactly that expected. The values in Figure 10.40 are averaged on the quantities obtained in both scenario 2 and 3, because the distance between the coordinators does not affect the de-association time.

Denoting by re-association time the time spent by the device to associate to another PAN coordinator after a de-association, Figure 10.41 shows the re-association times related to the scenarios 2 and 3. The information on the two scenarios have been shown separated, to compare the two results.

If we denote by $S_{\mathrm{LB}}$ the value of the threshold of the Leaky Bucket algorithm, starting from the second scenario we can observe in Figure 10.41 that for $S_{\mathrm{LB}} = 4$ the re-association time is lower than in the case $S_{\mathrm{LB}} = 2$. This is due to the fact that with $S_{\mathrm{LB}} = 4$, the device's de-association time is greater, therefore when it tries to re-associate, it will be probably closer to the second coordinator than in the case $S_{\mathrm{LB}} = 2$, and consequently it will experience a high quality of the channel and connect quickly to the coordinator. This does not happen in the third scenario, where the coordinators are very close, so that if $S_{\mathrm{LB}} = 4$, when D is de-associated from the first coordinator probably it is too far from the second coordinator too, and the re-association fails. The same observation can be drawn if we compare the re-association times in the two scenarios, $S_{\mathrm{LB}}$ being equal. Analyzing Figure 10.41, we can then conclude that $S_{\mathrm{LB}}$ strongly impacts system performance, but there is no an optimum value for that threshold; $S_{\mathrm{LB}}$ needs to be chosen in accordance with the cellular planning.

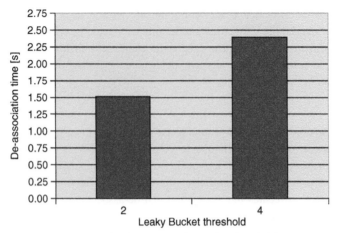

**Figure 10.40**   *De-association time for different Leaky Bucket thresholds*

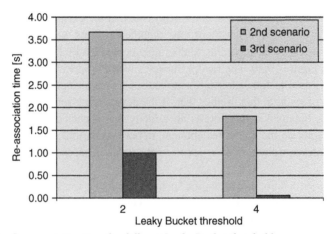

**Figure 10.41**   *Re-association time for different Leaky Bucket thresholds*

Concerning the data throughput and the PER, the results have been kept separated for the different scenarios and different $S_{LB}$. Moreover, the performance related to the two coordinators have been kept separated. In Figure 10.42 it is possible to observe that the throughput related to the first coordinator (meaning by first alternately C1 or C2 depending on the vehicle direction) does not change significantly, and it is similar to that in the case of one coordinator; this is obvious because the presence of a second coordinator, independent of the first and working on a different channel, does not generate interference. However, we can notice that in the $S_{LB} = 4$ case, the throughput is slightly lower than in the $S_{LB} = 2$ case, in particular in scenario 2; this is due to the fact that increasing $S_{LB}$, the time during which D remains connected to the first coordinator increases, the number of corrected delivered packet being equal, therefore the throughput decreases. The figure shows also that the performance related to the second PAN coordinator varies

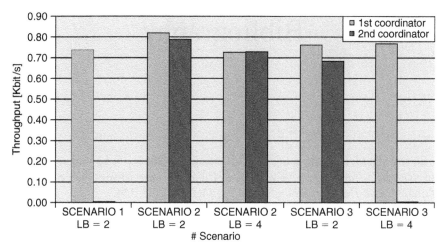

**Figure 10.42** *Throughput for different scenarios and Leaky Bucket thresholds*

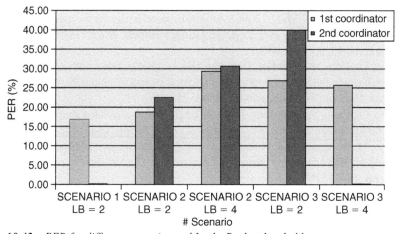

**Figure 10.43** *PER for different scenarios and Leaky Bucket thresholds*

for different $S_{LB}$; as a matter of fact, in the second scenario the throughput for $S_{LB} = 2$ is larger than in the case $S_{LB} = 4$. This suggests that when the link quality decreases, the faster D de-associates from the first coordinator, the better is the performance; this behaviour is confirmed if we observe the third scenario case, where the distance between the coordinators is smaller.

Finally, Figure 10.43 shows the system PER. As expected, the Figure substantially confirms the conclusions drawn from the throughput graph; as a matter of fact, the two quantities are strictly related, and the Figures carry the same results in a different way.

# Bibliography

AboElFotoh, H. M. F., Iyengar, S. S. & Krishnendu, C. (2005). Computing reliability and message delay for cooperative wireless distributed sensor networks subject to random failures. *IEEE Transactions on Reliability*, **54**(1), pp. 145–154.

Akyildiz, I. F., Su, W., Sankarasubramaniam & Cayirci, E. (2002). A survey on sensor networks. *IEEE Communications Magazine*, **40**(8), pp. 102–114.

Al-Nuaimi, M. O. & Stephens, R. B. L. (1998). Measurements and prediction model optimization for signal attenuation in vegetation media at centimetre wave frequencies. In *Proceedings of the IEEE on Microwaves, Antennas and Propagation*, Vol. 145, pp. 201–206.

Alouini, M. S. & Simon, M. K. (2000). An MGF-based performance analysis of generalized selection combining over Rayleigh fading channels. *IEEE Transactions on Communications*, **48**, pp. 401–415.

Andrisano, O., Conti, A., Dardari, D. & Pasolini, G. (2003). Bluetooth and IEEE 802.11b coexistence: Analytical performance evaluation in fading channels. *IEEE Journal on Selected Areas in Communications*, **21**, pp. 259–269.

Andrisano, O., Tralli, V. & Verdone, R. (1998). Millimetre waves for short-range multimedia communication systems. *Proceedings of IEEE*, **86**, pp. 1383–1401.

Andrisano, O., Verdone, R. & Nakagawa, M. (2000). Intelligent transportation systems: The role of third generation mobile radio networks. *IEEE Communications Magazine*, **38**(9), pp. 144–151.

Annamalai, A., Deora, G. & Tellambura, C. (2002). Unified error probability analysis for generalized selection diversity in Rician fading channels. In *Proceedings of the IEEE Vehicular Technology Conference (VTC '02-Spring)*, Vol. 4, pp. 2042–2046.

Arias-de Reyna, E., D'Amico, A. A. & Mengali, U. (2006). UWB energy detection receivers with partial channel knowledge. In *Proceedings of the IEEE International Conference on Communications*, Vol. 10, pp. 4688–4693.

Arici, T., Gedik, B., Altunbasak, Y. & Liu, L. (2003). Pinco: A pipelined in network compression scheme for data collection in wireless sensor networks. In *Proceedings of the 12th International Conference on Computer Communications and Networks*, pp. 539–544.

Bahl, P. & Padmanabhan, V. N. (2000). Radar: An in-building RF-based user-location and tracking system. In *Proceedings of the IEEE INFOCOM 2000, Tel Aviv, Israel*, pp. 775–784.

Batariere, M. D., Blankenship, T. K., Kepler, J. F. & Krauss, T. P. (2004). Seasonal variations in path loss in the 3.7 GHz band. In *Proceeding of the IEEE Radio and Wireless Conference*, 2004, pp. 399–402.

Bektas, F., Vondra, B., Veith, P. E., Faltin, L., Pohl, A. & Scholtz, Al. (2003). Bluetooth communication employing antenna diversity. In *Proceedings of the IEEE International Symposium on Computers and Communications (ISCC 2003)*, Vol. 1, pp. 652–657.

Benzair, K. (1995). Measurements and modelling of propagation losses through vegetation at 1–4 GHz. In *Proceedings of the Ninth International Conference on Antennas and Propagation, ICAP '95*, Vol. 2, pp. 54–59.

Bettstetter, C. H. (2005). Connectivity of wireless multihop networks in a shadow fading environment. *Wireless Networks*, **11**(5), pp. 571–579.

Bettstetter, C. H. & Zangl, J. (2002). How to achieve a connected ad hoc network with homogeneous range assignment: An analytical study with consideration of border effects. In *Proceedings of the 4th International Workshop on Mobile and Wireless Communications Network*, 2002, pp. 125–129.

Biaz, S. & Welch, J. (2001). Closed form bounds for clock synchronization under simple uncertainty assumptions. *Information Processing Letters*, **80**(3), pp. 151–157. http://www.iteseer.ist.psu.edu/biaz01closed.html

Bletsas, A., Shin, H., Win, M. Z. & Lippman, A. (2006). Cooperative diversity with opportunistic relaying. In *Proceedings of the IEEE Wireless Commununications and Networking Conference 2006, Las Vegas, NE.*

Bluetooth™ (2004). Specificaton of the Bluetooth System, 0-3 (Version 2.0 +EDR). http://www.standards.ieee.org/getieee802/802.15.html.

Blum, R. S., Kassam, S. A. & Poor, H. V. (1997). Distributed detection with multiple sensors – Part II: Advanced topics. *Proceedings of IEEE*, **85**(1), pp. 64–79.

Booth, L., Bruck, J., Cook, M. & Franceschetti, M. (2003). Ad hoc wireless networks with noisy links. In *Proceedings of the IEEE ISIT 2003, Yokohama, Japan*, 29 June–4 July 2003, p. 386.

Bluetooth, several information on this can be found on the website: http://www.bluetooth.com

Braginsky, D. & Estrin, D. (2002). Rumor routing algorithm for sensor networks. In *Proceedings of the First Workshop on Sensor Networks and Applications, WSNA'02 Atlanta, CA, USA*, September 28, 2002, pp. 22–37.

Brennan, S. M., Mielke, A. M. & Torney, D. C. (2005). Radioactive source detection by sensor networks. *IEEE Transactions on Nuclear Science*, **52**(3), pp. 813–819.

Bulusu, N., Heidemann, J. & Estrin, D. (2000). GPS-less low-cost outdoor localization for very small devices. *IEEE Personal Communications*, **7**(5), pp. 28–34 [*see also IEEE Wireless Communications*].

Buratti, C., Giorgetti, A. & Verdone, R. (2005). Cross-layer design of an energy-efficient cluster formation algorithm with carrier sensing multiple access for wireless sensor networks. *EURASIP Journal on Wireless Communications and Networking*, **5**, pp. 672–685.

Cardell-Oliver, R., Smettem, K., Kranz, M. & Mayer, K. (2004). Field testing a wireless sensor network for reactive environmental monitoring. In *Proceedings of the Intelligent Sensors, Sensor Networks and Information Processing Conference*, December 14–17, pp. 7–12.

Cassioli, D., Win, M. Z. & Molisch, A. F. (2002). The ultra-wide bandwidth indoor channel: From statistical model to simulations. *IEEE Journal on Selected Areas in Communications*, **20**(6), pp. 1247–1257.

Cassioli, D., Win, M. Z., Vatalaro, F. & Molisch, A. F. (2007). Low-complexity Rake receivers in ultra-wideband channels. *IEEE Transactions on Wireless Communications*, **6**(4), pp. 1265–1275.

Cetin, M., Chen, L., Fisher, J. W. III, Ihler, A. T., Moses, R. L., Wainwright, M. J. & Willsky, A. S. (2006). Distributed fusion in sensor networks [a graphical models perspective]. *IEEE Signal Processing Magazine*, **23**(4), pp. 42–55.

Chamberland, J.-F. & Veeravalli, V. V. (2003). Decentralized detection in sensor networks. *IEEE Transactions on Signal Processing*, **51**(2), pp. 407–416.

Chao, Y.-L. & Scholtz, R. A. (2005). Ultra-wideband transmitted reference systems. *IEEE Transactions on Vehicular Technology*, **54**(5), pp. 1556–1569.

Chazan, D., Zakai, M. & Ziv, J. (1975). Improved lower bounds on signal parameter estimation. *IEEE Transactions on Information Theory*, **21**(1), pp. 90–93.

Chen, B., Jamieson, K., Balakrishnan, H. & Morris, R. (2001). Span: An energy efficient coordination algorithm for topology maintenance in ad hoc wireless networks. *Proceedings of MobiCom*, pp. 70–80.

Chen, B., Tong, L. & Varshney, P. K. (2006). Channel-aware distributed detection in wireless sensor networks. *IEEE Signal Processing Magazine*, **23**(4), pp. 16–26.

Chen, B., Jiang, R., Kasetkasem, T. & Varshney, P. K. (2004). Channel-aware decision fusion in wireless sensor networks. *IEEE Transactions on Signal Processing*, **52**(12), pp. 3454–3458.

Chen, Y. & Zhao, Q. (2005). On the lifetime of wireless sensor networks. *IEEE Communication Letters*, **9**(11), pp. 976–978.

Cheong, P., Rabbachin, A., Montillet, J. P., Yu, K. & Oppermann, I. (2005). Synchronization, TOA and position estimation for low-complexity LDR UWB devices. In *Proceedings of the IEEE International Conference on Utra-Wideband, Zurich, Switzerland*, pp. 480–484.

Chiani, M., Conti, A. & Verdone, R. (2001). Partial compensation signal-level-based up-link power control to extend terminal battery duration. *IEEE Transactions on Vehicular Technology*, **50**(4), pp. 1125–1131.

Chiasserini, C.-F. & Rao, R. R. (1999). A model for battery pulsed discharge with recovery effect. *WCNC'99, New Orleans, USA*, September.

Chiasserini, C.-F. & Rao, R. R. (2000). Stochastic battery discharge in portable communication devices. *Proceedings of the 15th Annual Battery Conference on Applications and Advances, Long Beach, USA*, January.

Chipcon Products from Texas Instruments, URL: http://www.chipcon.com

Choi, J. D. & Stark, W. E. (2002). Performance of ultra-wideband communications with suboptimal receivers in multipath channels. *IEEE Journals on Selected Areas in Communications*, **20**(9), pp. 1754–1766.

Chong, C.-C. & Yong, S. K. (2005). A generic statistical-based UWB channel model for high-rise apartments. *IEEE Transactions on Antennas and Propagation*, **53**(8), pp. 2389–2399.

Chong, C.-Y. & Kumar, S. P. (2003). Sensor networks: evolution, opportunities, and challenges. *Proceedings of IEEE*, **91**(8), pp. 1247–1256.

Conti, A. & Dardari, D. (2004). The effects of node spatial distribution on the performance of wireless sensor networks. In *Proceedings of the IEEE of Vehicular Technology Conference (VTC 2004-Spring)*, Vol. 5, pp. 2724–2728.

Conti, A., Win, M. Z., Chiani, M. & Winters, J. H. (2003). Bit error outage for diversity reception in shadowing environment. *IEEE Communication Letters*, **7**, pp. 15–17.

Cox, D., Jovanov, E. & Milenkovic, A. (2005). Time synchronization for ZigBee networks. In *Proceedings of the Thirty-Seventh Southeastern Symposium on System Theory, SSST '05*, pp. 135–138.

Cramer, R. J., Scholtz, R. A. & Win, M. Z. (2002). An Evaluation of the Ultra-Wideband Propagation Channel. *IEEE Transactions on Antennas and Propagation*, **50**(5), pp. 561–570.

Crossbow (2003). *User Manual: MPR Mote Processor Radio Board, MIB Mote Interface, Programming Board User 's Manual MPR500CA, MPR510CA, MPR520CA, MPR400CB, MPR410CB, MPR420CB, MPR300CA, MPR310CA, MIB300CA, MIB500CA, MIB510CA*, Revised edition. Crossbow.

CRUISE, WP112, D112.1: Report on WSN applications, their requirements, application-specific WSN issues and evaluation metrics, December 2006, website: http://www.ist-cruise.eu.

CRUISE, WP112, D112.2: Update on WSN applications, their requirements, application-specific WSN issues and evaluation metrics, December 2006, website: http://www.istcruise.eu

Culler, D., Estrin, D. & Srivastava, M. (2004). Overview of sensor networks. *IEEE Computer*, **37**(8), pp. 41–49.

Dalley, J. E. J., Smith, M. S. & Adams, D. N. (1999). Propagation losses due to foliage at various frequencies. In *Proceedings of the IEE National Conference on Antennas and Propagation, 1999*, pp. 267–270.

Dardari, D. (2007). On the connected nodes position distribution in ad hoc wireless networks with statistical channel models. *IEEE Transactions on Wireless Communications*, Submitted.

Dardari, D. (2007). On the connected nodes position distribution in ad hoc wireless networks with statistical channel models. In *Proceedings of the IEEE International Conference on Communications, ICC 2007, Glasgow, UK*.

Dardari, D. & Conti, A. (2004). A sub-optimal hierarchical maximum likelihood algorithm for collaborative localization in ad-hoc networks. In *Proceedings of the First IEEE International Conference on Sensor and Ad hoc Communications and Networks, SECON 2004, Santa Clara, CA*, pp. 425–429.

Dardari, D., Conti, A., Buratti, C. & Verdone, R. (2007). Mathematical evaluation of environmental monitoring estimation error through energy-efficient wireless sensor networks. *IEEE Transactions on Mobile Computing*, **6**(7).

Dardari, D., Chong, C.-C. & Win, M. Z. (2006*a*). Analysis of threshold-based ToA estimators in UWB channels. In *Proceedings of the European Signal Processing Conference, EUSIPCO 2006, Florence, Italy*, Invalid paper.

Dardari, D., Chong, C.-C. & Win, M. Z. (2006*b*). Improved lower bounds on time-ofarrival estimation error in realistic UWB channels. In *Proceedings of the IEEE International Conference on Ultra-Wideband, ICUWB 2006, Waltham, MA, USA*.

Dardari, D., Giorgetti, A., Chiani, M. & Win, M. Z. (2006). A stop-and-go transmitted-reference UWB receiver. In *Proceedings of the IEEE International Conference on Ultra-Wideband, ICUWB 2006, Waltham, MA, USA*.

Dardari, D., Giorgetti, A. & Win, M. Z. (2007). Time-of-arrival estimation in the presence of narrow and wide bandwidth interference in UWB channels. In *Proceedings of the IEEE International Conference on Ultra-Wideband, ICUWB 2007, Singapore*, Invited paper.

Dardari, D. & Win, M. Z. (2006). Threshold-based time-of-arrival estimators in UWB dense multipath channels. In *Proceedings of the IEEE International Conference on Communications, ICC 2006, Istanbul, Turkey*.

Dardari, D., Chong, C.-C. & Win, M. Z. (2008). Threshold based time-of-arrival estimators in UWB dense multipath channels. *IEEE Transactions on Communications*, 56, to be published.

Dardari, D., Fabbri, F. & Verdone, R. (2008). Mathematical characterization of connectivity in wireless networks with arbitrary channel models. *IEEE Transactions on Wireless Communications*, Submitted.

Dardari, D., Conti, A., Ferner, U., Giorgetti, A. & Win, M. Z. (2008). Ranging in UWB channels: From Theory to Practice. *Proceedings of IEEE, Special Issue on UWB Technology and Emerging Applications*.

Dardari, D., Conti, A. & Lien, J. (2008). The effect of cooperation in UWB-based positioning systems using experimental data. EURASIP Journal on Advances in Signal Processing, Special Issue on Cooperative Localization in Wireless Ad Hoc and Sensor Networks, Submitted.

De Pedri, A., Zanella, A. & Verdone, R. (2003). An energy-efficient protocol for wireless ad hoc sensor networks. In *Proceedings of the IEEE International Symposium on Autonomous Intelligent Networks and Systems (AINS '03), Menlo Park, CA*, June 30–July 1, 2003.

Dijkstra, E. W. (1959). A note on two problems in connexion with graphs. *Numerische Mathematik*, **1**, pp. 269–271.

Doherty, L., Pister, K. S. & Ghaoui, L. E. (2001). Convex position estimation in wireless sensor networks. In *Proceedings of the Twentieth Annual Joint Conference of the IEEE Computer and Communications Societies*, INFOCOM 2001, Vol. 3, pp. 1655–1663.

Dousse, O., Thiran, P. & Hasler, M. (2002). Connectivity in ad hoc and hybrid networks. In *Proceedings of the 21st Annual Joint Conference of the IEEE Computer and Communications Societies*, INFOCOM02, *New York, USA*, Vol. 2, pp. 1079–1088.

Dyck, R. E. V. (2001). Detection performance in self-organized wireless sensor networks. In *Proceedings of the Conference on Information Science and Systems, Baltimore, ML, USA*.

ECMA368 ecma-international (2006). High rate ultra-wideband PHY and MAC standard. http://www.ecma-international.org/publications/files/ECMA-ST/ECMA368.pdf.

Ekahau (n.d.). Ekahau positioning system. Available online: www.ekahau.com

Elnahrawy, E., Xiaoyan, L. & Martin, R. P. (2004). The limits of localization using signal strength: A comparative study. In *IEEE Proceedings of the Sensor and Ad Hoc Communications and Networks (SECON '04)*, pp. 406–414.

Elson, J. & Romer, K. (2003). Wireless sensor networks: A new regime for time synchronization. *SIGCOMM Computer Communication Review*, **33**(1), pp. 149–154.

Elson, J. & Estrin, D. (2001). Time synchronization for wireless sensor networks. In *Proceedings of the 15th International Parallel and Distributed Processing Symposium*, pp. 1965–1970.

Elson, J., Girod, L. & Estrin, D. (2002). Fine-grained network time synchronization using reference broadcasts. *SIGOPS Operating Systems Review*, **36**(SI), pp. 147–163.

Ermis, E. B., Alanyali, M. & Saligrama, V. (2006). Search and discovery in an uncertain networked world [how signal processing algorithms should be designed for this application]. *IEEE Signal Processing Magazine*, **23**(4), pp. 107–118.

e-SENSE, WP1, D1.2.1: Scenarios and audio visual concepts, September 2006, website: http://www.ist-esense.org

Esseghir, M., Bouabdallah, N. & Pujolle, G. (2005). A novel approach for improving wireless sensor network lifetime. In *Proceeding of PIMRC, Berlin*, September 2005, Vol. 4, pp. 11–14.

European Commission (2007). Commission decision of 21 February 2007 on allowing the use of the radio spectrum for equipment using ultra-wideband technology in a harmonised manner in the community. *Official Journal of the European Union*, C(2007), p. 522.

Falsi, C., Dardari, D., Mucchi, L. & Win, M. Z. (2006). Time of arrival estimation for UWB localizers in realistic environments. *EURASIP Journal on Applied Signal Processing (Special Issue on Wireless Location Technologies and Applications)*, 2006, Article ID 32082, 13.

Fan, R. & Lynch, N. (2004). Gradient clock synchronization, In *PODC04, Proceedings of the Fifth Annual ACM Symposium on Principles of Distributed Computing, St John's, Newfoundland, Canada.*

Fanimokun, A. & Frolik, J. (2003). Effects of natural propagation environments on wireless sensor network coverage area. In *Proceedings of the 35th Southeastern Symposium on System Theory*, 2003, pp. 16–20.

Federal Communications Commission (adopted February 14, 2002, released April 22, 2002). Revision of Part 15 of the Commission's rules regarding ultra-wideband transmission systems, First Report and Order (ET Docket 98–153).

Ferrari, G., Tonguz, O. K. & Bhatt, M. (2004). Impact of receiver sensitivity on the performance of sensor networks. In *Proceedings of the IEEE 59th Vehicular Technology Conference, 2004, VTC 2004-Spring*, Vol. 3, pp. 1350–1354.

Foh, C. H. & Lee, B. S. (2004). A closed form network connectivity formula one-dimensional MANETs. In *Proceedings of the IEEE International Conference on Communications, ICC 2004, Paris, France*, Vol. 6, pp. 3739–3742.

Fox, D., Hightower, J., Liao, L., Schulz, D. & Borriello, G. (2003). Bayesian filtering for location estimation. *IEEE Pervasive Computing*, 2(3), pp. 24–33.

Fu, H. & Kam, P. Y. (2005). Performance comparison of selection combining schemes for binary DPSK on non-selective Rayleigh-fading channels with interference. *IEEE Transactions on Wireless Communications*, 4, pp. 192–201.

Ganeriwal, S., Kumar, R. & Srivastava, M. (2003). Timing synch protocol for sensor networks. In *Proceedings of the ACM SenSys, Los Angeles, CA.*

Gardner, W. A. (1990). *Introduction to Random Processes with Applications to Signals and Systems*, 2nd edn. New York: McGraw-Hill.

Gastpar, M., Vetterli, M. & Dragotti, P. L. (2006). Sensing reality and communicating bits: A dangerous liaison [Is digital communication sufficient for sensor networks?]. *IEEE Signal Processing Magazine*, 23(4), pp. 70–83.

Gezici, S., Tian, Z., Giannakis, G. B., Kobayashi, H., Molisch, A. F., Poor, H. V. & Sahinoglu, Z. (2005). Localization via ultra-wideband radios: A look at positioning aspects for future sensor networks. *IEEE Signal Processing Magazine*, 22, pp. 70–84.

Gifford, W. M. & Win, M. Z. (2004). On transmitted-reference UWB communications. In *Proceedings of the 38th Asilomar Conference on Signals, Systems and Computers, Pacific Grove, CA*, pp. 1526–1531, Invited Paper.

Giorgetti, A., Chiani, M. & Win, M. Z. (2005). The effect of narrowband interference on wideband wireless communication systems. *IEEE Transactions on Communications*, 53(12), pp. 2139–2149.

Golmie, N., Chevrollier, N. & Rebala, O. (2003). Bluetooth and WLAN coexistence: Challenges and solutions. *IEEE Wireless Communications*, **10**(Article ID 78954), pp. 22–29.

Guo, C., Zhong, L. C. & Rabaey, J. M. (2001). Low power distributed MAC for ad hoc sensor radio networks. In *Proceedings of the Globecom 2001*, Vol. 5, pp. 2944–2948.

Guo, Y., Corke, P., Poulton, G., Wark, T., Bishop-Hurley, G. & Swain, D. (2006). Animal behaviour understanding using wireless sensor networks, ICN. *Proceedings, 2006, 31st IEEE Conference on Local Computer Networks*, 2006, pp. 607–614.

Gupta, P. & Kumar, P. R. (1998). Critical power for asymptotic connectivity. In *Proceedings of the 37th IEEE Conference on Decision and Control, Tampa, Florida, USA*, Vol. 1, 16–18 December, 1998, pp. 1106–1110.

Gupta, P. & Kumar, P. R. (2000). The capacity of wireless networks. *IEEE Transactions on Information Theory*, **46**(2), pp. 388–404.

Guvenc, I. & Sahinoglu, Z. (2005a). Threshold-based TOA estimation for impulse radio UWB systems. In *Proceedings of the IEEE International Conference on Utra-Wideband, Zurich, Switzerland*, pp. 420–425.

Guvenc, I. & Sahinoglu, Z. (2005b). TOA estimation with different IR-UWB transceiver types. In *Proceedings of the IEEE International Conference on Utra-Wideband, Zurich, Switzerland*, pp. 426–431.

Haschberger, P., Bundschuh, M. & Tank, V. (1996). Infrared sensor for the detection and protection of wildlife. *Optical Engineering*, **35**(3), pp. 882–889.

Hawkes, M. & Nehorai, A. (2003). Wideband source localization using a distributed acoustic vector-sensor array. *IEEE Transactions on Signal Processing*, **51**(6), pp. 1479–1491.

He, T., Huang, C., Blum, B. M., Stankovic, J. A. & Abdelzaher, T. (2003). Range-free localization schemes for large-scale sensor networks. In *Proceedings of the ACM MobiCom 2003, San Diego, CA*.

Hedetniemi, S. & Liestman, A. (1988). A survey of gossiping and broadcasting in communication networks. *Networks*, **18**(4), pp. 319–349.

Heinzelman, W., Chandrakasan, A. & Balakrishnan, H. (2000). Energy-efficient communication protocol for wireless microsensor networks. In *Proceedings of the 33rd Hawaii International Conference on System Sciences*, Vol. 2.

Heinzelman, W. B., Chandrakasan, A. P. & Balakrishnan, H. (2002). An application-specific protocol architecture for wireless microsensor networks. *IEEE Transactions on Wireless Communications*, **1**, pp. 660–670.

Hightower, J. & Borriello, G. (2001). Location systems for ubiquitous computing. *Computer*, **34**(8), pp. 57–66.

Hong, Y.-W. & Scaglione, A. (2005). A scalable synchronization protocol for large-scale sensor networks and its applications. *IEEE Journal on Selected Areas in Communications*, **23**(5), pp. 1085–1099.

Hong, Y.-W., Scaglione, A. & Varshney, P. K. (2005). A communication architecture for reaching consensus in decision for a large network. In *Proceedings of the IEEE Workshop on Statistical Signal Processing, Paris, France*, pp. 1220–1225.

Hu, Z. & Li, B. (2004). On the fundamental capacity and lifetime limits of energy-constrained wireless sensor networks. In *Proceedings of IEEE RTAS 2004, Toronto*, 25–28 May 2004, pp. 2–9.

IEEE (2006). *Signal Processing Magazine Special Issue*. Distributed signal processing in sensor networks.

IEEE 802.15.4 Standard (2006). Part 15.4: Wireless medium access control (Mac) and physical layer (Phy) specifications for low-rate wireless personal area networks (lr-wpans). http://www.standards.ieee.org/getieee802/802.15.html

IEEE 802.15.4a Standard (2007). Part 15.4: Wireless MAC and PHY specifications for low-rate wireless personal area networks (LR-WPANs): Amendment to add alternate PHY., August 2007.

IEEE 802.15.4 website: http//www.eee802.org/15/pub/TG4.html

IEEE 802.15.1 website: http://www.ieee802.org/15/pub/TG1.html

Intanagonwiwat, C., Govindan, R. & Estrin, D. (2000). Directed diffusion: A scalable and robust communication paradigm for sensor networks. In *Proceedings of the ACM Mobi-Com*, pp. 56–67.

Jakes, W. C. (1995). *Microwave Mobile Communications*, Classic reissue edition. Piscataway, NJ: IEEE Press.

Jourdan, D. B., Dardari, D. & Win, M. Z. (2006*a*). Position error bound and localization accuracy outage in dense cluttered environments. In *Proceedings of the IEEE International Conference on Ultra-Wideband, ICUWB 2006, Waltham, MA, USA*.

Jourdan, D., Dardari, D. & Win, M. Z. (2006*b*). Position error bound for UWB localization in dense cluttered environments. In *Proceedings of the IEEE International Conference on Communications, ICC 2006, Istanbul, Turkey*.

Jourdan, D. B., Dardari, D. & Win, M. Z. (2007). Position error bound for UWB localization in dense cluttered environments. *IEEE Transactions on Aerospace and Electronic Systems*, **43**, 2007, to be published.

Jourdan, D. B., Deyst, J. J., Win, M. Z. & Roy, N. (2005). Monte-Carlo localization in dense multipath environments using UWB ranging. In *Proceedings of the IEEE International Conference on Utra-Wideband, Zürich, Switzerland*, pp. 314–319.

Kimura, N. & Latifi, S. (2005). A survey on data compression in wireless sensor networks. In *Proceedings of the International Conference on Information Technology: Coding and Computing (ITCC '05)*, Vol. 2, pp. 8–13.

Kingsland, S. (1982). The refractory model: The logistic curve and the history of population ecology. *Quarterly Review of Biology*, **57**, pp. 29–52.

Kulik, J., Heinzelman, W. & Balakrishnan, H. (2002). Negotiation-based protocols. *Wireless Networks*, **8**, pp. 169–185.

Kumar, S., Lai, T. H. & Balogh, J. (2004). On-coverage in a mostly sleeping sensor network. In *Proceedings of the Tenth Annual International Conference on Mobile Computing and Networking (ACM MobiCom), Philadelphia, PA*, pp. 144–158.

Kumar, S., Arora, A. & Lai, T. H. (2005). On the lifetime analysis of always-on wireless network applications. In *Proceeding of MASS 2005, Washington, DC*, November 7–10.

Laneman, J. N., Tse, D. N. C. & Wornell, G. W. (2004). Cooperative diversity in wireless networks: Efficient protocols and outage behavior. *IEEE Transactions on Information Theory*, **50**(12), pp. 3062–3080.

Laneman, J. N. & Wornell, G. W. (2003). Distributed space–time coded protocols for exploiting cooperative diversity in wireless networks. *IEEE Transactions on Information Theory*, **59**, pp. 2415–2525.

Langendoen, K. & Reijers, N. (2003). Distributed localization in wireless sensor networks: A quantitative comparison. *Computer Networks*, **43**, pp. 499–518.

Lee, J.-Y. & Scholtz, R. A. (2002). Ranging in a dense multipath environment using an UWB radio link. *IEEE Journal on Selected Areas in Communications*, **20**(9), pp. 1677–1683.

Levanon, N. (2000). Lowest GDOP in 2-D scenarios. *IEEE Proceedings, Radar, Sonar, Navigation*, **147**(3), pp. 149–155.

Lindsey, S. & Raghavendra, C. (2002 September). Pegasis: Power-efficient gathering in sensor information systems. In *Proceeding of the IEEE Aerospace Conference*, Vol. 3, pp. 1125–1130.

Liu, X. & Srikant, R. (2004). An information-theoretic view of connectivity in wireless sensor networks. In *Proceedings of the First Annual IEEE Communications Society Conference on Sensor and Ad Hoc Communications and Networks, 2004, SECON*, pp. 508–516.

Lu, G., Krishnamachari, B. & Raghavendra, C. S. (2004). An adaptive energy efficient and low-latency MAC for data gathering in wireless sensor networks. In *Proceedings of the 18th International Parallel and Distributed Processing Symposium*.

Lucchi, M., Giorgetti, A. & Chiani, M. (2005). Cooperative diversity in wireless sensor networks. In *Proceedings of the 8th International Symposium on Wireless Personal Multimedia Communications (WPMC'05), Aalborg, DK*, September 2005, pp. 1738–1742.

Lundelius-Welch, J. & Lynch, N. A. (1984). An upper and lower bound for clock synchronization. *Information and Control*, **62**, pp. 190–204.

Luo, Z-Q., Gastpar, M., Liu, J. & Swami, A. (2006). Distributed signal processing in sensor networks [from the guest editors]. *IEEE Signal Processing Magazine*, **23**(4), pp. 14–15.

Ma, Y. & Chai, C. C. (2000). Unified error probability analysis for generalized selection combining in nakagami fading channels. *IEEE Journal on Selected Areas in Communications*, **18**, pp. 2198–2210.

Manjeshwar, A. & Agarwal, D. P. (2001). Teen: A routing protocol for enhanced efficiency in wireless sensor networks. In *Proceedings of the 15th International Workshop on Parallel and Distributed Processing Symposium*, pp. 2009–2015.

Maroti, M., Kusy, B., Simon, G. & Ledeczi, A. (2004). The flooding time synchronization protocol. In *Proceedings of the Second ACM Conference on Embedded Networked Sensor Systems (SenSys), Baltimore, Maryland, USA*, pp. 39–49.

Marron, P. J., Minder, D. & Embedded WiSeNts Consortium (2006). *Embedded WiseNts Research Roadmap*. Berlin, Germany, Information Society Technologies.

Marvasti, F. (1986). Signal recovery from nonuniform samples and spectral analysis on random nonuniform samples. In *Proceedings of the IEEE International Conference on Acoustic, Speech, and Signal Processing*, Vol. 11, pp. 1649–1652.

Masini, B., Conti, A., Dardari, D. & Pasolini, G. (2006). Exploiting diversity for coverage extension of Bluetooth-based mobile services. *EURASIP Journal on Wireless Communications and Networking*, **20**(Article ID 78954), pp. 1–9.

Masini, B., Dardari, D., Conti, A. & Pasolini, G. (2006*a*) Selection diversity for Bluetooth in the presence of IEEE802.11g interference. In *Proceedings of the IEEE Personal, Indoor and Mobile Radio Communications, 2006 (PIMRC '06)*, pp. 1–5.

Masini, B., Dardari, D., Conti, A. & Pasolini, G. (2006*b*). Selection diversity for Bluetooth in the presence of IEEE802.11g interference. In *Proceedings of the IEEE Vehicular Technology Conference, 2006 (VTC'06)*, Fall, pp. 1–5.

Masini, B. M., Conti, A., Pasolini, G. & Dardari, D. (2007). On the benefits of diversity schemes for Bluetooth coverage extension in the presence of IEEE802.11g interference. *Wireless Communications and Mobile Computing*, Vol. 7. Wiley InterService, pp. 1–11.

Masry, E. (1978). Poisson sampling and spectral estimation of continuous-time processes. *IEEE Transactions on Information Theory*, **IT-24**, pp. 173–183.

Mills, D. L. (1991). Internet time synchronization: The network time protocol. *IEEE Transactions on Communications*, 39(10), pp. 1482–1493.

Molisch, A. F., Balakrishnan, K., Cassioli, D., Chong, C-C., Emami, S., Fort, A., Karedal, J., Kunisch, J., Schantz, H., Schuster, U. & Siwiak, K. (2005). IEEE 802.15.4a channel model–Final report.

Molisch, A. F., Cassioli, D., Chong, C.-C., Emami, S., Fort, A., Kannan, B., Karedal, J., Kunisch, J., Schantz, H., Siwiak, K. & Win, M. Z. (2006). A comprehensive standardized model for ultra-wideband propagation channels. *IEEE Transactions on Antennas and Propagation*, 54(11), pp. 3151–3166. Special Issue on Wireless Communications.

Motley, A. J. & Keenan, J. M. P. (1988). Personal communication radio coverage in buildings at 900 MHz and 1700 MHz. *Electronics Letters*, 24(12), pp. 763–764.

Muthukrishnan, K., Lijding, M. & Havinga, P. J. M. (2005). Towards smart surroundings: Enabling techniques and technologies for localization. In Strang, T. & Linnhoff-Popien, C. (eds), *Proceedings of the First International Workshop on Location and Context-Awareness (LoCA 2005)*, *Berlin*, Vol. 3479, pp. 350–362.

Nemati, M. A., Mitra U. & Scholtz, R. A. (2006). Optimum integration time for UWB transmitted reference and energy detector receivers. In *Proceedings of the MILCOM 2006*, pp. 1–7.

Nemzek, R., Dreicer, J., Torney, D. C. & Warnock, T. (2004). Distributed sensor networks for detection of mobile radioactive sources. *IEEE Transactions on Nuclear Science*, 51(4), pp. 1693–1700.

Niculescu, D. & Nath, B. (2001). Ad hoc positioning system (APS). In *Proceedings of the Global Telecommunications Conference, 2001 (GLOBECOM '01. IEEE)*, Vol. 5, pp. 2926–2931.

Niu, R., Chen, B. & Varshney, P. (2006). Fusion of decisions transmitted over Rayleigh fading channels in wireless sensor networks. *IEEE Transactions on Signal Processing*, 54(3), pp. 1018–1026.

Niu, R. & Varshney, P. K. (2005). Distributed detection and fusion in a large wireless sensor network of random size. *EURASIP Journal on Wireless Communications Network*, 5(4), pp. 462–472.

Orriss, J. & Barton, S. K. (2003). Probability distributions for the number of radio transceivers which can communicate with one another. *IEEE Transactions on Communications*, 51(4), pp. 676–681.

Pados, D. A., Halford, K. W., Kazakos, D. & Papantoni-Kazakos, P. (1995). Distributed binary hypothesis testing with feedback. *IEEE Transactions on Systems, Man, and Cybernetics*, 25(1), pp. 21–42.

Parsons, J. D. (1992). *The Mobile Radio Propagation Channel*, 1st edn. New York: John Wiley.

Patwari, N., Ash, J. N., Kyperountas, S., III Hero, A. O., Moses, R. L. & Correal, N. S. (2005). Locating the nodes: Cooperative localization in wireless sensor networks. *IEEE Signal Processing Magazine*, 22(4), pp. 54–69.

Pavani, T., Costa, G., Mazzotti, M., Conti, A. & Dardari, D. (2006). Experimental results on indoor localization techniques through wireless sensors network. In *Proceedings of the IEEE of Vehicular Technology Conference (VTC 2006, Spring)*, Vol. 2, pp. 663–667.

Pavani, T., Costa, G., Mazzotti, M., Dardari, D. & Conti, A. (2006). Experimental results on indoor localization technique through wireless sensors network. In *Proceedings of the IEEE Vehicular Technology Conference (VTC 2006-Spring), Melbourne, Australia*.

Pavani, T., Dardari, D., Conti, A. & Andrisano, O. (2007). Localization and immersive guide in indoor mobile environments through wireless sensor networks: A support for emergency situations. In *Proceedings of the IEEE Conference on Wireless Rural and Emergency Communications, WRECOM 2007', Rome, Italy*, 6 pp.

Penrose, M. D. (1997). The longest edge of the random minimal spanning tree. *Annals of Applied Probability*, **7**(2), pp. 340–361.

Penrose, M. D. (1999). A strong law for the longest edge of the minimal spanning tree. *Annals of Applied Probability*, **27**, pp. 246–260.

Perur, S. & Iyer, S. (2006). Characterization of a connectivity measure for sparse wireless multihop networks. In *Proceedings of the 26th IEEE International Conference on Distributed Computing Systems, 2006 ICDCS*, pp. 80–80.

Petrović, D., Shah, R. C., Ramchandran, K. & Rabaey, J. (2003). Data funneling: Routing with aggregation and compression for wireless sensor networks. In *Proceedings of the First IEEE International Workshop on Sensor Network Protocols and Appplications*, pp. 156–162.

Philips, T. K., Panwar, S. S. & Tantawi, A. N. (1989). Connectivity properties of a packet radio network model. *IEEE Transactions on Information Theory*, **35**(5), pp. 1044–1047.

Pinto, P. C. & Win, M. Z. (2006*a*). Communication in a Poisson field of interferers. In *Proceedings of the Conference on Information Science and Systems, Princeton, NJ*, March, pp. 432–437.

Pinto, P. C. & Win, M. Z. (2006*b*). Design of covert military networks: A spectral outage-based approach. In *Proceedings of the Military Commununication & Conference, Washington, DC*, October, pp. 1–6.

Pinto, P. C., Chong, C.-C., Giorgetti, A., Chiani, M. & Win, M. Z. (2006*c*). Narrowband communication in a Poisson field of ultrawideband interferers. In *Proceedints of the IEEE International Conference on Utra-Wideband, Waltham, MA*, pp. 387–392. Best Student Paper Award.

Pinto, P. C. & Win, M. Z. (2006*d*). Spectral outage due to cumulative interference in a Poisson field of nodes. In *Proceedings of the IEEE Global Telecommunication Conference, San Francisco, CA*, pp. 1–6.

Pinto, P. C. & Win, M. Z. (2007 December). Spectral characterization of wireless networks. *IEEE Wireless Communications Magazine*, Special Issue on Wireless Sensor Networking.

Pottie, J. K. & Kaiser, W. J. (2000). Wireless integrated network sensors. *Communications of ACM*, **43**(5), pp. 51–58.

Pradham, S. S., Kusuma, J. & Ramchandran, K. (2002). Distributed compression in a dense microsensor network. *IEEE Signal Processing Magazine*, **19**, pp. 51–60.

Predd, J. B., Kulkarni, S. R. & Poor, H. V. (2006). Distributed learning in wireless sensor networks: Applications issues and the problem of distributed inference. *IEEE Signal Processing Magazine*, **23**(4), pp. 56–69.

Premkumar, K. & Srinivasan, S. H. (2005). Diversity techniques for interference mitigation between IEEE 802.11 WLANs and Bluetooth. In *Proceedings of the IEEE Personal, Indoor and Mobile Radio Communications, 2005 (PIMRC'05)*, Vol. 3, pp. 1468–1472.

Priyantha, N. B., Balakrishnan, H., Demaine, E. & Teller, S. (2003). Anchor-free distributed localization in sensor networks, Tech Report 892, MIT Laboratory for Computer Science. Available at: http://www.nms.lcs.mit.edu/cricket.

Proakis, J. G. (2001). *Digital Communications*, 4th edn. New York: McGraw-Hill.

Puri, R., Majumdar, A., Ishwar, P. & Ramchandran, K. (2006). Distributed video coding in wireless sensor networks; exploiting the spatiotemporal redundancy in broadband networks. *IEEE Signal Processing Magazine*, **23**(4), pp. 94–106.

Quek, T. Q.-S. & Win, M. Z. (2004). Performance analysis of ultrawide bandwidth transmitted-reference communications. In *Proceedings of the IEEE Semiannual Vehicular Technology Conference, Italy, Milan*, Vol. 3, pp. 1285–1289.

Quek, T. Q.-S. & Win, M. Z. (2004). Ultrawide bandwidth transmitted reference signaling. In *Proceedings of the IEEE International Conference on Communications, Paris, France*, Vol. 6, pp. 3409–3413.

Quek, T. Q.-S. & Win, M. Z. (2005 September). Analysis of UWB transmitted reference communication systems in dense multipath channels. *IEEE Journal on Selected Areas in Communications*, **23**(9), pp. 1863–1874.

Quek, T. Q.-S., Win, M. Z. & Dardari, D. (2005 September). UWB transmitted-reference signaling schemes – Part I: Performance analysis. In *Proceedings of the IEEE International Conference on Utra-Wideband, Zürich, Switzerland*, pp. 587–592.

Quek, T. Q.-S., Win, M. Z. & Dardari, D. (2005 September). UWB transmitted-reference signaling schemes – Part II: Narrowband interference analysis. In *Proceedings of the IEEE International Conference on Utra-Wideband, Zürich, Switzerland*, pp. 593–598.

Quek, T. Q.-S., Dardari, D. & Win, M. Z. (2006a). Cooperation in bandwidth-constrained wireless sensor networks. In *Proceedings of the IEEE International Workshop Wireless Ad Hoc Sensor Networks, New York, NY, USA*.

Quek, T. Q.-S., Dardari, D. & Win, M. Z. (2006b). Energy efficiency of cooperative dense wireless sensor networks. In *Proceedings of the International Wireless Communications and Mobile Computing Conference, IWCMC 2006, Vancouver, Canada*, pp. 1323–1330.

Quek, T. Q.-S., Dardari, D. & Win, M. Z. (2006c), Energy efficiency of dense wireless sensor networks: To cooperate or not to cooperate. In *Proceedings of the IEEE International Conference on Communications, ICC 2006, Istanbul, Turkey*, Vol. 10, pp. 4479–4484.

Quek, T. Q.-S., Dardari, D. & Win, M. Z. (2007). Energy efficiency of dense wireless sensor networks: To cooperate or not to cooperate. *IEEE JSAC (Special Issue on Cooperative Communications and Networking)*, **25**(2), pp. 459–470.

Quek, T. Q.-S. & Win, M. Z. (2005). Analysis of UWB transmitted-reference communication systems in dense multipath channels. *IEEE Journal on Selected Areas in Communications*, **23**(9), pp. 1863–1874.

Quek, T. Q.-S., Win, M. Z. & Dardari, D. (2007). Analysis of UWB transmitted-reference schemes with narrowband interference. *IEEE Transactions on Wireless Communications*, **6**(5).

Quek, T. Q.-S., Win, M. Z. & Dardari, D. (2007 June). Unified analysis of UWB transmitted-reference schemes in the presence of narrowband interference. *IEEE Transactions on Wireless Communications*, **6**(6), pp. 2126–2139.

Rabbachin, A., Oppermann, I. & Denis, B. (2006). ML time-of-arrival estimation based on low complexity UWB energy detection. In *Proceedings of the IEEE International Conference on Utra-Wideband (ICUWB), Waltham, MA*, pp. 599–604.

Raghavendra, C. S., Sivalingam, K. & Znati, T. (2004). *Wireless Sensor Networks*. Kluwer Springer.

Rajendran, V., Obraczka, K. & Garcia-Luna-Aceves, J. J. (2003). Energy-efficient, collision-free medium access control for wireless sensor networks. *Proceedings of the ACM SenSys*, **03**, pp. 181–192.

Rappaport, T. S. (1996). *Wireless Communications*, 1st edn. Upper Saddle River, NJ: Prentice Hall.

Ridolfi, A. & Win, M. Z. (2006). Ultrawide bandwidth signals as shot-noise: a unifying approach. *IEEE Journal on Selected Areas in Communications*, **24**(4), pp. 899–905.

Rodoplu, V. & Meng, T. H. (1999). Minimum energy mobile wireless networks. *IEEE Journal on Selected Areas in Communication*, **17**(8), pp. 1333–1344.

Romer, K. (2003). Temporal message ordering in wireless sensor networks. *IFIP Mediterreranean Workshop on Ad-Hoc Networks 2003, Mahdia, Tunisia*, June 2003, pp. 131–142.

Rugin, R. & Mazzini, G. (2004). A simple and efficient MAC-routing integrated algorithm for sensor network. *Proceedings of the ICC*, **6**, pp. 3499–3503.

Sahin, M. E., Guvenc, I. & Arslan, H. (2005). Optimization of energy detector receivers for UWB systems, In *Proceedings of the IEEE 61st ' Vehicular Technology Conference, 2005 (VTC 2005)*, Spring, Vol. 2, pp. 1386–1390.

Salbaroli, E. & Zanella, A. (2006). A connectivity model for the analysis of a wireless ad-hoc network of finite area. In *Proceedings of the 3rd Annual IEEE Communications Society on Sensor and Ad Hoc Communications and Networks, 2006 SECON '06*, Vol. 3, pp. 756–760.

Saleh, A. & Valenzuela, R. A. (1987). A statistical model for indoor multipath propagation. *IEEE Journal on Selected Areas in Communication*, **5**(2), pp. 128–137.

Santi, P. (2005). *Topology Control in Wireless Ad Hoc and Sensor Networks*. Chichester, UK: John Wiley.

Santoni, T. A., Santucci, J. F., De Gentili, E. & Costa, B. (2006). Using wireless sensor network for wildfire detection. A discrete event approach of environmental monitoring tool. *Proceeding of ISEIMA 06*, pp. 115–120.

Savarese, C., Rabaey, J. M. & Beutel, J. (2001). Location in distributed ad-hoc wireless sensor networks. In *Proceedings of the 2001 IEEE International Conference on Acoustics, Speech, and Signal Processing, 2001 (ICASSP '01), Salt Lake City, Utah*, May 2001, Vol. 4, pp. 2037–2040.

Savvides, A., Park, H. & Srivastava, M. (2002). The bits and flops of the N-hop multilateration primitive for node localization problems. In *Proceedings of the First ACM Workshop on Wireless Sensor Networks and Application, (WSNA), Atlanta, GA*, pp. 112–121.

Savvides, A., Garber, W. L., Moses, R. L. & Srivastava, M. B. (2005). An analysis of error inducing parameters in multihop sensor node localization. *IEEE Transactions on Mobile Computing*, **4**(6), pp. 567–577.

Sayed, A. H., Tarighat, A. & Khajehnouri, N. (2005). Network-based wireless location: Challenges faced in developing techniques for accurate wireless location information. *IEEE Signal Processing Magazine*, **22**(4), pp. 24–40.

Schurgers, C. & Srivastava, M. B. (2001). Energy-efficient routing in wireless sensor networks. *IEEE Military Communications Conference*, **1**, pp. 357–361.

Scott, J. C. (1998). Applications of telecommunications and information technology for humanitarian health initiatives. *Proceedings of the Medical Technology Symposium*, 17–20 August, pp. 197–204.

Sendonaris, A., Erkip, E. & Aazhang, B. (2003*a*). User cooperation diversity – Part II: Implementation aspects and performance analysis. *IEEE Transactions on Communications*, **51**(11), pp. 1939–1948.

Sendonaris, A., Erkip, E. & Aazhang, B. (2003*b*). User cooperation diversity – Part I: System description. *IEEE Transactions on Communications*, **51**(11), pp. 1927–1938.

Severi, S., Liva, G., Chiani, M. & Dardari, D. (2007). A new low-complexity user tracking algorithm for WLAN-based positioning systems. In *Proceedings of the 16th IST Mobile and Wireless Communications Summit, Budapest, Hungary*, July 2007, pp. 1–5.

Severi, S. & Dardari, D. (2008). Performance limits of time synchronization in wireless sensor networks. In *IEEE International Conference on Communications, ICC 2008*, Beijing, China, May 2008.

Shainoglu, Z. & Guvenc, I. (2006). Multiuser interference mitigation in noncoherent UWB ranging via nonlinear filtering. *EURASIP Journal as Wireless Communications and Networking*, pp. 1–10.

Shen, Y. & Win, M. Z. (2007 March). Fundamental limits of wideband localization accuracy via Fisher information. In *Proceedings of the IEEE Wireless Communications and Networking Conference, Hong Kong*.

Shen, Y., Wymeersch, H. & Win, M. Z. (2007 March). Fundamental limits of wideband cooperative localization via Fisher information. In *Proceedings of the IEEE Wireless Communications and Networking Conference, Hong Kong*, Best Paper Award.

Shen, Y., Lien, J., Wymeersch, H. & Win, M. Z. (2007 May). Location, location, location: Performance bounds and algorithms for cooperative localization in UWB networks. *IEEE Communication Theory Workshop (CTW 2007), Sedona, AZ*.

Shnayder, V., Hempstead, M., Chen, B., Allen, G. W. & Welsh, M. (2004). Simulating the power consumption of large scale sensor network applications. In *Proceedings of the Second ACM Conference on Embedded Networked Sensor Systems (ACM SenSys), Baltimore, MD*.

Sichitiu, M. L. & Veerarittiphan, C. (2003). Simple, accurate time synchronization for wireless sensor networks. In *Proceedings of the IEEE Wireless Communications and Networking, 2003, WCNC 2003*, Vol. 2, pp. 1266–1273.

Sihavaran, T., Blair, G., Friday, A., Wu, M., Limon, H. D., Okanda, P. & Sorensen, C. F. Cooperating sentient vehicles for next generation automobiles. In *Proceeding of MobiSys 2004, 1st ACM Workshop on Applications of Mobile Embedded Systems*, June 2004.

Simon, M. K. & Alouini, M. S. (2004). *Digital Communication over Fading Channels: A Unified Approach to Performance Analysis*, 2nd edn. New York: John Wiley.

Simon Wong, K. F., Tsang, I. W., Cheung, V., Gary Chan, S.H., Kwok, J. T. (2005). Position estimation for wireless sensor networks. In *Proceedings of IEEE Globecom 2005, St. Louis*, November 28, December 2, 2005.

Singh, S. & Raghavendra, C. S. (1998). Pamas: Power aware multi-access protocol with signalling for ad hoc networks. *ACM Computer Communication Review*, **28**(3).

Sivrikaya, F. & Yener, B. (2004). Time synchronization in sensor networks: A survey. *IEEE Network*, **18**(4), pp. 45–50.

Slepian, D. & Wolf, J. (1973). Noiseless coding of correlated information sources. *IEEE Transactions on Information Theory*, **19**, pp. 471–480.

Sohrabi, K., Gao, J., Ailawadhi, V. & Pottie, G. J. (2000). Protocols for self-organization of a wireless sensor network. *IEEE Personal Communications*, **7**, pp. 16–27.

Soltanian, A. & Van Dyck, R. E. (2001). Performance of the Bluetooth system in fading dispersive channels and interference. In *Proceedings of the IEEE Global Communications Conference (GCC '01)*, Vol. 6, pp. 3499–3503.

Souza, R. D., Moore, C., Galvin, D. & Randall, D. (2006). Global connectivity from local geometric constraints for sensor networks with various wireless footprints. In *Proceedings of the 5th International Conference on Information Processing in Sensor Networks, 2006 IPSN*, pp. 19–26.

Spilker, J. (1978). GPS signal structure and performance characteristics. *Journal of the Institute of Navigation*, **25**(2), pp. 121–146.

Srikanth, T. K. & Toueg, S. (1987). Optimal clock synchronization. *Journal of the ACM*, **34**(3), pp. 626–645.

Stankovic, J. A., Cao, Q., Doan, T., Fang, L., He, Z., Kiran, R., Lin, S., Son, S., Stoleru, R. & Wood, A. (2005). Wireless sensor networks for in-home healthcare: Potential and challenges. *Proceedings of High Confidence Medical Device Software and Systems (HCMDSS) Workshop, Philadelphia, PA*, June 2–3, 2005.

Su, W. & Akyildiz, I. F. (2005). Time-diffusion synchronization protocol for wireless sensor networks. *IEEE/ACM Transactions on Networking*, **13**(2), pp. 384–397.

Sundararaman, B., Buy, U. & Kshemkalyani, A. D. (2005). Clock synchronization for wireless sensor networks: A survey. *Ad Hoc Networks*, **3**(3), pp. 281–323.

Sung, Y., Tong, L. & Swami, A. (2005). Asymptotic locally optimal detector for large-scale sensor networks under the Poisson regime. *IEEE Transactions on Signal Processing*, **53**(6), pp. 2005–2017.

Suwansantisuk, W. & Win, M. Z. (2005 March). Fundamental limits on spread spectrum signal acquisition. In *Proceedings of the Conference on Information Science and System, Baltimore, MD*.

Suwansantisuk, W. & Win, M. Z. (2005 March). Optimal search procedures for spread spectrum acquisition. In *Proceedings of the Conference on Information Science and Systems, Baltimore, MD*.

Suwansantisuk, W. & Win, M. Z. (2005 September) Optimal search strategies for ultrawide bandwidth signal acquisition. In *Proceedings of the IEEE International Conference on Utra-Wideband, Zürich, Switzerland*, 349–354.

Suwansantisuk, W., Win, M. Z. & Shepp, L. A. (2005 May). Properties of the mean acquisition time for wide-bandwidth signals in dense multipath channels. In *Proceedings of the 3rd SPIE International Symposium on Fluctuation and Noise in Communication Systems, Austin, TX*, Vol. 5847 pp. 121–135.

Suwansantisuk, W. & Win, M. Z. (2006 February). On the asymptotic performance of multi-dwell signal acquisition in dense multipath channels. In *Proceedings of the*

*IEEE International Zurich Seminar on Communications, Zürich, Switzerland*, Invited Paper.

Suwansantisuk, W. & Win, M. Z. (2006 February). Fundamental limits of multipath aided serial search acquisition. In *Proceedings Information of the Theory and Applications – Inaugural Workshop, La Jolla, CA*, Invited Paper.

Suwansantisuk, W. & Win, M. Z. (2007). Multipath aided rapid acquisition: Optimal search strategies. *IEEE Transactions on Information Theory*, **52**(1), pp. 174–193.

Suwansantisuk, W., Win, M. Z. & Shepp, L. (2005 September). On the performance of wide-bandwidth signal acquisition in dense multipath channels. *IEEE Transactions on Vehicular Technology*, **54**(5), pp. 1584–1594.

Svensoon, A. (2005). Executive Summary – The Safe Traffic Project, February.

Swaszek, P. & Willett, P. (1995). Parley as an approach to distributed detection. *IEEE Transactions on Aerospace and Electronic Systems*, **31**(1), pp. 447–457.

Szewczyk, R., Osterweil, E., Polastre, J., Hamilton, M., Mainwaring, A. & Estrin, D. (2004). Habitat monitoring with sensor networks. *Communications of the ACM*, **47**(6), pp. 34–40.

Tam, W. K. & Tran, V. N. (1995). Propagation modeling for indoor wireless communication. *Electronics and Communication Engineering Journal*, **7**(5), pp. 221–228.

Time-domain, httpwww.timedomain.com

Tseng, Y.-C., Ni, S.-Y., Chen, Y.-S. & Sheu, J.-P. (2002). The broadcast storm problem in a mobile ad hoc network. *Wireless Networks*, **8**(2/3), pp. 153–167.

Tsitsiklis, J. N. (1993). Decentralized detection. *Advances in Statistical Signal Processing*, **2**, pp. 297–344.

Van Dick, R. E., Miller, L. E. & Gaithersburg, M. (2001). Distributed sensor processing over an ad hoc wireless network: Simulation framework and performance criteria. In *Proceedings of the IEEE Military Communications Conference (MILCOM '01), Washington, DC, USA*, Vol. 2, 28–31 October 2001, pp. 894–898.

Trees, V. & Harry, L. (1968). *Detection, Estimation, and Modulation Theory*, 1st edn. New York: John Wiley.

Verdone, R. (2005). A statistical analysis of wireless connectivity in three dimensions. In *Proceedings of the IEEE International Conference on Communications, Seoul, Korea*, Vol. 5, pp. 3207–3210.

Verdone, R. (2004). An energy-efficient decentralized communication protocol for a network of uniformly distributed sensors polled by a wireless transceiver. In *Proceedings of the IEEE International Conference on Commununications, Paris, France*, Vol. 6, pp. 3491–3498.

Varshney, P. K. (1997). *Distributed Detection and Data Fusion*. New York: Springer.

Wang, Y. & Harder, A. (2006). A toolkit-based approach to indoor localization. In *Proceedings of the International Conference on Parallel Processing (ICPP '06)*, pp. 229–238.

Weissberger, M. A. (1981). An initial critical summary of models for predicting the attenuation of radio waves by foliage. In *ECACTR-81-101*. Annepolis, MD: Electromagnetic Compatibility Analysis Center.

Werner-Allen, G., Johnson, J. B. Ruiz, M. Lees, J. & Welsh, M. (2005). Monitoring volcanic eruptions with a wireless sensor networks. In *Proceedings of the Second European Workshop on Wireless Sensor Networks, EWSN 2005, Istanbul, Turkey*, 31 January–2 February 2005, pp. 108–120.

Williams, B. & Camp, T. (2002). Comparison of broadcasting techniques for mobile ad hoc networks. In *Proceedings of the ACM Symposium on Mobile Ad Hoc Networking and Computing, Lausanne, Switzerland*, pp. 194–205.

WiMedia Alliance (2005), Wimedia. http://www.wimedia.org

Win, M. Z. (2003). On the monotonicity of matched-filter bounds for diversity combining receivers. In *Proceedings of the IEEE Global Telecommunications Conference, San Francisco, CA*, Vol. 3, pp. 1684–1688.

Win, M. Z. (2002*a*). Spectral density of random time-hopping spread-spectrum UWB signals. *IEEE Communication Letters*, **6**(12), pp. 526–528.

Win, M. Z. (2002*b*). A unified spectral analysis of generalized time-hopping spread-spectrum signals in the presence of timing jitter. *IEEE Journal on Selected Areas in Communications*, **20**(9), pp. 1664–1676.

Win, M. Z. & Kostic, Z. A. (1999*a*). Virtual path analysis of selective Rake receiver in dense multipath channels. *IEEE Communication Letters*, **3**(11), pp. 308–310.

Win, M. Z. & Kostic, Z. A. (1999*b*). Impact of spreading bandwidth on Rake reception in dense multipath channels. *IEEE Journal on Selected Areas in Communications*, **17**(10), pp. 1794–1806.

Win, M. Z. & Scholtz, R. A. (2000). Ultra-wide bandwidth time-hopping spread-spectrum impulse radio for wireless multiple-access communications. *IEEE Transactions on Communications*, **48**(4), pp. 679–691.

Win, M. Z., Scholtz, R. A. & Fullerton, L. W. (1996). Time-hopping SSMA techniques for impulse radio with an analog modulated data subcarrier. In *Proceedings of the IEEE Fourth International Symposium on Spread Spectrum Techniques and Applications, Mainz, Germany*, pp. 359–364.

Win, M. Z., Qiu, X., Scholtz, R. A., Victor, O. & Li, K. (1999 May). ATM-based THSS-MA network for multimedia PCS. *IEEE Journal on Selected Areas in Communications*, **17**(5), pp. 824–836.

Win, M. Z., Chrisikos, G. & Sollenberger, N. R. (2000). Effects of chip rate on selective Rake combining. *IEEE Communication Letters*, **4**(7), pp. 233–235.

Win, M. Z., Chrisikos, G. & Sollenberger, N. R. (2000). Performance of Rake reception in dense multipath channels: Implications of spreading bandwidth and selection diversity order. *IEEE Journal on Selected Areas in Communications*, **18**(8), pp. 1516–1525.

Win, M. Z. & Scholtz, R. A. (1998). On the robustness of ultra-wide bandwidth signals in dense multipath environments. *IEEE Communication Letters*, **2**(2), pp. 51–53.

Win, M. Z. & Scholtz, R. A. (1998). On the energy capture of ultra-wide bandwidth signals in dense multipath environments. *IEEE Communication Letters*, **2**(9), pp. 245–247.

Win, M. Z. & Scholtz, R. A. (1998). Impulse radio: How it works. *IEEE Communication Letters*, **2**(2), pp. 36–38.

Win, M. Z. & Wen, Y. (2005). Extreme power dispersion profiles for Nakagami-*m* fading channels with maximal-ratio diversity. *IEEE Communication Letters*, **9**(5), pp. 385–387.

Win, M. Z., Mallik, R. K. & Chrisikos, G. (2003). Higher order statistics of antenna subset diversity. *IEEE Transactions on Wireless Communications*, **2**, pp. 871–875.

Win, M. Z. & Scholtz, R. A. (2002*a*). Characterization of ultra-wide bandwidth wireless indoor communications channel: A communication theoretic view. *IEEE Journal Selected Areas in Communications*, **20**(9), pp. 1613–1627.

Win, M. Z. & Wen, Y. (2005). Extreme power dispersion profiles for Nakagami-*m* fading channels with maximal-ratio diversity. *IEEE Communication Letters*, **9**(5), pp. 385–387.

Win, M. Z. & Winters, J. H. (2001). Virtual branch analysis of symbol error provavility for hybrid selection/maximal-ratio combining in rayleigh fading. *IEEE Transactions on Communications*, **49**, pp. 1926–1934.

Win, M. Z., Pinto, P. C., Giorgetti, A., Chiani, M. & Shepp, L. A. (2006 August). Error performance of ultrawideband systems in a Poisson field of narrowband interferers. In *Proceedings of the IEEE International Symposium on Spread Spectrum Techniques and Applications, Manaus, Brazil*, pp. 410–416, Invited Paper.

Winters, J. H. (1987). On the capacity of radio communication systems with diversity in Rayleigh fading environment. *IEEE Journal Selected Areas in Communications*, **5**(5), pp. 871–878. SAC

Woo, A. & Culler, D. (2001). A transmission control scheme for media access in sensor network. In *ACM Proceedings of the International Conference on Mobile Computing and Networking (MobiCom '01), Rome, Italy*, pp. 272–286.

Xiao, J.-J., Ribeiro, A., Luo, Z.-Q. & Giannakis, G. B. (2006). Distributed compression-estimation using wireless sensor networks. *IEEE Signal Processing Magazine*, **23**(4), pp. 27–41.

Xu, Y., Heidemann, J. & Estrin, D. (2001). Geography-informed energy conservation for ad-hoc routing. In *Proceedings of the Seventh Annual ACM/IEEE International Conference on Mobile Computing and Networking, Rome, Italy*, pp. 70–84.

Xue, Q. & Ganz, A. (2004). Maximizing sensor network lifetime: Analysis and design guidelines. *Proceedings of MILCOM*.

Yang, L. & Alouini, M. S. (2006). Performance comparison of different selection combining algorithms in presence of co-channel interference. *IEEE Transactions on Vehicular Technology*, **55**, pp. 559–571.

Yang, L. & Giannakis, G. B. (2004). Ultra-wideband communications: An idea whose time has come. *IEEE Signal Processing Magazine*, **21**(6), pp. 26–54.

Ye, W., Heidemann, J. & Estrin, D. (2002). An energy-efficient MAC protocol for wireless sensor networks. In *Proceedings of the IEEE INFOCOM*, Vol. 3, pp. 1567–1576.

Yen, Y.-S., Hong, S., Chang, R.-S. & Chao, H.-C. (2007). An energy efficient and coverage guaranteed wireless sensor network. In *Proceedings of IEEE WCNC 2007, Hong Kong*, 11–15 March 2007, pp. 2923–2928.

Yu, Y., Estrin, D. & Govindan, R. (2001). Geographical and energy-aware routing: A recursive data dissemination protocol for wireless sensor networks. In *UCLA Computer Science Department Technical Report*.

Zhang, F. & Deng, G. Y. (2005). Probabilistic time synchronization in wireless sensor networks. In *Proceedings of the International Conference on Wireless Communications, Networking and Mobile Computing*, Vol. 2, pp. 980–984.

Zhao, F., Liu, J., Guibas, L. & Reich, J. (2003). Collaborative signal and information processing: An information-directed approach. *Proceedings of IEEE*, **91**(8), pp. 1199–1209.

Zhao, Q., Swami, A. & Tong, L. (2006). The interplay between signal processing and networking in sensor networks. *IEEE Signal Processing Magazine*, **23**(4), pp. 84–93.

Zhong, L. C., Shah, R., Guo, C. & Rabaey, J. (2001). An ultra-low power and distributed access protocol for broadband wireless sensor networks. In *Proceedings of the IEEE Broadband Wireless Summit, Las Vegas, NY*, May 2001.

ZigBec Website: http//www/zigbec.org/en/index.asp

Zorzi, M. & Rao, R. R. (2003*a*). Geographic random forwarding (geraf) for ad hoc and sensor networks: Energy and latency performance. *IEEE Transactions on Mobile Computing*, **2**(4), pp. 349–365.

Zorzi, M. & Rao, R. R. (2003*b*). Geographic random forwarding (geraf) for ad hoc and sensor networks: multihop performance. *IEEE Transactions on Mobile Computing*, **2**(4), pp. 337–348.

## Further Reading

Cayirci, E. (2003 August). Data aggregation and dilution by modulus addressing in wireless sensor networks. *IEEE Communication Letters*, **7**(8), pp. 355–357.

Chang, J.-H. & Tassiulas, L. (2004). Maximum lifetime routing in wireless sensor networks. *IEEE/ACM Transactions on Networking*, **12**(4), pp. 609–619.

Chiasserini, C. F. & Rao, R. R. (2001 July). Improving battery performance by using traffic shaping techniques. *IEEE Journal on Selected Areas in Communications*, **19**(7), pp. 1385–1394.

Cui, S., Goldsmith, A. J. & Bahai, A. (2004 August). Energy-efficient of MIMO and cooperative MIMO techniques in sensor networks. *IEEE Journal on Selected Areas in Communications*, **22**(6), pp. 1089–1098.

Guerin, R. & Orda, A. (2002 October). Computing shortest paths for any number of hops. *IEEE/ACM Transactions on Networking*, **10**(5), pp. 613–620.

Hara, S., Zhao D. Z., Yanagihara, K. Taketsugu, J. Fukui, K. Fukunaga, S. Kitayama, K. (2005). Propagation characteristics of IEEE 802.15.4 radio signal and their application for location estimation. In *Proceedings of IEEE Vehicular Technology Conference, (VTC-2005-Spring)*, Vol. 1, pp. 97–101.

Lindsey, S., Raghavendra, C. & Sivalingam, K. M. (2002 September). Data gathering algorithms in sensor networks using energy metrics. *IEEE Transactions on Parallel and Distributed Systems*, **13**(9), pp. 924–935.

Mukhrjee, S. & Avidor, D. (2005). On the probability distribution of the minimal number of hops between any pair of nodes in a bounded wireless ad-hoc network subject to fading. In *Proceedings of International Workshop on Wireless Ad-Hoc Networks, IWWAN 2005, London, UK*, 23–26 May, 2005.

Santi, P. & Blough, D. M. (2003). The critical transmitting range for connectivity in sparse wireless ad hoc networks. *IEEE Transactions on Mobile Computing*, **2**(1), pp. 25–39.

Schurgers, C., Tsiatsis, V., Ganeriwal, S. & Srivastava, M. (2002). Optimizing sensor networks in the energy-latency-density design space. *IEEE Transactions on Mobile Computing*, **1**(1), pp. 70–80.

Schurgers, C., Kulkarni, G. & Srivastava, M. B. (October 2002). Distributed on-demand address assignment in wireless sensor networks. *IEEE Transaction on Parallel and Distributed Systems*, **13**(10), pp. 1056–1065.

Verdone, R. (2008). Wireless sensor networks. *Proceedings of the 5th European Conference, EWSN 2008, Bologna, Italy*, January/February 2008, LNCS 4913, Springer-Verlag, Berlin, January 2008.

Xue, F. & Kumar, P. R. (2004). The number of neighbors needed for connectivity of wireless networks. *Wireless Networks*, **10**(2), pp. 169–181.

# Index

Printed and bound by CPI Group (UK) Ltd, Croydon, CR0 4YY

03/10/2024

01040314-0014